Probability and Stochastic Processes with Applications to Communications, Systems and Networks, 2nd Edition

Probability and Stochastic Processes with Applications to Communications, Systems and Networks, 2nd Edition

Guest Editors

Gurami Tsitsiashvili
Alexander Bochkov

MDPI

Basel • Beijing • Wuhan • Barcelona • Belgrade • Novi Sad • Cluj • Manchester

Guest Editors
Gurami Tsitsiashvili
Institute for Applied Mathematics
Far Eastern Branch of Russian Academy Sciences
Vladivostok
Russia

Alexander Bochkov
Scientific and Technical Complex Research and Design Institute for Information Technology, Signaling and Telecommunications in Railway Transportation – NIIAS, JSC
Moscow
Russia

Editorial Office
MDPI AG
Grosspeteranlage 5
4052 Basel, Switzerland

This is a reprint of the Special Issue, published open access by the journal *Mathematics* (ISSN 2227-7390), freely accessible at: https://www.mdpi.com/si/mathematics/Probab_Stoch_Process_Appl_Commun_Syst_Netw_II.

For citation purposes, cite each article independently as indicated on the article page online and as indicated below:

Lastname, A.A.; Lastname, B.B. Article Title. *Journal Name* **Year**, *Volume Number*, Page Range.

ISBN 978-3-7258-3663-5 (Hbk)
ISBN 978-3-7258-3664-2 (PDF)
https://doi.org/10.3390/books978-3-7258-3664-2

Contents

About the Editors

Gurami Tsitsiashvili

Gurami Tsitsiashvili, Dr. Science, is currently a Professor at the Institute for Applied Mathematics, the Far Eastern Branch of the Russian Academy of Sciences, Russia. He received an M. Sc. in Mathematics and Physics from Moscow Physic Technical Institute in 1972 and a Ph.D. in Mathematics and Physics from the Institute for Applied Mathematics, the Far Eastern Branch of the Russian Academy of Sciences in 1976. In 1992, he achieved Dr. Sci. in Mathematics and Physics with a Rehabilitation Thesis "Decomposition Analysis of Complex Systems". In 1972, he was stationed in the Far East division of the Russian Academy of Sciences, where he has worked until now, starting as a trainee-researcher and later becoming the Deputy Director of Science at the Institute of Applied Mathematics of the RAS. He is the author of more than 200 works, including four monographs, one textbook and more than 110 articles in international peer-reviewed magazines. He studies complex systems, mass service system stability, cooperativity in multiple stochastic systems and multiplicative theorems for queueing networks.

Alexander Bochkov

Alexander Bochkov, Dr. Science, is currently a Scientific Secretary at the Research and Design Institute for Information Technology, Signaling and Telecommunications in Railway Transportation—NIIAS, JSC, Co-founder of Gnedenko-Forum. Alexander Bochkov graduated from the faculty of the Moscow Aviation Institute (MAI) in 1994. After graduation, he worked as a researcher at the Scientific-Technological Complex System Analysis, Russian Research Center "Kurchatov Institute", Scientific Institutes and organizations of Gazprom. In 2000, he produced a dissertation for the degree of Ph.D. (candidate of technical sciences), and in 2019, he produced a dissertation for the degree of Dr. Science. Dr. Bochkov participated in many Russian and international conferences and scientific schools and has published in Russian and foreign journals more than 150 scientific papers and eight monographs. His research interests are life-support systems, water purification electrochemical systems, non-stationary processes, risk assessment, analysis and management, system analysis and operation research and vulnerability and survivability of large-scale systems.

mathematics

Editorial

Preface to the Special Issue on Probability and Stochastic Processes with Applications to Communications, Systems and Networks, 2nd Edition

Alexander Bochkov [1,*] and Gurami Tsitsiashvili [2]

1 Scientific and Technical Complex Research and Design Institute for Information Technology, Signaling and Telecommunications in Railway Transportation—JSC NIIAS, 109029 Moscow, Russia
2 Institute for Applied Mathematics, Far Eastern Branch of Russian Academy of Sciences, 690041 Vladivostok, Russia; guram@iam.dvo.ru
* Correspondence: a.bochkov@gmail.com or a.bochkov@vniias.ru

Citation: Bochkov, A.; Tsitsiashvili, G. Preface to the Special Issue on Probability and Stochastic Processes with Applications to Communications, Systems and Networks, 2nd Edition. *Mathematics* **2025**, *13*, 922. https://doi.org/10.3390/math13060922

Received: 4 March 2025
Accepted: 9 March 2025
Published: 11 March 2025

Citation: Bochkov, A.; Tsitsiashvili, G. Preface to the Special Issue on Probability and Stochastic Processes with Applications to Communications, Systems and Networks, 2nd Edition. *Mathematics* **2025**, *13*, 922. https://doi.org/10.3390/math13060922

Probability theory and stochastic processes are fundamental mathematical tools that play a critical role in understanding and modeling randomness and uncertainty in various fields, including communications, systems engineering, and network design. These concepts provide the foundation for analyzing and designing systems that operate in unpredictable environments, such as wireless communication networks, internet traffic, and signal processing systems.

The Special Issue is the continuation of the work initiated in [1], representing a reprint of the second edition of the Special Issue "Probability and Stochastic Processes with Applications to Communications, Systems and Networks, 2nd Edition". The authors' geographical distribution is shown in Table 1; the 25 authors are from ten different countries. Note that it is usual for a paper to be written by more than one author and for authors to collaborate with authors with different or multiple affiliations.

Table 1. Geographic distribution of authors by country.

Country	Number of Authors
Republic of Korea	5
India	6
Russia	4
France	2
Italy	1
Iran	1
Saudi Arabia	3
Tunisia	1
Canada	1
Egypt	1

In Contribution 1, the authors explore diversity selection in three-user interference-aligned (IA) MIMO channels, emphasizing reliability enhancement through diversity order (DO) analysis. While degrees of freedom (DoF) have traditionally been prioritized for bandwidth optimization in IA, diversity order has received less attention for improving error performance. This paper introduces the concept of conditional diversity order and proposes a beamforming vector selection method that achieves a conditional diversity order of $M^2/2$, where M is the number of antennas per transceiver. The proposed two-stage decoding approach integrates zero-forcing for interference suppression with maximum likelihood (ML) decoding and orthogonalization techniques to recover the desired signal. The authors demonstrate that their scheme outperforms traditional IA methods in both error probability and conditional DO, particularly in scenarios with a large number of antennas. These findings offer valuable insights into optimizing IA for enhanced reliability and lay the groundwork for future research on advanced decoding and beamforming strategies.

In Contribution 2, a new linear transceiver optimization problem is addressed for correlated MIMO interference channels in the presence of channel state information (CSI) errors. This scenario is more realistic and practical compared to previous studies focusing on uncorrelated MIMO interference channels. The optimization problem is formulated as minimizing the mean square error (MSE) under general power constraints. Since the objective function for precoders and receive filters is not jointly convex, the problem is decomposed into two convex subproblems, which are solved iteratively to obtain linear precoders and receive filters. The proposed algorithm is guaranteed to converge to a local minimum. Numerical results show that the algorithm significantly reduces sensitivity to CSI errors compared to existing robust schemes in correlated MIMO interference channels.

In Contribution 3, the authors extend the k-out-of-n: G reliability system to a multi-server queueing model with N repair policies. The system consists of n servers with identically and exponentially distributed service times. Servers fail at an exponential rate, and repairs follow N distinct policies. Although servers operate independently, maintenance is only performed if at least k servers are operational. The steady-state model is analyzed using a matrix-analytic method, and an optimization problem is formulated. Performance measures are evaluated and the results are presented, providing insights into the system's behavior under various repair policies.

In Contribution 4, a two-server queueing system is analyzed, where the arrival and service processes are interdependent. The system's evolution is governed by transitions on the product space of three Markov chains, which describe the arrival and service processes. Transitions follow a semi-Markov rule, with dwell times exponentially distributed. The authors derive the stability condition for the system and compute the stationary distribution at equilibrium. Several performance measures are evaluated, and numerical illustrations are provided to validate the model's effectiveness.

In Contribution 5, positive recurrence is established for a single-server queueing system under generalized service intensity conditions. Unlike previous studies, the authors do not assume the existence of a service distribution density function but instead use a lower bound of integral type as a sufficient condition. Positive recurrence ensures the existence of an invariant distribution and guarantees slow convergence to it in the full variation metric.

In Contribution 6, weak convergence results are presented for a family of parameter-dependent Gibbs measures. The authors show that the limiting distribution is centered on the set of global minima of the limiting Gibbs potential. An explicit calculation of the limit distribution is provided, and the method is applied as an alternative to Lyapunov function or direct convergence approaches for stationary probabilities. The theoretical findings are illustrated with a numerical example, including an application to the repairman problem.

In Contribution 7, the redundancy allocation problem (RAP) is studied under the assumption that component failure times follow an Erlang distribution with a time-dependent rate parameter. Redundancy choices for each subsystem include none, active, standby, or mixed. A genetic algorithm is employed to solve the optimal allocation problem. Numerical examples are analyzed to investigate the impact of time dependence, and a case study from the literature is examined. The results highlight that the time dependence of the mean time between failures (MTBF) distribution parameters can significantly influence optimal redundancy allocation.

In Contribution 8, the authors address the problem of adaptive event filtering for a class of nonlinear discrete-time systems modeled using interval type-2 fuzzy models. The system is subject to Markov switching and susceptible to cheating attacks. To reduce unnecessary signal transmissions over the communication channel, an improved event-triggering mechanism is proposed. Additionally, an extended dissipativity specification is introduced to quantify the transient dynamics of filtering errors. Using the linear matrix inequalities (LMI) approach and leveraging information from upper and lower membership functions, the authors establish sufficient conditions for the existence of a desired filter. This ensures the root mean square (RMS) stability and extended dissipativity of the augmented filtering system. An optimization algorithm based on particle swarm optimization (PSO) is proposed to compute the filter gains with optimal efficiency. The effectiveness of the developed scheme is validated through numerical experiments on a single-link robotic arm and a lower limb system.

In Contribution 9, the authors study the formation of a particle flow that replicates an image and estimate the distance between the copy and the original using a specialized probabilistic metric. The ability of a ball flow to cover a surface during ball grinding is analyzed using stochastic geometry formulas. The recovery of inhomogeneous Poisson flow characteristics from imprecise observations is examined using the point-coloring theorem of Poisson flow. The dependence of the Poisson distribution parameter on the peak load generated by an inhomogeneous input Poisson flow is evaluated in a queueing system with an infinite number of servers and deterministic service time. The models consist of an inhomogeneous Poisson flow of points, each labeled with attributes such as mass, area, volume, observability, and service time. The authors establish an asymptotic step dependence between the objective functions of the model and the parameters of label splitting. The results have potential applications in nanotechnology, powder metallurgy, ecology, and smart city initiatives, and are supported by real-world observations and experiments.

In Contribution 10, the authors explore the Hawkes process, a self-excited point process with a clustering effect, where the jump rate depends on the entire history of the process. While the Hawkes process is generally non-Markovian, it can exhibit Markovian properties under specific conditions, such as when the excitation function is exponential or a sum of exponentials. The intensity of the Hawkes process is determined by the sum of a base intensity and terms dependent on the process's history. In this study, the authors consider a linear Hawkes model with a randomly determined baseline intensity and investigate large deviation principles for this new model. The random baseline intensity Hawkes processes have broad applications in insurance, finance, queueing theory, and statistics.

In Contribution 11, the authors analyze the waiting time distribution for a single-server queueing system with finite buffer capacity \(N\). Clients arrive according to a Poisson process, and the server operates under a fast service rule. The service times follow a general distribution independent of the arrival process. The authors propose an alternative approach to derive the probability distribution of the queue length at the epoch following batch departures. Using an embedded Markov chain, Markov renewal theory, and semi-Markov processes, they obtain the queue length distribution at a random epoch. The waiting time distribution for a random customer is determined by establishing a functional relationship between the probability-generating function of the queue length distribution and the Laplace–Stieltjes transform (LST) of the waiting time distribution. The authors derive the probability density function of the waiting time and provide numerical results.

In Contribution 12, the authors address a critical issue in power grids arising from the high energy consumption of mining devices in bitcoin-based financial trading systems (BFTSs). This poses a significant risk, as cybercriminals may exploit vulnerabilities to disrupt optimal energy management. The paper introduces a new type of cyber-attack, termed "miner-misuse", and proposes an online anomaly detection approach based on reinforcement learning (RL) to counteract such attacks. Due to incomplete system information, the authors implement an Observable Markov Decision Process (OMDP) within the RL framework to block miner attacks. The proposed method is shown to perform optimally and accurately when tuning parameters are correctly set during training. A hybrid mechanism combining optimization and learning ensures both optimal solutions and fast convergence.

Additionally, the authors propose an Intelligent Priority Selection (IPS) algorithm integrated with the RL method to enhance the detection of mining attacks. To validate the approach, they model the energy consumption of mining devices as a function of the hashing rate in the BFTS. Uncertain fluctuations in energy consumption are addressed using the Unscented Transform (UT) method, which accurately models uncertain parameters with high probability. The results demonstrate that the proposed anomaly detection method outperforms other approaches in detecting real-time miner attacks, as evidenced by the F-score value and the probability of successful attack detection.

Thus, the collection covers a wide range of advanced topics in communication systems, queueing theory, stochastic processes, and optimization, with applications ranging from wireless networks to power grids and nanotechnology. Taken together, all of the papers emphasize the importance of advanced mathematical and computational tools for solving complex problems in communication systems, queueing theory, and stochastic processes. The integration of optimization algorithms, machine learning, and probabilistic models enables the development of robust and efficient solutions for real-world applications ranging from improving the reliability of wireless communications to securing power grids and optimizing industrial processes. This work not only advances theoretical knowledge, but also provides practical foundations for solving emerging problems in technology and infrastructure.

Probability theory and stochastic processes are powerful tools for understanding and managing randomness in communications, systems, and networks. From modeling network traffic to designing robust communication systems, these concepts enable engineers and researchers to tackle real-world challenges in the presence of uncertainty. As technology continues to evolve, the importance of probability and stochastic processes will only grow, making them essential knowledge for anyone working in these fields.

This Special Issue serves as a starting point for exploring the rich and fascinating world of probability and stochastic processes. Whether you are a student, researcher, or practitioner, mastering these concepts will open doors to innovative solutions and deeper insights into the behavior of complex systems.

Conflicts of Interest: The authors declare no conflicts of interest.

List of Contributions

1. Jin, D.; Jin, X. Study on Selection Diversity for MIMO 3-User Interference Channel with Interference Alignment. *Mathematics* **2024**, *12*, 3877. https://doi.org/10.3390/math12243877.
2. Kang, J.-M.; Lim, D.-W. Robust Transceiver Design for Correlated MIMO Interference Channels in the Presence of CSI Errors under General Power Constraints. *Mathematics* **2024**, *12*, 801. https://doi.org/10.3390/math12060801.
3. Krishnamoorthy, A.; Joshua, A.N.; Mathew, A.P. The k-out-of-n:G System Viewed as a Multi-Server Queue. *Mathematics* **2024**, *12*, 210. https://doi.org/10.3390/math12020210.

4. Sindhu, S.; Krishnamoorthy, A.; Kozyrev, D. A Two-Server Queue with Interdependence between Arrival and Service Processes. *Mathematics* **2023**, *11*, 4692. https://doi.org/10.3390/math11224692.
5. Veretennikov, A. On Positive Recurrence of the Mn/GI/1/∞ Model. *Mathematics* **2023**, *11*, 4514. https://doi.org/10.3390/math11214514.
6. Chetouani, H.; Limnios, N. Gibbs Distribution and the Repairman Problem. *Mathematics* **2023**, *11*, 4120. https://doi.org/10.3390/math11194120.
7. Zio, E.; Gholinezhad, H. Redundancy Allocation of Components with Time-Dependent Failure Rates. *Mathematics* **2023**, *11*, 3534. https://doi.org/10.3390/math11163534.
8. Kchaou, M.; Regaieg, M.A. Event-Triggered Extended Dissipativity Fuzzy Filter Design for Nonlinear Markovian Switching Systems against Deception Attacks. *Mathematics* **2023**, *11*, 2064. https://doi.org/10.3390/math11092064.
9. Tsitsiashvili, G.; Osipova, M. Asymptotic Relations in Applied Models of Inhomogeneous Poisson Point Flows. *Mathematics* **2023**, *11*, 1881. https://doi.org/10.3390/math11081881.
10. Seol, Y. Large Deviations for Hawkes Processes with Randomized Baseline Intensity. *Mathematics* **2023**, *11*, 1826. https://doi.org/10.3390/math11081826.
11. Chaudhry, M.; Datta Banik, A.; Barik, S.; Goswami, V. A Novel Computational Procedure for the Waiting-Time Distribution (in the Queue) for Bulk-Service Finite-Buffer Queues with Poisson Input. *Mathematics* **2023**, *11*, 1142. https://doi.org/10.3390/math11051142.
12. Almalaq, A.; Albadran, S.; Mohamed, M.A. An Adoptive Miner-Misuse Based Online Anomaly Detection Approach in the Power System: An Optimum Reinforcement Learning Method. *Mathematics* **2023**, *11*, 884. https://doi.org/10.3390/math11040884.

Reference

1. Tsitsiashvili, G.; Bochkov, A. (Eds.) *Probability and Stochastic Processes with Applications to Communications, Systems and Networks*; MDPI Books: Basel, Switzerland, 2023.

mathematics

MDPI

Article

An Adoptive Miner-Misuse Based Online Anomaly Detection Approach in the Power System: An Optimum Reinforcement Learning Method

Abdulaziz Almalaq [1,*], Saleh Albadran [1] and Mohamed A. Mohamed [2,*]

1 Department of Electrical Engineering, Engineering College, University of Ha'il, Ha'il 55476, Saudi Arabia
2 Electrical Engineering Department, Faculty of Engineering, Minia University, Minia 61519, Egypt
* Correspondence: a.almalaq@uoh.edu.sa (A.A.); dr.mohamed.abdelaziz@mu.edu.eg (M.A.M.)

Abstract: Over the past few years, the Bitcoin-based financial trading system (BFTS) has created new challenges for the power system due to the high-risk consumption of mining devices. Briefly, such a problem would be a compelling incentive for cyber-attackers who intend to inflict significant infections on a power system. Simply put, an effort to phony up the consumption data of mining devices results in the furtherance of messing up the optimal energy management within the power system. Hence, this paper introduces a new cyber-attack named miner-misuse for power systems equipped by transaction tech. To overwhelm this dispute, this article also addresses an online coefficient anomaly detection approach with reliance on the reinforcement learning (RL) concept for the power system. On account of not being sufficiently aware of the system, we fulfilled the Observable Markov Decision Process (OMDP) idea in the RL mechanism in order to barricade the miner attack. The proposed method would be enhanced in an optimal and punctual way if the setting parameters were properly established in the learning procedure. So to speak, a hybrid mechanism of the optimization approach and learning structure will not only guarantee catching in the best and most far-sighted solution but also become the high converging time. To this end, this paper proposes an Intelligent Priority Selection (IPS) algorithm merging with the suggested RL method to become more punctual and optimum in the way of detecting miner attacks. Additionally, to conjure up the proposed detection approach's effectiveness, mathematical modeling of the energy consumption of the mining devices based on the hashing rate within BFTS is provided. The uncertain fluctuation related to the needed energy of miners makes energy management unpredictable and needs to be dealt with. Hence, the unscented transformation (UT) method can obtain a high chance of precisely modeling the uncertain parameters within the system. All in all, the F-score value and successful probability of attack inferred from results revealed that the proposed anomaly detection method has the ability to identify the miner attacks as real-time-short as possible compared to other approaches.

Keywords: reinforcement learning; FDI attack; mining device; power system; UT method; attack detection

MSC: 49-11

Citation: Almalaq, A.; Albadran, S.; Mohamed, M.A. An Adoptive Miner-Misuse Based Online Anomaly Detection Approach in the Power System: An Optimum Reinforcement Learning Method. *Mathematics* **2023**, *11*, 884. https://doi.org/10.3390/math11040884

Academic Editors: Gurami Tsitsiashvili and Alexander Bochkov

Received: 8 January 2023
Revised: 6 February 2023
Accepted: 7 February 2023
Published: 9 February 2023

1. Introduction

Nowadays, on account of widely utilizing the smart varied business tools for social welfare enhancement, there is a need to develop an environmentally friendly financing bed that can guarantee financial affairs in a risk-free and reliable way. Meanwhile, launching a financial trading network based on the platform of blockchain technology such as the Bitcoin systems needs to catch the remarkable energy consumption through the mining devices. Likewise, such technology being a scale-great load demand is technically known as another major difficulty for the power systems. This results in the electrical grid facing optimal energy scheduling for not being aware of the consumed power and seating within the grid configuration. Recently, energy management based on blockchain became an attention center for researchers. However, the deployment of blockchain tech does not entrust a straightforward choice owing to the challenging obstacles in the way. Thus, the current study tried to settle the issue that blockchain technology may be successful in optimal energy management, whereas blockchain tech was introduced as an energy consumer for the power system. One aspect of the problem that has yet to be considered is the cyber-injurious intrusion in the form of making spurious mining devices and false data injection (FDI) into the defined financial trading platform, aiming to irregularly enlarge the load demands of the electrical grid [1]. This issue can emerge new inescapable challenges facing energy management that need to be dealt with. In the literature, various cyber-attack detection approaches are investigated. Looking over the above-mentioned considerations, the literature of the relevant works is addressed in the following.

1.1. Cyber-Attack Detection in the Power System

For the last few years, cyber-threats focused on the electrical grid's performance for the sake of depending on industrial technologies in every country. Hence, the cyber-attacks of FDI type are one of the most efficient ways to launch widespread sabotage in the varied fields of the cyber-physical system, e.g., the control system [2–5], energy markets [6], electrical grid [7,8], and so forth, in the sense that FDI attacks may bring irrecoverable injuries involving the electrical grid versus the modest cyber cost involved by the hackers. Indeed, FDI attacks tend to manipulate sensing and measuring info with the aim of improperly reading into the monitoring system [9]. Broadly speaking, most attackers, known as denial of service (DOS) attacks [10–12], aim to mislead the data estimation process to get into the all-embracing destruction, e.g., the wide-area blackouts in the power systems. The DOS attacks are able to decline the chance of having access to the system for meter devices [13–15], whereas, jamming attacks manipulate the process of data measurement by means of applying noise [16]. On the basis of blockchain tech, it is utilized in many applications, such as communication structures, smart contracts, the Internet of Things, electronic voting, and so on. Despite these all applications, blockchain is vulnerable to various cyber-attacks classified under the following: (1) attacks related to the used mathematical techniques (stale and orphaned blocks), (2) attacks of peer-to-peer structure (51% attack, distributed DoS attack), and (3) attacks of the application context [17]. In the literature, many reports have expressed the increase in data security risks for blockchain tech. For instance, a hacker in June 2016 could manage to steal USD 50 million from "The DAO" [18]. As mentioned already, on account of the intricate looping structure of the electrical network, any subversion or failure in the grid's vulnerable points lead to cascading damages over the grid for a short period of time. Keeping these in mind, it seems that timely anomaly detection is critical once the grid continually suffers cyber-attacks. In this case, quickly performing the detection method is of crucial significance in assessing the cyber-attack detection methods. In a more proper way, when a variation occurs in the sensing data at an unknown hour, the goal is to trace the data variation as soon as possible with emphasis on lowering the false alarm for the measuring devices that is available over time. Another matter of great concern that needs urgent attention is that the detection speed and detection accuracy are in conflict with each other once launching a cyber-attack. Conversely, it is very significant to find a way to reconcile these two conflicting views. The reinforcement learning (RL) concept is defined with the aim of controlling stochastic environments. For this reason, the observable Markov decision process (OMDP) problem may be solvable by inspiration from the RL concept [19,20]. Indeed, learning the underlying OMDP problems can be achieved by a model-based RL algorithm [21–25].

1.2. The Mining Devices

Satoshi Nakamoto proposed a Bitcoin-based cash trading system guaranteed by blockchain tech in 2008 [26]. A blockchain network is defined based on the ledgers maintained by the wide nodes in the form of a distributed structure [27]. Blockchain has the needed ability to prove transaction security and trust in wide usages, e.g., smart healthcare, real estate, the logistics industry, and so forth. The consensus algorithm shows that the proof of work (PoW) is the most widely and commonly used in the blockchain and is able to check the bilateral trust among nodes within the network [28]. Indeed, the PoW mechanism brings high security with the help of working the mining devices while consuming lots of power during the hashing computation. For instance, the literature indicates that the bitcoin-based financial trading system (BFTS) takes an electrical consumption of about 1–32 TWh per year, whilst the power consumption of Denmark is estimated to be up to 32 TWh. This means that BFTS, notwithstanding providing a safe mutual fiscal bed, can turn into a serious menace to the electrical grid as a large-scale load demand. Hence, in [29], the authors tried to indicate a new consensus algorithm called proof of stake (PoS) for application in blockchain tech. All in all, new investigations are working on and starting to eliminate this concern.

1.3. Motivation and Contributions

In view of the fact, it is possible that the data monitoring center is overshadowed by cyber-attacks aimed at making counterfeit data instead of documenting it. Thus, it seems that the monitoring data need to be checked before being employed by the decision-maker. As noted, despite that the blockchain technology could guarantee data security for the power system through distributed data management, it can be vulnerable and can be a target for cyber attackers thanks to the presence of high-consumption miners. Accordingly, having a close look at the data detection mechanism to counter it is a must. This paper aimed at, first, introducing a new attack based on the consensus structure of blockchain and proposing an adaptive detection mechanism to deal with it. A cyber nascence attack called miner misuse was introduced here for energy systems utilizing the blockchain security structure. Indeed, the adversary's intent of deliberately increasing the miner's energy is problematic for the energy systems secured by blockchain tech. So to speak, a cyber attacker can often lead astray the monitoring operator by way of false data injection related to the miner energy. This work is possible in making misprision decisions about energy management. As such, to counter the aforementioned concern, the monitoring data related to the miners' energy consumption must be put into a checking way. Meanwhile, this paper captured a miner data detection method inspired by the reinforcement learning concept to raise the chance of trustiness for energy systems. Inasmuch as the learning mechanisms are in essence unable to catch the optimal solution, they need to be vouched by the optimization approaches. To this end, this paper followed another goal based on introducing an Intelligent Priority Selection (IPS) optimization algorithm melded with the learning mechanism. Hence, to differentiate this paper compared to others, Table 1 briefly indicates the main differences.

Table 1. Categorization of the different approaches.

	Block Chain	Cyber Attack	RL Mechanism	Optimization Method	Uncertainty
[17]	✓	✓			
[30,31]	✓	✓			✓
[32–34]			✓	✓	
[35–37]				✓	
Proposed model	✓	✓	✓	✓	✓

Therefore, the underlying contributions and characteristics of this paper concerning previous publications are categorized as follows:

- Introducing and mathematically modeling a new cyber nascence attack based on the miner misuse of the blockchain mechanism whose aim is data security within the smart energy system.
- Evolving an effective and adoptive IPS-RL-based online anomaly detection method to pinpoint the unknown malicious intrusions in the power system based on BFTSs called the miner misuse.
- Presenting the UT-based uncertainty approach to conjure up the high-risk energy consumption of the mining devices and check up on their effects on energy management.

The rest of the paper is provided such that the mathematical formulations of the studied model are described in Section 2. Section 3 expresses the proposed detection method. The solution process is represented in Section 4. Section 5 explains the UT-based uncertainty method. The effectiveness of the studied model proposed in this paper is evaluated and validated in Section 6, and Section 7 provides the conclusion.

2. The Modeling Structure of the Smart Grid Based on BFTS

2.1. The Basic Framework of the Power System

Recently, there is a considerable effort employing smart modern tools in the daily life of people in a way that social welfare would be on the rise. Developing the Bitcoin-based financial trading system (BFTS) can be a manifest case of the above considerations. For all these reasons, there is a tendency toward growing load demands in the power system, as shown in Figure 1. For instance, BFTSs need to employ mining devices for securing goals while taking the significantly consumed energy for hashing and blocking information into the well-known blockchain process. Developing BFTSs along with increasing the number of mining devices causes unavoidable challenges facing modern power systems to emerge. Ignoring the energy consumption of the mining device in energy scheduling management makes it impost the irrefutable limitations such as the high operation cost, line congestion, uncompromising power generation, and so forth on the power systems [38]. To realize the effects of the mining devices on energy management, a closer look at the energy operating framework of the electrical grids is needed. Hence, this section is dedicated to explaining how to meet the load demands by the generation units with an emphasis on satisfying the relevant constraints and the objective function followed by the power system. Keeping this in mind, the fossil fuel-based traditional generation units are operated by an energy control center aimed at following the relevant objective function with an emphasis on power limitations as defined by Equations (1)–(13). In this case, the generation scheduling goal described in (1) indicates that these generation units can catch in/out permits and the generated power in line with minimizing the operation cost and the relevant shot-down/start-up costs. Indeed, all generation units would be operated with an emphasis on the price/generation curve fulfilled by the electricity market, but this goal is satisfied by looking at the transmission grid limitations. Meanwhile, one of the significant issues that needs urgent attention concerns the power limitations related to power generation and ramping. These constraints are modeled by Equations (2)–(5) in the operating framework. It may be true that the most important factor for holding the grid power stability is regularly checking the generated/consumed power variation during the grid operation. Hence, Equation (2) defines balancing the swirling active power among the resources, the flowing power of lines, and load demands, including the public loads and the consumed power of the mining devices. Conversely, this elucidation can be accurately inferred for the reactive power as shown in Equation (3). Equations (4) and (5) are the limitations of the generated powers. This means that the generation units should adhere to keep their generated powers to a minimum/maximum level due to the technical conditions. Of course, each generation unit is allowed to ramp up the output power with the aim of covering the critical load fluctuations. Thus, we modeled the ramping power limitations of generation units based on Equations (6) and (7). It is a right assertion that the grid lines are connected in the form of a looping structure with the goal of coupling the generation units and loads [38]. With this description, there is a need to model in which line l the power of bus n is injected/received to/from. In this regard, Equations (8) and (9) illustrate the active/reactive power injection based on the bus voltage and the technical properties of lines. Needless to say, one underlying concern for grid operators is the overloading of lines once raising the period of peak load. To counter this problem, the restriction of power flow is observed by constraints in Equations (12) and (13). Finally, to achieve voltage stability and optimal operation, it is necessary that the bus voltages and angles are restricted up to a permissible level based on Equations (10) and (11).

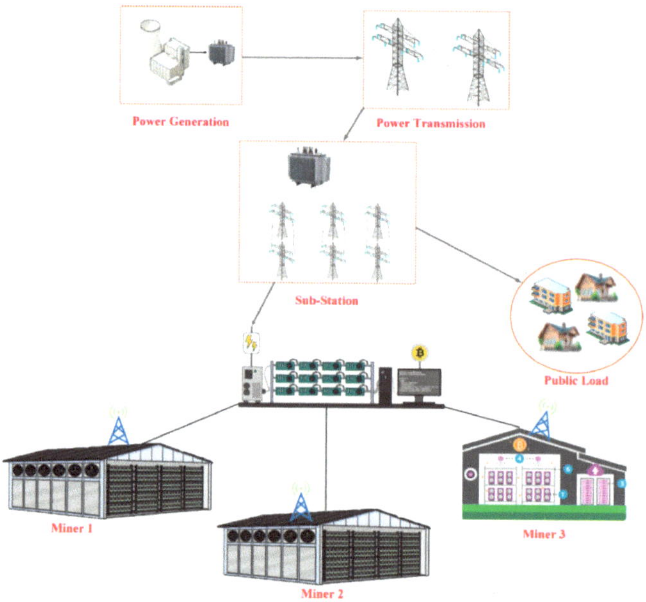

Figure 1. Illustration of the electrical grid based on the mining devices.

$$min\ TC^{grid} = \sum_t \sum_g \left[f^c\left(P_{t,g}\right) + S_{t,g}^{up} + D_{t,g}^{down} \right] \tag{1}$$

$$min\ TC^{grid} = \sum_t \sum_g \left[f^c\left(P_{t,g}\right) + S_{t,g}^{up} + D_{t,g}^{down} \right] \tag{2}$$

$$\sum_g \left(P_{t,g}\right) - \sum_l \left(P_{t,l}^L\right) = P_{n,t}^L + P^{mining} \qquad \forall t \in \Omega^T, \forall n \in \Omega^n \tag{3}$$

$$\sum_g Q_{t,g} + \sum_l \left(Q_{t,l}^L\right) = Q_{n,t}^L \qquad \forall t \in \Omega^T, \forall n \in \Omega^n \tag{4}$$

$$P^{min} z_{g,t} \leq P_{t,g} \leq P^{max} z_{g,t} \qquad \forall t \in \Omega^T, \forall g \in \Omega^g \tag{5}$$

$$Q^{min} z_{g,t} \leq Q_{t,g} \leq Q^{max} z_{g,t} \qquad \forall t \in \Omega^T, \forall g \in \Omega^g \tag{6}$$

$$P_{t,g} - P_{t-1,g} \leq D_G^+ z_{g,t-1} \qquad \forall t \in \Omega^T, \forall g \in \Omega^g \tag{7}$$

$$P_{t-1,g} - P_{t,g} \leq D_G^- z_{g,t} \qquad \forall t \in \Omega^T, \forall g \in \Omega^g \tag{8}$$

$$P^L{}_l = (V_n - V_m) R_l' - X_l' \eta_n \qquad \forall n \in \Omega^n, \forall m \in \Omega^m, \forall l \in \Omega^l \tag{9}$$

$$V^{min} \leq V_{n,t} \leq V^{max} \qquad \forall t \in \Omega^T, \forall n \in \Omega^n \tag{10}$$

$$\eta_n^{min} \leq \eta_{n,t} \leq \eta_n^{max} \qquad \forall t \in \Omega^T, \forall n \in \Omega^n \tag{11}$$

$$-P_l^{L\_max} \leq P_{l,t}^L \leq P_l^{L\_max} \qquad \forall t \in \Omega^T, \forall l \in \Omega^l \tag{12}$$

$$-Q_l^{L\_max} \leq Q_{t,l}^{L} \leq Q_l^{L\_max} \qquad \forall t \in \Omega^T, \forall l \in \Omega^l \tag{13}$$

2.2. The Mathematical Definition of the Mining Device

As recently mentioned, supplying the power involved in BFTSs, thanks to the mining devices, is a significant concern in energy management that needs to be dealt with. Indeed, mining devices are made up of a set of central processing units (CPUs), all of which have a duty in line with finding a solution for the targeted hashing puzzle. Broadly speaking, the needed power in this process is practically referred to as the computing power of these CPUs for cyclically solving the specified puzzle. For effective realization, this section aims to mathematically develop the mining device's behavior. To this end, the CPU's energy consumption was calculated based on the frequency and voltage associated with Equation (14) in the first place.

$$p^{mining} = Su^2 f \tag{14}$$

where:

- p^{mining}: The consumed power.
- S: A constant value related to the computing device.
- u: Voltage.
- f: Frequency.

In (14), the operating frequency is a variable taking into account the cycle number required for the hashing process per second. Meanwhile, assume the frequency and power consumption are in a direct way; as a result, Equation (14) with a slight change can substitute into Equation (15):

$$p^{mining} = qf \rightarrow q = Su^2 \tag{15}$$

By having considered $f = hw$, the aforementioned formulation is updated as:

$$p^{mining} = hq\omega \tag{16}$$

Additionally, the share in percentage related to CPU i for a mining device is achieved by the following:

$$\eta = \frac{P_i^{mining}}{\sum_{j \in N} P_j^{mining}} \tag{17}$$

Looking over the process, each CPU i takes a remarkable chance of solving the targeted puzzle per second. By adopting the Poisson distribution to the mining devices, the successful likelihood of CPU i can be obtained based on Equation (18):

$$\underset{i}{\mathrm{P(r)}} = \eta e^{-\lambda \beta t r_i} \tag{18}$$

In (18), obtaining a successful CPU in the solution is subject to be altered, with an emphasis on the number of transactions inserted into the data block and constant ψ. It is worth pointing out that mining devices are encouraged to catch gains as tax in the form of public service if rapidly puzzling out compared to others. Nutshell, the final benefit arising from each mining device based on the power consumption is earned as follows:

$$bif_{B_i}\left(P_i^{mining}\right) = \frac{TP_i^{mining} e^{-\lambda \beta t r_i}}{\sum_{j \in N} P_j^{mining}} - y_i P_i^{mining} \tag{19}$$

3. The Miner-Misuse-Based Detection Approach

In the previous section, it was expressed how the mining device's performance is in line with the energy scheduling for the electrical grids. By having a look at the high-risk energy consumption related to these mining devices, one obtains a practical realization of badly overshadowing energy management if the abnormally growing trend in employing mining devices is ignored. This concern can be turned into a high chance of imposing cyber-demolitions within the power system structure. Indeed, an adroit hacker in an effective effort can misuse this opportunity and can launch a bogus mining device called miner-misuse as the load demand in the electrical grid. For this reason, it seems that developing the BFTS-based electrical grid structure relying on a misuse detection approach to prevent any attack launch is a must. Hence, this section is dedicated to presenting the proposed RL-based online detection method aimed at meeting the miner-attack alarum.

3.1. The Definition of the Proposed RL Concept-Based Detection Approach

Recently, the learning machine in the field of cyber security illustrates has been utilized effectively in various industrial applications. Needless to say, learning machine problems are mainly defined in different forms of the learning concept, i.e., (1) supervised learning, (2) reinforcement learning (RL), and (3) unsupervised learning. In the literature, the RL method is introduced as an efficient and reliable security policy due to the learnable agent's employment in cooperating with the environment. Indeed, it was proven that the learning process could be more improvable with the chance of having environmental data. So to speak, the stepping learnedness concerning the penalty/reward approach makes the agent more reliable for realizing the environment as well as possible. Thus, as the agent receives the environmental data, it gives the corresponding retroaction in line with the relevant goal as well. Afterward, obtaining feedback on the agent's performance can be the best and most far-sighted solution for allotting the penalty/reward coefficients. An agent takes rewards on account of offering acceptable output within the environment. In contrast to gaining a reward, an agent would be fined due to misstating the fact of the environment. This work can help the agent recognize the environment where it is put into. To realize how the RL method works in action, Figure 2 shows briefly the RL approach's performance [32].

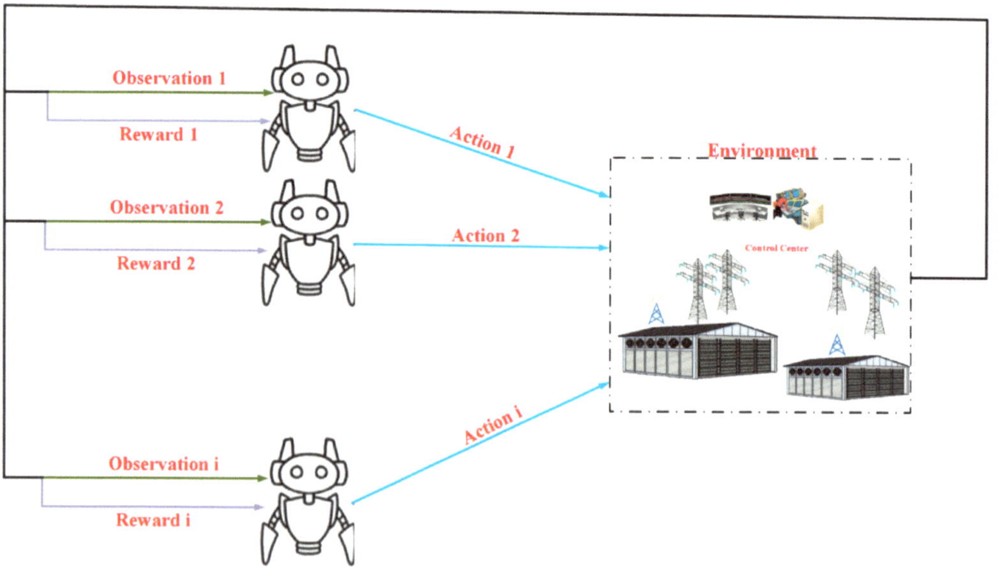

Figure 2. The reinforcement learning structure.

As can be seen, the launching of the RL method is defined by two spaces: (1) environment and (2) agent. Additionally, in order to declare anomalies in the environment, the RL method is carried out in reliance during the learning phase and the online detection phase. The learning phase needs to be described in the first place. In this step, the agent learns which environment observation each appropriate action belongs to. the agent takes a reward from the environment as a result of the action selection, which guides the agent on how to modify its selection process versus the environmental observations. This learning phase goes toward enabling the agent to pick out the best action under an unknown environment based on uncertain states. Generally, in this step, the RL elements' performance is respectively served below:

The agent: (1) receives the environmental observation; (2) chooses the equivalent action; (3) takes a reward for the environment.

The environment: (1) indicates the observation; (2) examinates the relevant action; (3) releases a reward in line with the agent's action.

As mentioned, such an RL method is aimed at merely learning the agent by means of the unknown environmental observations that it needs to delineate in detail. This concept is inspired by the OMDP idea in which the environment is invisibly defined. Having said this, the elements made up of an OMDP problem are marked out based on the concealed states (s), environmental observations (o), the transition probability of the agent from a state to another one (T), the reward (r), and the agent's action (a). In such a problem, firstly, the agent deploys into the unknown environment and receives the relevant observations, and then it selects a related but not optimal action. By doing so, the environment obtains a reward related to the agent's action and the current state at t.

Now, after defining an OPMP problem, there is a need to express how such a problem can be mapped for the attack detection method. Hence, Figure 3 briefly illustrates the OMDP-based detection problem framework that is proposed in this paper. In this way, assume that a hacker could obtain access to data and launch an anomaly in reliance on the unknown method in the system at $t = k$. Due to the nameless attack strategy, let us suggest the environmental conditions in varied states based on 'prior-attack', 'later-attack', and 'terminal'. In each state, the agent can choose one of the 'continue' and 'stop' actions with an emphasis on the environmental observation. If the agent has the accurate election considering the current state, it receives reward = 0 from the environment and moves toward the 'terminal' state, which means the best choice. If so, the agent takes reward = 1 or reward = b concerning the current state due to the evil detection that needs amendment. Keeping these in mind, the goal followed by the agent in the approach is functioned by Equation (20):

$$minR^{pen} = E^{\kappa}\left[(re_t | t_s < k_t) + \sum_{t=\kappa}^{\infty} b | t_s > k_t \right] \tag{20}$$

where t_s and R^{pen} indicate the stopping time and the relevant reward emitted by the environment. It is important to say that the reward is fixed based on the detection time once an attack is launched on the system at $t = k$. For future realization, the learning process is explained in detail. Assume the agent's state is the 'prior-attack' state in the first place. Meanwhile, if the agent selected the 'continue' action, it would take *reward* = b owing to the ineptness in describing the anomaly at $t_s > k$. Conversely, *reward* = *1* will be emitted for the agent if it badly adopts the 'stop' action when it is exposed to the 'later-attack' state at $t_s > k$. Therefore, in the learning phase, for each action-observation pair, it is suggested to learn a $P(o, a)$ value by using the RL algorithm. All in all, the learning phase trend is briefly stepped by:

(1) Choose the arbitrary $P(o^t, a^t)$;
(2) Check the stopping time and alarm based on the above explanations;
(3) Collect the measurement X_t;
(4) Obtain the observation signal (o^{t+1}) and choose the action a^{t+1} based on the ε-greedy policy for which it is determined by the minimal $P(o^{t+1}, a^{t+1})$;
(5) Update the $P(o^t, a^t)$ value based on $P(o^t, a^t) = P(o^t, a^t) + \alpha (r + P(o^{t+1}, a^{t+1}) - P(o^t, a^t)$.

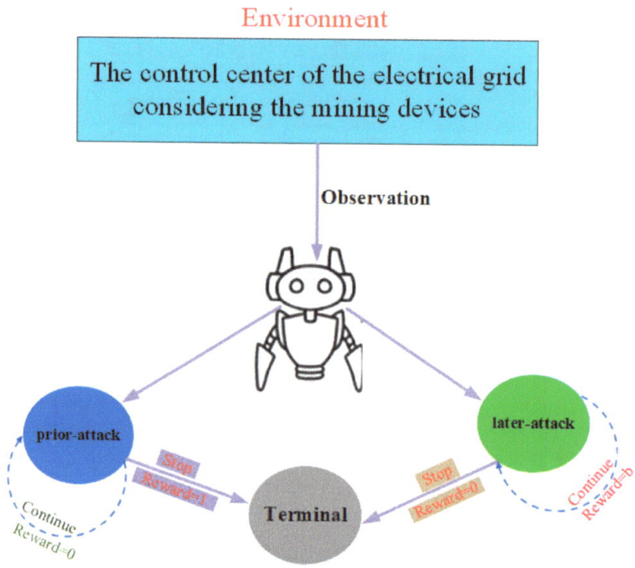

Figure 3. The anomaly detection method.

The previous phase, the agent needs the ability to detect every possible anomaly if it occurs in the system. In the detection phase, assume the agent observes signal o by relying on the 'prior-attack' state and chooses the 'continue' action. In the next step, the agent repeats steps 3–5 of the learning phase and opts for the best corresponding action (a). Here, there is a key point that needs an explanation. In step 5 of the learning phase, the updating process related to the p-value is dependent on the coefficient α, which is effective for being optimal in the learning phase. Hence, this paper suggests a coefficient Intelligent Priority Selection (IPS)-based optimization algorithm to obtain the appropriate coefficient α.

3.2. Intelligent Priority Selection Based Optimization Algorithm

The learning methods, especially the RL method, are unable to find the optimal solution. By keeping this in mind, this part tries to cover a concern of the learning phase in the RL-based attack detection algorithm with the help of an effective optimization method. Indeed, one main goal of this paper is to find a hybrid optimum detection method. In the literature, varied approaches have been proposed to obtain an efficient strategy for optimization issues through mathematical solving methods and artificial intelligence. However, the fairly low authenticity and the high time-passing and time-consumption are significant concerns involved in these methods. Hence, the proposed algorithm aimed to cover them. In the statistics, computing out of n from set N is shown below:

$$\binom{N}{n} = \frac{N!}{(n!).(N-n)!} \tag{21}$$

This results from above show that the problem space for finding an optimal solution may be so much that it leads to a high time passage. This method guarantees the solution to be more reliable. To overcome this problem, the proposed IPS approach has the ability to mitigate the time-consumption as much as possible. This method was inspired by the smart random permutation concept to find out the best solution based on the candidate's answers [39]. Hence, the proposed method follows the following steps:

Step1: Let us assume P is defined as the possible candidate for solving a problem. Additionally, the initial solutions are randomly chosen and stored in matrix K. The rest of the solutions (P-K) are saved in matrix W. The probable solution sets result from substituting the members of matrix W into matrix K, named KT as indicated in (25). Next, it is necessary to calculate the best solution of the objective function arising from the outcomes of the matrix KT indicated as $F^{best}_{W_1,K'_n}$ in (26). Equation (27) shows the sorted matrix related to the objective function calculation. Of course, the optimal solution is achieved when opting for the best answer among candidates [39].

$$P = [p_1, \ldots, p_N] \tag{22}$$

$$K = [k_1, \ldots, k_n] \tag{23}$$

$$W = [w_1, \ldots, w_m] \tag{24}$$

$$KT = \left[\begin{array}{c} \left.\begin{array}{cccc} \begin{array}{c} k/1 = k'_1 \\ \uparrow \\ w_1 \end{array} & k_2 & \cdots & k_n \\ k_1 & \begin{array}{c} k/2 = k'_2 \\ \uparrow \\ w_1 \end{array} & \cdots & k_n \\ k_1 & k_2 & \cdots & \begin{array}{c} k/n = k'_n \\ \uparrow \\ w_1 \end{array} \end{array}\right\} H_{w_1} \cdots \left.\begin{array}{cccc} \begin{array}{c} k/1 = k'_1 \\ \uparrow \\ w_m \end{array} & k_2 & \cdots & k_n \\ & & \vdots & \\ k_1 & k_2 & \cdots & \begin{array}{c} k/n = k'_n \\ \uparrow \\ w_m \end{array} \end{array}\right\} H_{w_m} \\ \begin{array}{cc} \downarrow & \downarrow \\ F(H_{w_1}) = F^{best}_{w_{1,k''_1}} & F(H_{w_m}) = F^{best}_{w_{m,k''_m}} \end{array} \end{array} \right], \tag{25}$$

$$H_{w_i} = [H_{w_1}, H_{w_2}, \ldots, H_{w_m}] \qquad \forall i \in \Omega^i$$
$$k''_M = [k'_1, k'_2, \ldots, k'_n] \qquad \forall M \in \Omega^M$$

On the basis of the objective function matrix (27), the relevant elements of the matrix W are arranged and saved in the matrix W'_j as illustrated in (28). This process is reiterated for set K''_j (29). Finally, the first level of the matrix F^{best_sort} is considered the best answer (see (30)).

$$F^{best}_m = \begin{bmatrix} F^{best}_{w_{1,k''_1}} \\ \vdots \\ F^{best}_{w_{m,k''_m}} \end{bmatrix} \qquad \forall m \in \Omega^m \tag{26}$$

$$F^{best\_sort} = \begin{bmatrix} F^{best}_{w'_1 \to k''_1} \\ \vdots \\ F^{best}_{w'_m \to k''_m} \end{bmatrix} \tag{27}$$

$$w'_j = [w'_1, \ldots, w'_m] \forall m \in \Omega^m \tag{28}$$

$$k''_j = [k''_1, \ldots, k''_m] \forall m \in \Omega^m \tag{29}$$

$$F = F^{best}_{w'_1 \to k''_1} \tag{30}$$

Step2: In this step, the matrix KT is updated by the new matrix KT^{new}_r. In this way, updating the W_j matrix is conducted by (31) based on w'_{j+1} in the next nitration. Then, the new matrix $K1^{new}_j$ is computed by eliminating the w'_j from the matrix K_j (32). Merging matrixes KT^{new}_r and w'_j is in the next stage based on Equation (33), in which ψ_r shows the coming together of the matrices and r is 1 to $m - j$. Value m is the length of matrix W. For each element of the matrix ψ_r, the relevant objective function is calculated, and the best answer ($F1^{Best}$) is obtained. As a result, the corresponding component of the matrix ψ_r related to $F1^{Best}$ is inserted in matrix K, as illustrated in (34)–(36).

$$W_j = w'_{j+1} \qquad \forall j \in \Omega^j \tag{31}$$

$$K1_j^{new} = \left\{ x \middle| x \in K_j, x \neq k''_j, x \neq w'_j \right\} \qquad \forall j \in \Omega^j \tag{32}$$

$$\psi_r = KT_r^{new} \cup w'_j \qquad \forall j \in \Omega^j, \forall r \in \Omega^r = [1, 2, \ldots, m-j] \tag{33}$$

$$F1_r = f(\psi_r) \tag{34}$$

$$F_j = F1^{Best} \qquad \forall j \in \Omega^j \tag{35}$$

$$K_j = \psi^{Best} \qquad \forall j \in \Omega^j \tag{36}$$

Step3: In this step, the most optimal solution is obtained through the last answer based on the objective function.

$$F^{best\_total} = F^{Best} \tag{37}$$

The step-by-step structure of the proposed IPS algorithm is shown in Figure 4 [40].

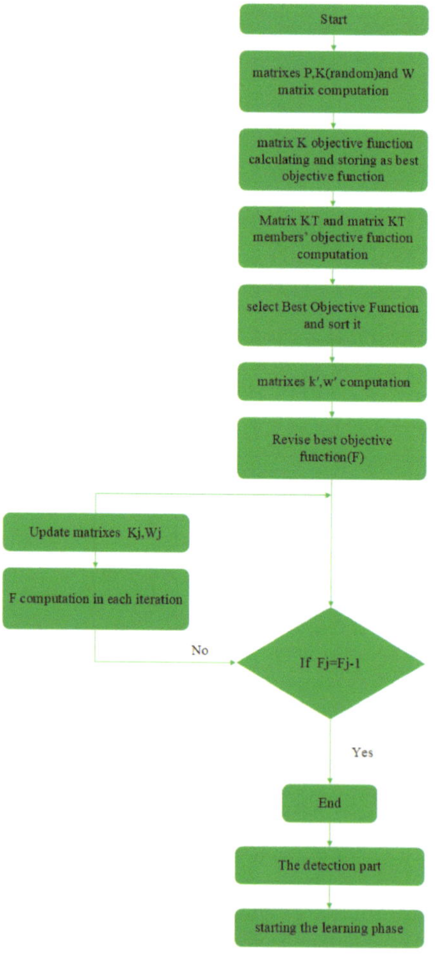

Figure 4. The framework of the IPS algorithm.

4. Stochastic Quantization Approach

Taking a close look at uncertainty effects on energy management can help to make the energy scheduling more faithful, more controllable, and so forth in the power system. To this end, this section takes inspiration from the UT approach-based uncertainty concept for modeling the relevant parameters such as the load demands and the consumed power of the mining devices in BFTSs. To recapitulate briefly, the UT method was designed as such that the points randomly generated are turned from the probability density function into the discreet distribution function [41]. Assume that the uncertain outcome of vector Q is obtained by the nonlinear function $T = \hat{f}(Q)$, in which Q comprises the number of parameters p with mean value z and covariance matrix A. The solving steps of the UT method for points $2p + 1$ are mainly explained as follows:

Step 1: In the first place, all points are calculated based on (38)–(40).

$$Q^0 = z \tag{38}$$

$$Q^k = z + \left(\sqrt{\frac{p}{1 - W^0} A_{aa}}\right)_k \quad k = 1, 2, \ldots, p \tag{39}$$

$$Q^{k+c} = z - \left(\sqrt{\frac{p}{1 - W^0} A_{aa}}\right)_k \quad k = 1, 2, \ldots, p \tag{40}$$

where A_{aa} is defined as the covariance matrix and $\overline{R} = z$.

Step 2: Computing the point Weight is achieved by (41):

$$W^k = \frac{1 - W^0}{2p} \quad k = 1, 2, \ldots, 2p \tag{41}$$

It is true that the sum of the weights should be equal to 1.

Step 3: By inserting these points into the nonlinear function $T^k = \hat{f}\left(Q^k\right)$, the stochastic outcome is computed by substituting these points into the function $\hat{f}\left(Q^k\right)$.

$$\overline{T} = \sum_{k=0}^{2p} W^k T^k \tag{42}$$

$$P_{TT} = \sum_{k=1}^{2p} W^k \left(T^k - \overline{T}\right)\left(T^k - \overline{T}\right)^R \tag{43}$$

5. Simulation Results

Exposing a discussion on widely developing BFTSs in modern social businesses and their effect on energy management in power systems are the underlying concerns in this paper. Indeed, the needed energy related to the mining devices to seek a puzzle in blockchain-based security platforms creates outstanding challenges in the spheres of generation scheduling and cyber-anomalies for the power systems. To be more precise, this weak spot can be an Achilles' heel for the cyber-hackers who put an end to energy management by faking out the mining devices in the face of energy consumption on the rise. Hence, this section is dedicated to indicating whether the proposed detection approach would be useful to cover this concern or not. To this end, we tried to carry out an IEEE 24-bus test system-based electrical grid [42,43] aimed at satisfying the public loads and needed power in the BFTSs. Next, we considered that an attacker spies out into the system to create the bogus mining device called miner-misuse for which the load demands would be spuriously increased. In this way, the proposed cyber-anomaly detection method's effectiveness in divulging the miner misuses was evaluated. Finally, regarding the high-risk energy consumption related to mining devices, we examined the uncertain effects on the energy management in the last part. To briefly sum up, the relevant consequences are defined as follows:

Case I: Evaluating the energy management based on BFTSs;

Case II: Simulating the IPS-RL-based detection approach considering miner-misuses;

Case III: Modeling the uncertainty of mining devices and their effect on energy consumption.

6. Evaluating the Energy Management Based on BFTSs

As said already, BFTSs have the ability to safely trade off fiscal businesses while taking the great consumed energy insofar as is irrefutable in the generation schedule. For more realization, in the case study, we assumed the number of mining devices embedded in the BFTS was 24 devices, which are located in the varied areas of the grid. These devices provide the needed activity in response to users connected to the BFTS. Indeed, the mining devices assure the users the guarantee of true financial transactions among themselves by participating in a hard competitive scheme. Hence, the mining device owners may use varied computing tools to find the targeted puzzle that is victorious in this scheme. Thereby, every mining device has different energy consumption across a wide range. In light of this evidence, we carried out the studied case and illustrated the relevant simulation results in Figure 5 and Table 2. Figure 5 shows the consumed power for 24 mining devices, all of which were satisfied by the defined electrical grid in a 24 h horizon. As can be seen, the mining devices take the varied needed powers in the daytime due to the different activities in the way of finding the targeted puzzle's solution. Looking over the figure, it is inferred that device No.12 overtakes 338.9 kW in hour 22, which is the most consumption compared to the other devices. It is worth mentioning that the peak-energy consumption related to these devices is directly dependent on the hardship level of the targeted puzzle and may be at any time. Generally speaking, such fluctuations in the consumption profile exude outstanding challenges facing energy management. Hence, the results in Table 2 prove these elucidations. The total operation cost has an ascending change of almost 56.44% when considering the mining devices in energy management. This means that the effects of the presence of the BFTSs on the power system's performance are irrefutable.

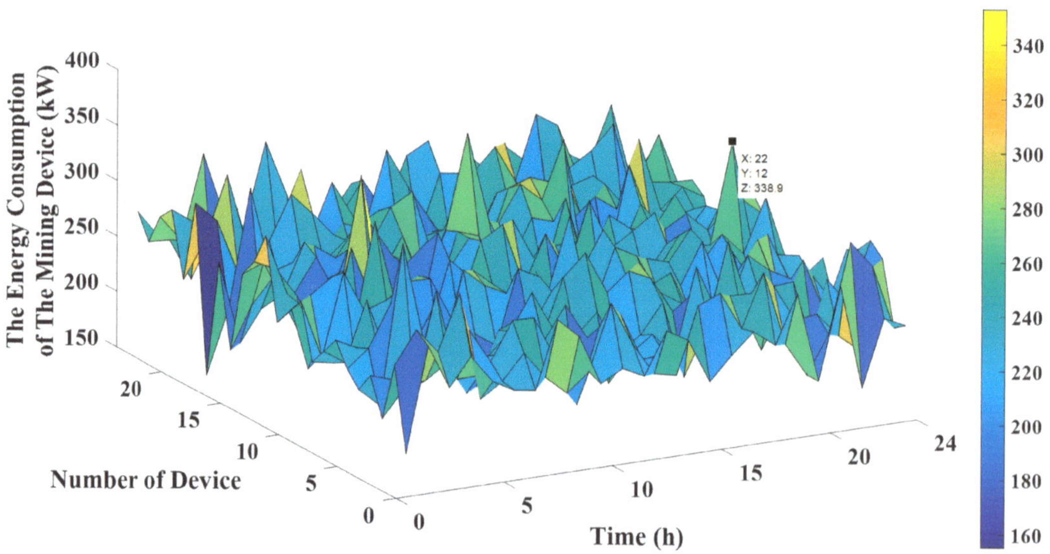

Figure 5. The energy consumption of the mining devices.

Table 2. Comparative results for the total cost.

Number of Case	Total Operation Cost (¢)
Considering the mining devices	$1.0793772038 \times 10^{10}$
Ignoring the mining devices	4.70×10^{9}

7. Simulating the IPS-RL Based Detection Approach Considering Miner-Misuses

As said already, in such a miner-based electrical grid, due to the high energy consumption of the miners, these devices can be known as a provocative factor for cyber-hackers who are able to bring to naught the energy scheduling process in power systems. Indeed, in this type of attack, intruders may carry out many bogus mining devices through deceitful data infusion into the control center, for which the energy consumption related to the miners takes an illusory tendency toward the ascending way. This assault is named the miner misuse in power systems based on BFTSs, and it needs to be dealt with. Hence, this section is dedicated to presenting how the proposed detection method would be worthwhile for tracking down the miner misuses in power systems based on BFTSs. To this end, we equipped the electrical grid with a security platform based on the proposed IPS-RL detection method. Meanwhile, we assume some hidden intrusions aimed at launching the miner misuse into the system have occurred. The absorbing consequences related to the proposed detection method's effectiveness in response to this attack are illustrated in Figures 6 and 7 and Table 3. In this way, Figure 6 demonstrates the results of monitoring the real energy consumption of the miner, which is compared to that manner being under attack. As can be seen, a cyber-hacker tried to launch the miner-attack at $t = 80$ s and could spuriously raise the energy consumption to a remarkable range of almost 25%. This means that the system stayed in the 'later-attack' state. In such a case, the IPS-RL platform was performed based on the defined phase 2 (see Section 3), and the agent unerringly chose the 'stop' action for which it could discover this anomaly at $t = 88$ s. The time-passing is calculated by the time delay between the start time and the end or the stopping time, which is 8 s. This shows that the proposed detection method not only is discrete and alarms anomalies, but it has the minimum stopping time. Besides that, we checked and evaluated this approach versus another attack at $t = 123$ s, as shown in Figure 6. In this situation, it could also detect the hidden intrusion for a lesser time delay of 5 s better than the other one. This advancement is obtained because the agent is learned by the prior anomaly and could provide superior performance. For more realization, Figure 7 indicates the energy consumption deviations for all attacks launched as well.

Table 3. Comparison among different trials.

%Number of Trials	Trials (1000)	Trials (1500)
False/Positive	%0	%0.09
True/Positive	%93.69	%91
False/Negative	%3.78	%4.84
True/Negative	%2.53	%4.07

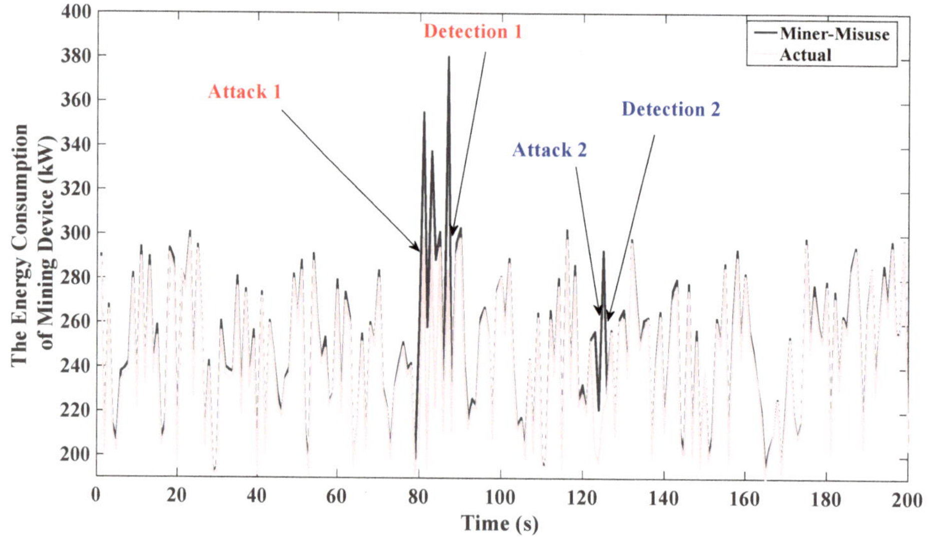

Figure 6. Illustration of the energy consumption under miner misuses.

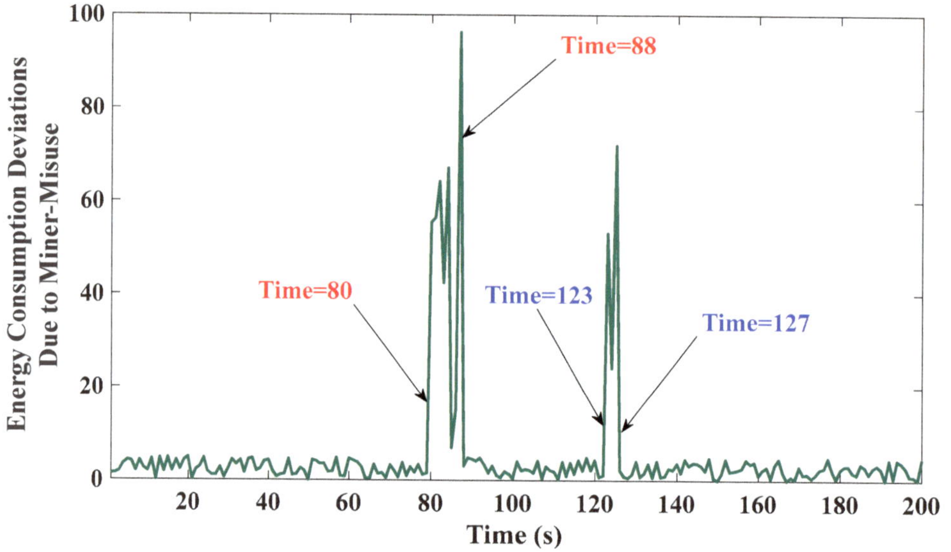

Figure 7. Illustration of the energy consumption deviations under miner misuses.

Speaking of the proposed detection method's effectiveness, it seems that the obtained results are not enough, and more tests and evaluations of this proposed method under many trials are needed for better decisions in assessing the security platform. To this end, we separately tested our approach for 1000 trials and 1500 trials here. The relevant consequences are provided in Table 3. In order to better perceive the performance, we tried to illustrate the platform's behavior for the true/false decisions in varied forms of false/positive, true/positive, false/negative, and true/negative. It can be seen that our method could accurately make the true/positive arbitration for 93.69% of trials when launching many malicious attacks on the system. On the other hand, this method unfortunately detected and erroneously stopped the system for 3.78% of trials when the system was clear, while it was 2.53% to take the true decision in the same condition. The second column of Table 2 indicates the same results for 1500 trials. Here, it is seen that the agent is successful and useful in the true decision for 91% of 1500 trials, which proves our detection method's effectiveness, as mentioned in the previous section. Another key factor that can make a profound impact on it being optimal and efficient in the RL-based detection method is setting the main parameters. Figure 8 shows the simulation result of the IPS-RL-based detection method considering the miner attack for the varied coefficients of α. Comparing the detection times obtained by the algorithm for α = 0.1, 0.5, 1, and 10, it proves that an optimal value α can remarkably be effective for mitigating the stopping time in the anomaly detection method's performance. As can be seen, the proposed algorithm based on α = 0.1 had a minimum time delay of 8 s, while it was obtained at 13 s for α = 10. This shows that the IPS optimization algorithm can optimally handle the learning procedure in the RL-based anomaly detection approach. It is worth saying that various optimization algorithms were suggested to catch the best solution in the literature. On the whole, we mainly found two underlying approaches that capture the optimal answer, including (1) mathematical methods and (2) meta-heuristic methods. On account of involving meta-heuristic methods in the local optimal solution for some studied cases, the first approach is referred to with an emphasis on accuracy. Meanwhile, we tried here to provide a mathematical optimization method named the IPS algorithm. To realize the effectiveness of the proposed algorithm, we plotted Figures 9 and 10 as the compared results of the convergence process for the various optimization methods. Our method experienced a shorter convergence process in comparison to other methods. Besides that, in order to assess the IPS optimization algorithm in the learning phase, we compared our detection method based on the IPS, PSO, and Bat optimization algorithms and indicated the F-score coefficient and the converging process for them. Figures 9–12 show the simulation results. It is important to say that all the above algorithms start with the same initial population to have an equitable comparison. According to this figure, the IPS algorithm not only first converged compared to the PSO and Bat algorithms, but it had a better F-score coefficient (indicating the ability to escape from the local optima). To validate the proposed method, this paper compared it to other detection approaches, i.e., the support vector machine (VSM) and reinforcement learning (RL). To this end, it is necessary that the F-score is calculated for trial numbers through the assessment of the precision versus and recall curves. Looking over Figure 12, it is inferred the IPS-RL method can be more sensitive to detecting the attack than the other models.

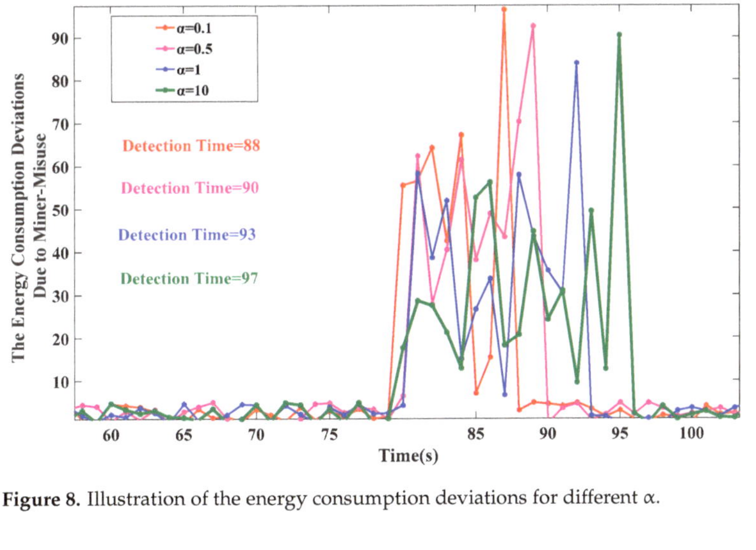

Figure 8. Illustration of the energy consumption deviations for different α.

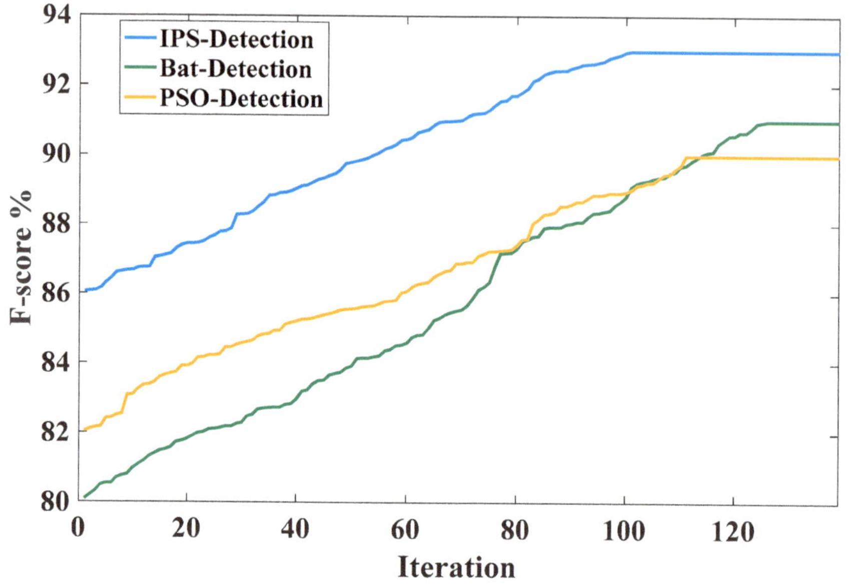

Figure 9. Evaluating the different optimization algorithms for F-score.

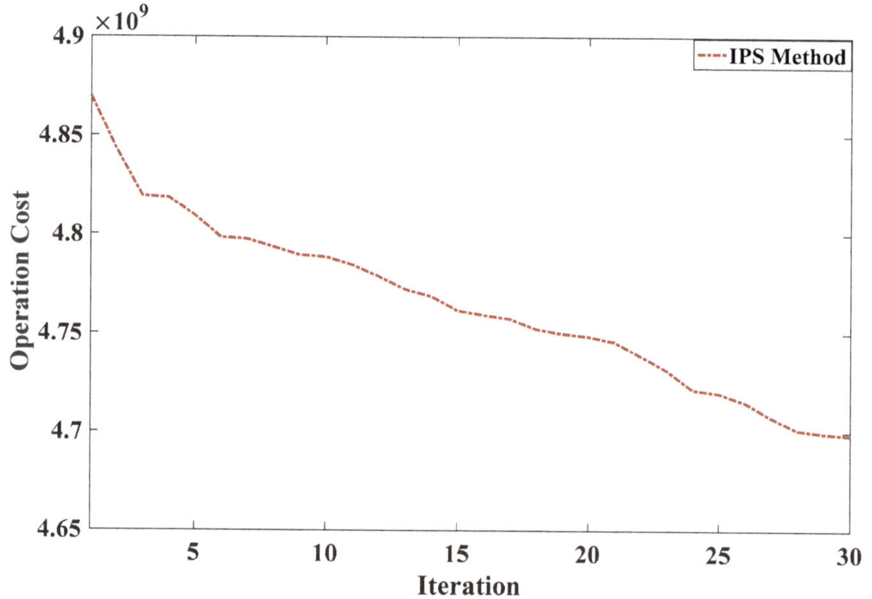

Figure 10. Convergence process of the IPS algorithm.

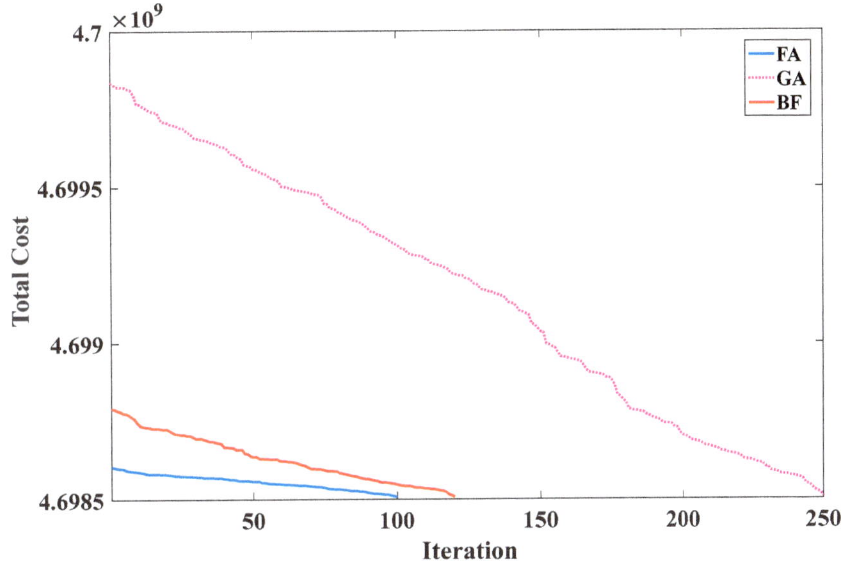

Figure 11. Comparison of algorithms for the convergence algorithm.

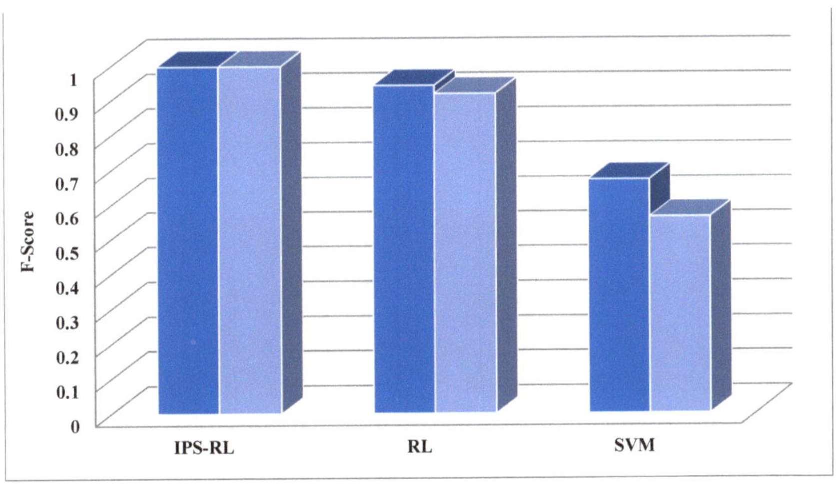

Figure 12. Illustration of the compared results for different detection approaches.

8. Modeling the Uncertainty of Mining Devices and Their Effect on the Energy Consumption

This section is dedicated to taking a close look at the effects of the uncertain energy consumption related to the mining devices on energy management within the electrical grid based on BFTSs. As stated already, the mining devices' behavior in the solving process of a targeted puzzle based on the blockchain tech causes them to make their energy consumption different and unstable. Hence, having considered the high energy consumption of these devices, modeling their uncertainty can remarkably increase the chance of having an optimal energy scheduling framework in the electrical grid. To clarify the truth of the matter, we modeled the uncertain parameters inspired by the UT concept and provided the simulation results in Figures 13 and 14. As can be seen in Figure 13, the average energy consumption of all mining devices takes a remarkable fluctuation of almost 33% for all times when modeling the uncertain parameters fixed into the grid structure. Additionally these changes can affect the total operation cost, which is on the rise as indicated in Figure 14. To put it in a nutshell, the energy management in the presence of the mining devices in the system needs to model the uncertainty for providing an optimal and accurate evaluation.

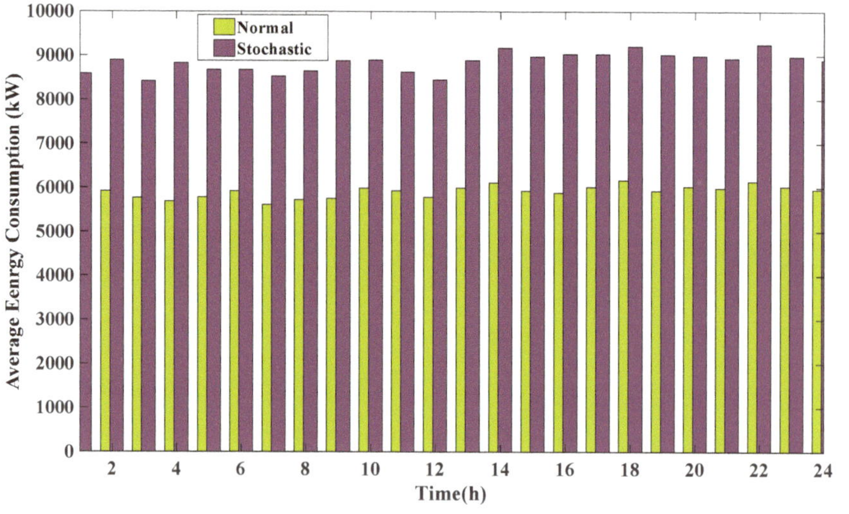

Figure 13. The average energy consumption of mining devices under uncertainty/normal conditions.

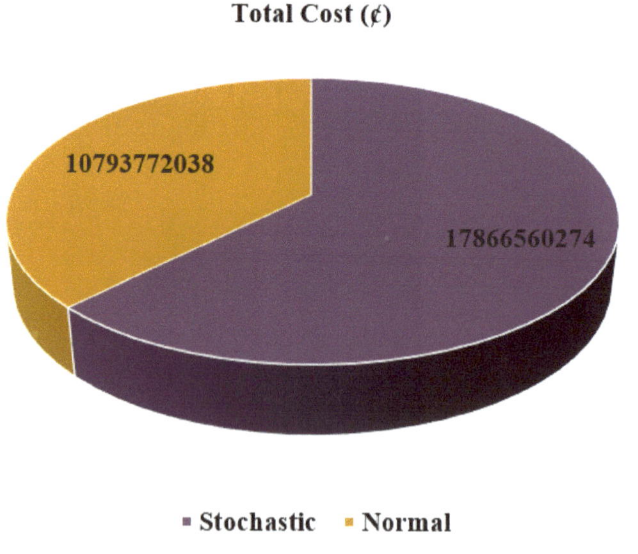

Figure 14. The total cost under uncertainty/normal conditions.

9. Conclusions

This article took an approach at an IPS-RL-based anomaly detection model to pinpoint the hidden malicious intrusions aimed to phony up the mining devices' behavior, which were named miner misuses within the power systems based on BFTSs. Indeed, in this type of anomaly, a cyber-hacker would try to make the bogus mining devices to create a spurious upward trend in the energy consumption, which is gobbled up by these devices to solve a targeted puzzle within BFTSs. To prove the truth of the matter, we carried out an IEEE 24-bus test system-based electrical grid in furtherance of bearing out the public loads and consumed power in the BFTSs. After that, we assumed an intruder could make many bogus miners through deceitful data infusion into the decision-making center of the electrical grid for which the energy consumption was spuriously increased. Hence, consequences emanating from the proposed detection model in response to this anomaly were provided in the result section. They indicated that the proposed method could precisely pick out the true/positive detection for 93.69% of 1000 trials when launching malicious anomalies on the grid, while the agent learned by the proposed method made false decisions only for 3.78% of trials. This result was 91% and 4.84% for 1500 trials in the same condition. Besides that, looking over the results, the IPS optimization algorithm could obtain an F-score coefficient of 92%, which indicates that it is more useful to help in the optimal learning process based on the RL method compared to the other algorithms. Finally, the results show that energy management in the grid based on BFTSs would be more accurate and flexible if precisely modeling the uncertainty of energy consumption in the mining devices.

Author Contributions: Conceptualization, A.A., S.A. and M.A.M.; methodology, A.A., S.A. and M.A.M.; software, A.A., S.A. and M.A.M.; validation, A.A., S.A. and M.A.M.; formal analysis, A.A., S.A. and M.A.M.; investigation, A.A., S.A. and M.A.M.; data curation, A.A., S.A. and M.A.M.; writing—original draft preparation, A.A., S.A. and M.A.M.; writing—review and editing, A.A., S.A. and M.A.M.; visualization, A.A., S.A. and M.A.M.; supervision, A.A., S.A. and M.A.M.; project administration, A.A. and S.A.; funding acquisition, A.A. and S.A. All authors have read and agreed to the published version of the manuscript.

Funding: This research has been funded by Deputy for Research & Innovation, Ministry of Education through Initiative of Institutional Funding at University of Ha'il, Saudi Arabia through project number IFP-22 099.

Institutional Review Board Statement: Not applicable.

Informed Consent Statement: Not applicable.

Data Availability Statement: Not applicable.

Conflicts of Interest: The authors declare no conflict of interest.

References

1. Alnowibet, K.; Annuk, A.; Dampage, U.; Mohamed, M.A. Effective energy management via false data detection scheme for the interconnected smart energy hub–microgrid system under stochastic framework. *Sustainability* **2021**, *13*, 11836. [CrossRef]
2. Almalaq, A.; Albadran, S.; Alghadhban, A.; Jin, T.; Mohamed, M.A. An effective hybrid-energy framework for grid vulnerability alleviation under cyber-stealthy intrusions. *Mathematics* **2022**, *10*, 2510.
3. Zhang, J.X.; Yang, G.H. Low-complexity tracking control of strict-feedback systems with unknown control directions. *IEEE Trans. Autom. Control* **2019**, *64*, 5175–5182.
4. Zhang, X.; Lewen, D. Image enhancement based on rough set and fractional order differentiator. *Fractal Fract.* **2022**, *6*, 214.
5. Zhang, J.X.; Yang, G.H. Fault-tolerant output-constrained control of unknown Euler–Lagrange systems with prescribed tracking accuracy. *Automatica* **2020**, *111*, 108606. [CrossRef]
6. Almalaq, A.; Albadran, S.; Mohamed, M.A. Deep machine learning model-based cyber-attacks detection in smart power systems. *Mathematics* **2022**, *10*, 2574.
7. Alsokhiry, F.; Annuk, A.; Kabanen, T.; Mohamed, M.A. A Malware Attack Enabled an Online Energy Strategy for Dynamic Wireless EVs within Transportation Systems. *Mathematics* **2022**, *10*, 4691.
8. Soliman, M.S.; Belkhier, Y.; Ullah, N.; Achour, A.; Alharbi, Y.M.; Al Alahmadi, A.A.; Abeida, H.; Khraisat, Y.S.H. Supervisory energy management of a hybrid battery/PV/tidal/wind sources integrated in DC-microgrid energy storage system. *Energy Rep.* **2021**, *7*, 7728–7740. [CrossRef]
9. Rahman, M.A.; Mohsenian-Rad, H. False data injection attacks with incomplete information against smart power grids. In Proceedings of the 2012 IEEE Global Communications Conference (GLOBECOM), Anaheim, CA, USA, 3–7 December 2012; pp. 3153–3158.
10. Liang, G.; Zhao, J.; Luo, F.; Weller, S.R.; Dong, Z.Y. A review of false data injection attacks against modern power systems. *IEEE Trans. Smart Grid* **2017**, *8*, 1630–1638.
11. Wang, W.; Lu, Z. Cyber security in the smart grid: Survey and challenges. *Comput. Netw.* **2013**, *57*, 1344–1371.
12. Yan, Y.; Qian, Y.; Sharif, H.; Tipper, D. A survey on cyber-security for smart grid communications. *IEEE Commun. Surv. Tutor.* **2012**, *14*, 998–1010.

13. Kurt, M.N.; Yilmaz, Y.; Wang, X. Distributed quickest detection of cyber-attacks in smart grid. *IEEE Trans. Inf. Forensics Secur.* **2018**, *13*, 2015–2030.
14. Asri, S.; Pranggono, B. Impact of distributed denial-of-service attack on advanced metering infrastructure. *Wirel. Pers. Commun.* **2015**, *83*, 2211–2223.
15. Zhang, Y.; Wang, L.; Sun, W.; Green, R.C.; Alam, M. Distributed intrusion detection system in a multi-layer network architecture of smart grids. *IEEE Trans. Smart Grid* **2011**, *2*, 796–808.
16. Kurt, M.N.; Yilmaz, Y.; Wang, X. Real-time detection of hybrid and stealthy cyber-attacks in smart grid. *IEEE Trans. Inf. Forensics Secur.* **2019**, *14*, 498–513.
17. Saad, M.; Spaulding, J.; Njilla, L.; Kamhoua, C.; Shetty, S.; Nyang, D.; Mohaisen, D. Exploring the attack surface of blockchain: A comprehensive survey. *IEEE Commun. Surv. Tutor.* **2020**, *22*, 1977–2008.
18. Siegel, D. Understanding the DAO Attack. Available online: https://www.coindesk.com/understandingdao-hack-journalists/ (accessed on 13 September 2019).
19. Ross, S.; Pineau, J.; Chaib-draa, B.; Kreitmann, P. A Bayesian approach for learning and planning in partially observable Markov decision processes. *J. Mach. Learn. Res.* **2011**, *12*, 1729–1770.
20. Doshi-Velez, F.; Wingate, D.; Roy, N.; Tenenbaum, J. Nonparametric Bayesian policy priors for reinforcement learning. *Adv. Neural Inf. Process. Syst.* **2010**, *1*, 532–540.
21. Jaakkola, T.; Singh, S.P.; Jordan, M.I. Reinforcement learning algorithm for partially observable Markov decision problems. *Adv. Neural Inf. Process. Syst.* **1994**, *7*, 345–352.
22. Perkins, T.J. Reinforcement learning for POMDPs based on action values and stochastic optimization. In Proceedings of the Eighteenth National Conference on Artificial Intelligence and Fourteenth Conference on Innovative Applications of Artificial Intelligence, Edmonton, AB, Canada, 28 July 28–1 August 2002; pp. 199–204.
23. Loch, J.; Singh, S.P. Using eligibility traces to find the best memoryless policy in partially observable Markov decision processes. In Proceedings of the 15th International Conference on Machine Learning, (ICML), Madison, WI, USA, 24–27 July 1998; pp. 323–331.
24. Lanzi, P.L. Adaptive agents with reinforcement learning and internal memory. In Proceedings of the Sixth International Conference on the Simulation of Adaptive Behavior (SAB2000), Paris, France, 17–18 October 2005; MIT Press: Cambridge, MA, USA, 2000; pp. 333–342.
25. Peshkin, L.; Meuleau, N.; Kaelbling, L.P. Learning policies with external memory. In Proceedings of the 16th International Conference on Machine Learning, San Francisco, CA, USA, 27–30 June 1999; pp. 307–314.
26. Yli-Huumo, J.; Ko, D.; Choi, S.; Park, S.; Smolander, K. Where is current research on blockchain technology?—A systematic review. *PLoS ONE* **2016**, *11*, e0163477.
27. Yuan, Y.; Wang, F.-Y. Parallel blockchain: Concept, methods and issues. *Acta Autom. Sin.* **2017**, *43*, 1703–1712.
28. Eyal, I.; Sirer, E.G. Majority is not enough: Bitcoin mining is vulnerable. In Proceedings of the International Conference on Financial Cryptography and Data Security, Christ Church, Barbados, 3–7 March 2014; pp. 436–454.
29. Kiayias, A.; Russell, A.; David, B.; Oliynykov, R. Ouroboros: A Provably Secure Proof-of-Stake Blockchain Protocol. Cryptology ePrint Archive. *Tech. Rep. 2016/889*. 2016. Available online: https://eprint.iacr.org/2016/889 (accessed on 6 December 2022).
30. Aien, M.; Hajebrahimi, A.; Fotuhi-Firuzabad, M. A comprehensive review on uncertainty modeling techniques in power system studies. *Renew. Sustain. Energy Rev.* **2016**, *57*, 1077–1089.
31. Sheikh, M.; Aghaei, J.; Chabok, H.; Roustaei, M.; Niknam, T.; Kavousi-Fard, A.; Shafie-Khah, M.; Catalão, J.P.S. Synergies between transportation systems, energy hub and the grid in smart cities. *IEEE Trans. Intell. Transp. Syst.* **2021**, *23*, 7371–7385. [CrossRef]
32. Chen, X.; Qu, G.; Tang, Y.; Low, S.; Li, N. Reinforcement learning for selective key applications in power systems: Recent advances and future challenges. *IEEE Trans. Smart Grid.* **2022**, *13*, 2935–2958. [CrossRef]
33. Cao, D.; Hu, W.; Zhao, J.; Zhang, G.; Zhang, B.; Liu, Z.; Chen, Z.; Blaabjerg, F. Reinforcement learning and its applications in modern power and energy systems: A review. *J. Mod. Power Syst. Clean Energy* **2020**, *8*, 1029–1042. [CrossRef]
34. Ruan, G.; Zhong, H.; Zhang, G.; He, Y.; Wang, X.; Pu, T. Review of learning-assisted power system optimization. *CSEE J. Power Energy Syst.* **2020**, *2*, 221–231.
35. Mohamed, M.A.; Awwad, E.M.; El-Sherbeeny, A.M.; Nasr, E.A.; Ali, Z.M. Optimal scheduling of reconfigurable grids considering dynamic line rating constraint. *IET Gener. Transm. Distrib.* **2020**, *14*, 1862–1871. [CrossRef]
36. Al Alahmadi, A.A.; Belkhier, Y.; Ullah, N.; Abeida, H.; Soliman, M.S.; Khraisat YS, H.; Alharbi, Y.M. Hybrid wind/PV/battery energy management-based intelligent non-integer control for smart DC-microgrid of smart university. *IEEE Access* **2021**, *9*, 98948–98961.
37. Makhadmeh, S.N.; Khader, A.T.; Al-Betar, M.A.; Naim, S.; Abasi, A.K.; Alyasseri, Z.A.A. Optimization methods for power scheduling problems in smart home: Survey. *Renew. Sustain. Energy Rev.* **2019**, *115*, 109362.
38. Roustaei, M.; Letafat, A.; Sheikh, M.; Sadoughi, R.; Ardeshiri, M. A cost-effective voltage security constrained congestion management approach for transmission system operation improvement. *Electr. Power Syst. Res.* **2022**, *203*, 107674. [CrossRef]
39. Mohamed, M.A.; Hajjiah, A.; Alnowibet, K.A.; Alrasheedi, A.F.; Awwad, E.M.; Muyeen, S.M. A secured advanced management architecture in peer-to-peer energy trading for multi-microgrid in the stochastic environment. *IEEE Access* **2021**, *9*, 92083–92100. [CrossRef]
40. Mohamed, M.A.; Mirjalili, S.; Dampage, U.; Salmen, S.H.; Obaid, S.A.; Annuk, A. A cost-efficient-based cooperative allocation of mining devices and renewable resources enhancing blockchain architecture. *Sustainability* **2021**, *13*, 10382. [CrossRef]

41. Zou, H.; Tao, J.; Elsayed, S.K.; Elattar, E.E.; Almalaq, A.; Mohamed, M.A. Stochastic multi-carrier energy management in the smart islands using reinforcement learning and unscented transform. *Int. J. Electr. Power Energy Syst.* **2021**, *130*, 106988.
42. Min, L.; Alnowibet, K.A.; Alrasheedi, A.F.; Moazzen, F.; Awwad, E.M.; Mohamed, M.A. A stochastic machine learning based approach for observability enhancement of automated smart grids. *Sustain. Cities Soc.* **2021**, *72*, 103071.
43. Chabok, H.; Aghaei, J.; Sheikh, M.; Roustaei, M.; Zare, M.; Niknam, T.; Lehtonen, M.; Shafi-khah, M.; Catalão, J.P.S. Transmission-constrained optimal allocation of price-maker wind-storage units in electricity markets. *Appl. Energy* **2022**, *310*, 118542.

mathematics

Article

A Novel Computational Procedure for the Waiting-Time Distribution (In the Queue) for Bulk-Service Finite-Buffer Queues with Poisson Input

Mohan Chaudhry [1,†], Abhijit Datta Banik [2,*,†], Sitaram Barik [2,†] and Veena Goswami [3,†]

1 Department of Mathematics and Computer Science, Royal Military College of Canada, STN Forces, P.O. Box 17000, Kingston, ON K7K 7B4, Canada
2 School of Basic Sciences, Indian Institute of Technology Bhubaneswar, Permanent Campus Argul, Jatni, Khurda 752050, India
3 School of Computer Applications, Kalinga Institute of Industrial Technology, Bhubaneswar 751024, India
* Correspondence: adattabanik@iitbbs.ac.in
† These authors contributed equally to this work.

Abstract: In this paper, we discuss the waiting-time distribution for a finite-space, single-server queueing system, in which customers arrive singly following a Poisson process and the server operates under (a, b)-bulk service rule. The queueing system has a finite-buffer capacity 'N' excluding the batch in service. The service-time distribution of batches follows a general distribution, which is independent of the arrival process. We first develop an alternative approach of obtaining the probability distribution for the queue length at a post-departure epoch of a batch and, subsequently, the probability distribution for the queue length at a random epoch using an embedded Markov chain, Markov renewal theory and the semi-Markov process. The waiting-time distribution (in the queue) of a random customer is derived using the functional relation between the probability generating function (pgf) for the queue-length distribution and the Laplace–Stieltjes transform (LST) of the queueing-time distribution for a random customer. Using LSTs, we discuss the derivation of the probability density function of a random customer's waiting time and its numerical implementations.

Keywords: Poisson input; batch service (a, b)-rule; finite-buffer queue; roots

MSC: 60K25; 68M20; 90B22

Citation: Chaudhry, M.; Datta Banik, A.; Barik, S.; Goswami, V. A Novel Computational Procedure for the Waiting-Time Distribution (In the Queue) for Bulk-Service Finite-Buffer Queues with Poisson Input. *Mathematics* **2023**, *11*, 1142. https://doi.org/10.3390/math11051142

Academic Editors: Gurami Tsitsiashvili, Alexander Bochkov, Jonathan Blackledge and Vladimir V. Rykov

Received: 8 January 2023
Revised: 19 February 2023
Accepted: 23 February 2023
Published: 24 February 2023

1. Introduction

Single-server queueing models have been extensively studied in the literature. However, many practical situations exist where customers need to be served in batches. Queueing models of this type are called general bulk-service (GBS) queues. Examples of bulk-service queues can be seen in the unloading and loading of cargoes at a sea port, in traffic-signal systems, and in a mass transportation system. The GBS rule is stated as follows. If n $(0 \leq n < a)$ customers are waiting after the completion of a batch service, then the server has to wait until the number reaches 'a' customers. However, the entire queue is served if the number of waiting customers in the queue is greater than or equal to a but less than or equal to b. The first 'b' customers will be taken into service if the number present in the queue is greater than b. This rule has been widely investigated in the literature, e.g., Chaudhry and Templeton [1] and Dshalalow [2,3]. An excellent survey of various types of infinite-buffer queues with bulk arrival, bulk service and vacations can be found in Dshalalow [4]. In general, the analysis of queueing models of this type is difficult, since their probability generating functions involve unknown probabilities in the numerators. Neuts [5] discussed the bulk-service queue using the theory of the semi-Markov process and obtained queue-length distribution. Borthakur [6] studied the bulk-service queue and derived the stationary probability distribution for the queue length directly as the sum of a constant multiplied by geometric terms, where the common ratios are the root of the associated characteristic equation. Following this, Medhi [7] obtained the waiting-time density function for $M/M^{(a,b)}/1$ queue in terms of roots. Later, Easton and Chaudhry [8] extended these results and discussed the waiting-time distribution for the $E_k/M^{(a,b)}/1$ queueing system. We can find an elaborate presentation of bulk-service queues in the books by Chaudhry and Templeton [1] and Medhi [9], where a complete list of references up to the early 1980s. One may note that it has been reported in the literature that the analysis of finite-buffer queueing models by solving simultaneous linear equations may pose some problems for some particular types of queueing models; see the remarks made by Powell ([10], pp. 64, 141). Chaudhry et al. [11] discussed the computational analysis of the $M/G^{(a,b)}/1$ and related non-bulk queues. Brière and Chaudhry [12] analyzed a batch-service queue with a random service batch size, i.e., $M/G^Y/1$ queue. They obtained the probability distribution for the queue length at different epochs by finding the roots of the associated characteristic equation. Waiting-time analysis in a finite-buffer $M/G/1$ queueing model was first investigated by Finch [13] in his pioneering work in this area of research. However, no analytical results have been reported in the literature on queues to analyze the waiting-time distribution of a finite-buffer $M/G/1$-type bulk-service queue. A few results related to stationary queue-length distributions in a finite-buffer single-server queue using the roots' method are addressed in the literature, e.g., see, Chaudhry et al. [14] and Chaudhry and Goswami [15]. Further, analysis of the finite-buffer bulk-service $M/G^Y/1$ queue was performed by Singh [16] and queue-length distributions were obtained. Later, Chaudhry and Gupta [17] performed the analytic analysis of the finite-buffer bulk-service $M/G^{(a,b)}/1/N$ queueing model and obtained the queue-length distributions at a post-departure and arbitrary epochs using the supplementary variable technique. However, no analytical results have been reported in the literature related to the analysis of the waiting-time distribution in a finite-buffer $M/G/1$-type bulk-service queue using roots.

We encounter finite-buffer bulk-service queueing systems in many real-world scenarios; see Gold and Tran-Gia [18] for applications related to manufacturing processes. In telecommunications networks, information units are stored in the system if a server is busy. If an arrival finds the buffer space full, then the arriving customer is blocked and considered to be lost. The main interest of the system designer is to provide sufficient buffer space so that the probability of blocking or loss is small. Thus, the blocking/loss probabilities are an important measure of concern for a system designer. In cloud computing, transmissions of user requests are of various types (e.g., data, voice, video, images). There are many applications which need bulk data to be transferred in the web which are accessed by a large number of clients at the same time. As per the service level agreement (SLA) all the requests received by the clients may be processed in bulk. The instances of the target web application running into virtual machines (VMs) act as service centers to process the requests in the queue. We figured out the process of client requests in a single web application with the following features: The inter-arrival times between two successive client requests may be identified by a Poisson process. When bulk requests are submitted in the queue, the requests are forwarded to the VM that is currently idle because of the lack of a sufficient number of requests. The requests are served in batches with a minimum of a and a maximum of b requests for this VM, where $1 \leq a \leq b \leq N$. The system under consideration contains multiple homogeneous VMs which render service in order of task request arrivals, i.e., first come, first served (FCFS). For a recent survey of the system model of bulk service on cloud, the readers are referred to Goswami et al. [19]. In recent years, considering batch-size-dependent service processes, the finite-buffer capacity $MAP/G_r^{(a,b)}/1/N$ and infinite-buffer $MAP/G_r^{(a,b)}/1$ queueing systems have been studied by Banerjee et al. [20], and Pradhan and Gupta [21], respectively.

This paper carries out the analytic analysis of the finite buffer $M/G^{(a,b)}/1/N$ queue using the method of the Markov renewal theory, the embedded Markov chain, the semi-Markov process and the roots. As stated earlier, the service to the queueing system is provided in batches of minimum size a and maximum size b $(1 \leq a \leq b)$. We obtain the steady-state, queue-length distribution at post-departure and random epochs. As stated above, using the functional relationship, we derive the probability density function for a random customer's waiting-time distribution (in the queue and in the system). Finally, to extend the work done by Chaudhry and Gupta [17] for the well-known model $M/G^{(a,b)}/1/N$, which deals with the number in the queue, this paper first gives a simple procedure to get the probability generation function (pgf) of a number in the queue by using the method of the Markov renewal theory, the embedded Markov chain and the semi-Markov process. Using the derived pgf, we deal with the probability density function for the waiting-time distribution (in queue as well as in the system) of a random customer.

2. Brief Description of the $M/G^{(a,b)}/1/N$ Queueing Model and Its Queue-Length Distributions

In this single-server queueing model, the arrival process of the customers is assumed to be a Poisson process with an average rate, λ. The queued customers are served in batches of minimum size, 'a', with a maximum threshold size, 'b'; however, if there are fewer than 'a' customers, the server must wait until the queue accumulates 'a' customers. The number 'N' denotes the buffer capacity of the queueing system, excluding the size of a batch with the server, and it is assumed that $1 \le a \le b \le N$. The service times of batches are independent identically distributed random variables (i.i.d.r.vs.) denoted by the notation S, where S has the probability density function (pdf) $b(t)$, cumulative distribution function (DF) $B(t)$, Laplace–Stieltjes transform (LST) $B^*(s)$ and mean service time $E(S) = b_1 = -B^{*(1)}(0)$, where $G^{*(1)}(0)$ is the derivative of $G^*(s)$ evaluated at $s = 0$. The mean service rate is $\mu = 1/b_1$ or $E(S) = 1/\mu$. The traffic intensity can be obtained as $\rho = \lambda b_1/b = \lambda/(\mu b)$. Let us define the conditional probability k_j $(j \ge 0)$, which denotes the probability that j customers arrive in the queueing system during a service time of a batch. Then,

$$k_j = \int_0^\infty \frac{e^{-\lambda v}(\lambda v)^j}{j!} dB(v).$$

The probability generating function (pgf) of $\{k_j\}_0^\infty$ is denoted by $K(z)$. It can be easily seen that $K(z) = B^*(\lambda - \lambda z)$; see Keilson and Servi [22] for a detailed derivation. Similarly, we denote the conditional probability $\widehat{k}_j$ $(j \ge 0)$, which represents the probability that j customers arrive in the queueing system during a stationary elapsed service time of a batch. This leads to:

$$\widehat{k}_j = \mu \int_0^\infty \frac{e^{-\lambda v}(\lambda v)^j}{j!}(1 - B(v))dv.$$

Similar to the above result, if we define the pgf of $\{\widehat{k}_j\}_0^\infty$ by $\widehat{K}(z)$, then it can be seen that $\widehat{K}(z) = \mu \frac{1-B^*(\lambda-\lambda z)}{\lambda - \lambda z}$. The probability distribution for the queue length at a random epoch can be obtained through relations among post-departure and random epoch probabilities using the embedded Markov chain, the Markov renewal theory and the semi-Markov processes. For this, consider the queue at the batch service completion epochs. Let the successive batch service completion epochs be marked in the time-axis as $t_0, t_1, t_2, \ldots$ The time epoch just after a service completion may be denoted by t_n^+. The state of the system at t_n^+ is defined as $\{N_q(t_n^+),\ n \ge 0\}$, where $N_q(t_n^+)$ is the number of customers $(0 \le N_q(t_n^+) \le N)$ in the queue after the n-th departure epoch of a batch. Thus, $\{N_q(t_n^+),\ n \ge 0\}$ becomes a Markov chain. This irreducible and aperiodic Markov chain becomes ergodic under any $\rho >=< 1$. In this discrete-time Markov chain, $N_q(t_{n+1}^+)$ is dependent on the random variable $N_q(t_n^+)$ and another random variable A_{n+1}, which is independent of $N_q(t_n^+)$. This gives:

$$N_q(t_{n+1}^+) = \min\left\{ \left(N_q(t_n^+) - b\right)^+ + A_{n+1},\ N \right\}, \tag{1}$$

where $\left(x\right)^{+} = \max\{0, x\}$ and A_{n+1} is the random number of arrivals during the $(n+1)$th batch service time. When $n \to \infty$, $P(A_{n+1} = j) = k_j\ (j \geq 0)$. The various elements of the transition probability matrix (TPM) of this Markov chain are given by:

$$P_{ij} = P\{N_q(t_{n+1}^+) = j | N_q(t_n^+) = i\} = \begin{cases} k_j, & 0 \leq i \leq b,\ 0 \leq j \leq N-1, \\ k_N^c, & 0 \leq i \leq b,\ j = N, \\ k_{j-(i-b)}, & j \geq i-b,\ b+1 \leq i \leq N,\ 0 \leq j \leq N-1, \\ k_{N-(i-b)}^c, & j \geq i-b,\ b+1 \leq i \leq N,\ j = N, \\ 0, & otherwise, \end{cases}$$

where $k_j^c = \sum_{i=j}^{\infty} k_i\ (j \geq 0)$. Thus, the limiting probabilities are defined as:

$$p^+(j) = \lim_{n\to\infty} P\{N_q(t_n^+) = j\}, \quad 0 \leq j \leq N.$$

Now $p^+(j)\ (0 \leq j \leq N)$ can be obtained by solving the system of equations $\boldsymbol{P}^+ = \boldsymbol{P}^+\boldsymbol{P}$, where $\boldsymbol{P}^+ = [p^+(0),\ p^+(1),\ p^+(2),\ \cdots,\ p^+(a-1),\ p^+(a),\ \cdots,\ p^+(N-1),\ p^+(N)]$, $\boldsymbol{P}$ is a $(N+1) \times (N+1)$ matrix whose elements are P_{ij}, and after simplification of $\boldsymbol{P}^+ = \boldsymbol{P}^+\boldsymbol{P}$, we get:

$$p^+(j) = \sum_{i=0}^{b} p^+(i)k_j + \sum_{r=1}^{\min\{j,\ N-b\}} p^+(b+r)k_{j-r},\ 0 \leq j \leq N-1, \tag{2}$$

$$p^+(N) = \sum_{i=0}^{b} p^+(i)k_N^c + \sum_{r=1}^{N-b} p^+(b+r)k_{N-r}^c. \tag{3}$$

Here, it may be noted that we can find the pgf of $\{p^+(n)\}_0^N$ using (2) and (3). Multiplying (2) by z^j and summing over j from 0 to $N-1$ and adding (3) after multiplying with z^N, we get after simplification the following expression:

$$p^{+*}(z) = \sum_{r=0}^{b} p^+(r)\Big[K(z) - \sum_{i=N+1}^{\infty} k_i z^i\Big] + \frac{1}{z^b}\sum_{j=b+1}^{N} p^+(j)z^j\Big[K(z) - \sum_{i=N+1}^{\infty} k_i z^i\Big], \tag{4}$$

where $p^{+*}(z) = \sum_{n=0}^{N} p^+(n)z^n$. Simplification of (4) by replacing $\sum_{j=b+1}^{N} p^+(j)z^j = p^{+*}(z) - \sum_{r=0}^{b} p^+(r)z^r$ in the right-hand side, we obtain:

$$p^{+*}(z) = \frac{\sum_{r=0}^{b-1}(z^b - z^r)p^+(r)K(z)}{z^b - K(z)} + \frac{\sum_{r=0}^{b-1}(z^r - z^b)p^+(r)\sum_{i=N+1}^{\infty} k_i z^i - p^{+*}(z)\sum_{i=N+1}^{\infty} k_i z^i}{z^b - K(z)}. \tag{5}$$

One can see that the numerator for the second term of Equation (5) has at least $(N+b)$ degree in z. Since $p^+(n)\ (0 \leq n \leq N)$ is the coefficient of z^n in the right-hand side of $p^{+*}(z)$ in Equation (5), then one may evaluate $p^+(n)$ using the following pgf:

$$p^{+*}(z) = \frac{\sum_{r=0}^{b-1}(z^b - z^r)p^+(r)K(z)}{z^b - K(z)}, \tag{6}$$

which is identical with the pgf of post-departure epoch probabilities in an infinite-buffer $M/G^{(a,b)}/1/\infty$ queueing model with the traffic intensity $\rho < 1$. The pgf of the infinite-buffer queueing model was mentioned earlier in the literature, e.g., see Equation (2.1) in [11] and also see Problem 11 in [1] (pp. 226–227)chaudhry1983first, where one may note that $K(z) = B^*(\lambda - \lambda z)$.

At this point, we intend to calculate the post-departure epoch probabilities in terms of the roots of the associated characteristic equation. Since rational LSTs (see Botta et al. [23]) cover a large number of distributions, we take $K(z) = B^*(\lambda - \lambda z) = B^*(s)$, where

$s = \lambda - \lambda z$ as a rational function in 's'. If the LST of the service-time distribution has a non-rational LST, then the LST can be approximated by rational functions using the Padé approximation technique.

In view of this, we consider those service-time distributions that have rational LST of the form $B^*(s) = P(s)/Q(s)$, where the degree of the polynomial $Q(s)$ is m and that of the polynomial $P(s)$ is y $(\leq m)$. Further, we make this assumption, as has been done by Botta et al. [23], since this assumption covers a large number of practical cases. Substituting $B^*(\lambda - \lambda z) = \frac{P(\lambda-\lambda z)}{Q(\lambda-\lambda z)} \equiv \frac{f(z)}{g(z)}$ in Equation (6), one can obtain the corresponding expression as given below,

$$p^{+*}(z) = \frac{\sum_{r=0}^{b-1}(z^b - z^r)p^+(r)f(z)}{z^b g(z) - f(z)}. \tag{7}$$

Since $K(z) = \frac{f(z)}{g(z)}$, $p^{+*}(z)$ in Equation (7) becomes a rational function with the denominator $(z^b g(z) - f(z))$ which is a polynomial of degree $(b+m)$. This polynomial $(z^b g(z) - f(z))$ has $(b+m)$ zeros in the whole complex plane. This leads to the following characteristic equation (CE):

$$z^b g(z) - f(z) = 0, \tag{8}$$

which has $(b+m)$ roots, namely, $\gamma_1 = 1$, γ_i $(i = 2, 3, \ldots, b+m)$.

Remark 1. *We could have avoided the derivation of pgf $p^{+*}(z)$, but since it gives a simpler procedure than what is available in the literature, we have included it here for completeness sake. This pgf is also used in deriving the waiting-time distribution. When software packages such as MAPLE and MATHEMATICA could not find (see Neuts [24], Kendall [25] and Kleinrock [26]) a large number of roots (they do now); Chaudhry responded to this (see Chaudhry [27,28]). In this connection, see also Chaudhry et al. [29]. Here is just one example. It can be solved in many ways, but we are giving just one method using the package MAPLE:*

```
> restart : with(RootFinding) :
> f := (x - 1) · (x - 2) · (x - 3)^4 · (x - 7)^3
                f := (x - 1)(x - 2)(x - 3)^4(x - 7)^3
> fsolve(f, x, complex)
                1., 2., 3., 3., 3., 3., 7., 7., 7.
```

The CE (8) has $b+m$ roots. As the queueing model under consideration has a finite buffer, the following three cases arise depending on ρ:

- Case 1 $(\rho < 1)$: Using Rouché's theorem, it can be shown that Equation (8) has b roots inside and on the unit circle $|z| = 1$ and the remaining m roots: γ_i $(i = b+1, b+2, \ldots, b+m)$ are in the region $|z| > 1$.
- Case 2 $(\rho = 1)$: As in Case 1, among the remaining m roots, one root (say, γ_{b+1}) is equal to one, and the other roots: γ_i $(i = b+2, b+3, \ldots, b+m)$ stay in the region $|z| > 1$.
- Case 3 $(\rho > 1)$: As above, among the remaining m roots, one root (say, γ_{b+1}) is inside the interval $(0, 1)$, and the other roots: γ_i $(i = b+2, b+3, \ldots, b+m)$ are in the region $|z| > 1$.

One may observe that as ρ (> 0) increases, from the remaining m roots of the CE (8), one positive real root gets closer to the origin from right to left. The above three cases have been discussed below.

- We first discuss the case when $\rho < 1$. For this case, when the traffic intensity of the queueing model under consideration is less than one, i.e., $\rho < 1$, one can use Equation (6) to find the post-departure epoch probabilities as described below. The unknown probabilities $\{p^+(r)\}_{r=0}^{b-1}$ on the right-hand side of the pgf in Equation (7) can be obtained by using the roots of Equation (8), i.e., $\gamma_1 = 1$ and $\gamma_2, \gamma_3, \ldots, \gamma_b$. This leads to:

$$\sum_{r=0}^{b-1} (\gamma_i^b - \gamma_i^r) p^+(r) f(\gamma_i) = 0, \quad i = 2, 3, \ldots, b, \tag{9}$$

and:

$$p^{+*}(1) = \frac{\left[\frac{d}{dz}\left[\sum_{r=0}^{b-1} (z^b - z^r) p^+(r) f(z)\right]\right]_{z=1}}{\left[\frac{d}{dz}[z^b g(z) - f(z)]\right]_{z=1}} = 1. \tag{10}$$

Therefore, for $\rho < 1$, using the partial fraction $p^{+*}(z)$ can be written as follows:

$$p^{+*}(z) = D \frac{\sum_{r=0}^{b-1} (z^b - z^r) p^+(r) f(z)}{z^b g(z) - f(z)} = D \sum_{i=b+1}^{b+m} \frac{\widehat{l}_i}{z - \gamma_i}, \tag{11}$$

where D is the normalizing constant and $\widehat{l}_i$'s are the constant coefficients of the partial fraction. Now, extracting the coefficients of z^n from the right-hand side of (11), one can obtain the post-departure epoch probabilities as:

$$p^+(n) = D \sum_{i=b+1}^{b+m} \frac{-\widehat{l}_i}{\gamma_i^{n+1}}, \qquad n = 0, 1, 2, \ldots, N, \tag{12}$$

where:

$$\widehat{l}_i = \frac{\sum_{r=0}^{b-1} (\gamma_i^b - \gamma_i^r) p^+(r) f(\gamma_i)}{\left[\frac{d}{dz}[z^b g(z) - f(z)]\right]_{z=\gamma_i}}, \quad i = b+1, b+2, \ldots, b+m.$$

One can obtain the unknown D in Equation (12) using the normalization condition as stated before and is given by:

$$D = \left(- \sum_{i=b+1}^{b+m} \frac{\widehat{l}_i}{\gamma_i} \frac{1 - \gamma_i^{-(N+1)}}{1 - \gamma_i^{-1}} \right)^{-1}, \quad \text{when } \rho < 1. \tag{13}$$

- For $\rho = 1$, we assume $\gamma_{b+1} = 1$, and in this case, Equations (11) and (12) are valid. The only exception is that $\widehat{l}_{b+1}$ may be obtained as follows:

$$\widehat{l}_{b+1} = \frac{\left[\frac{d^2}{dz^2}\left[\sum_{r=0}^{b-1} (z^b - z^r) p^+(r) f(z)\right]\right]_{z=1}}{\left[\frac{d^2}{dz^2}[z^b g(z) - f(z)]\right]_{z=1}} \to 1 \uparrow,$$

and:

$$D = \left(- \widehat{l}_{b+1}(N+1) - \sum_{i=b+2}^{b+m} \frac{\widehat{l}_i}{\gamma_i} \frac{1 - \gamma_i^{-(N+1)}}{1 - \gamma_i^{-1}} \right)^{-1}, \quad \text{when } \rho = 1. \tag{14}$$

It is also important to note that the normalization condition (10) does not work when the traffic intensity $\rho = 1$ as $z = 1$ is a repeated root of multiplicity two for the characteristic Equation (8). Thus, when $\rho = 1$, one may use Equations (16)–(18), which express the probabilities $\{p^+(j)\}_{j=b}^{N}$ in terms of the probabilities $\{p^+(r)\}_{r=0}^{b-1}$ and then one may use the normalization condition as $p^{+*}(1) = \sum_{j=0}^{N} p^+(j) = 1$. Alternatively, one can use the following normalization condition for $\rho = 1$:

$$p^{+*}(1) = \frac{\left[\frac{d^2}{dz^2}[\sum_{r=0}^{b-1}(z^b - z^r)p^+(r)f(z)]\right]_{z=1}}{\left[\frac{d^2}{dz^2}[z^b g(z) - f(z)]\right]_{z=1}} = 1. \quad (15)$$

- Lastly, we discuss the case when $\rho > 1$. As $p^{+*}(z)$ is convergent/analytic inside $|z| < \gamma_{b+1}$, the first b roots of the characteristic Equation (8) inside the region $|z| < \gamma_{b+1}$ should be the zeros of the numerator polynomial on the right-hand side of Equation (7). The unknown probabilities $\{p^+(r)\}_{r=0}^{b-1}$ on the right-hand side of the pgf in Equation (7) can be obtained by the set of linear Equations (9). This set of homogeneous linear equations is not sufficient to give non-trivial solution for the unknown probabilities $\{p^+(r)\}_{r=0}^{b-1}$ and thus one must use normalization condition along with the above system of simultaneous Equations (9) in the following way. Thus, when $\rho > 1$, one should use Equations (2) and (3) by expressing the probabilities $\{p^+(j)\}_{j=b}^{N}$ in terms of the probabilities $\{p^+(r)\}_{r=0}^{b-1}$ as follows:

$$p^+(b) = \frac{p^+(0)}{k_0} - \sum_{i=0}^{b-1} p^+(i), \quad (16)$$

$$p^+(b+j) = \frac{p^+(j) - \sum_{i=0}^{b} p^+(i)k_j - \sum_{r=1}^{j-1} p^+(b+r)k_{j-r}}{k_0}, \quad j = 1,\ 2,\ \ldots,\ N-b-1, \quad (17)$$

$$p^+(N) = \frac{\sum_{i=0}^{b} p^+(i)k_N^c + \sum_{r=1}^{N-b-1} p^+(b+r)k_{N-r}^c}{1 - k_b^c}, \quad (18)$$

and then one may use the normalization condition as follows:

$$\sum_{j=0}^{N} p^+(j) = 1, \quad (19)$$

where the probabilities $\{p^+(j)\}_{j=b}^{N}$ must be expressed in terms of the unknown probabilities $\{p^+(r)\}_{r=0}^{b-1}$ using the above Equations (16)–(18). The solution of the above system of linear Equations (9) and (19) gives the unknown probabilities $\{p^+(r)\}_{r=0}^{b-1}$ and from the above expressions (16)–(18) one can get the probabilities $\{p^+(j)\}_{j=b}^{N}$ in terms of the known probabilities $\{p^+(r)\}_{r=0}^{b-1}$. It may be noted that the above procedure of finding post-departure epoch probabilities works when $\rho <=> 1$. One may remark here that the above proposed recursive scheme (16)–(18) may sometimes be unstable, mainly when we are dealing with a very high buffer capacity $(N \to \infty)$ and due to the negative sign involved in the right-hand side of the recursive Equations (16) and (17). One may note here that it is possible to find post-departure epoch probabilities in closed form similar to the cases $\rho <= 1$ after getting the unknown probabilities $\{p^+(r)\}_{r=0}^{b-1}$ by using the above procedure. For this, we multiply both sides of Equation (7) by the factor $(z - \gamma_{b+1})$ and obtain the following equation:

$$p^{+*}(z)(z - \gamma_{b+1}) = \frac{\sum_{r=0}^{b-1}(z^b - z^r)p^+(r)f(z)(z - \gamma_{b+1})}{z^b g(z) - f(z)}. \quad (20)$$

Now, the left/right hand side of Equation (20) is convergent/analytic inside and on the unit circle $|z| = 1$. Thus, one may write Equation (20) in the following way:

$$p^{+*}(z)(z - \gamma_{b+1}) = \frac{\sum_{r=0}^{b-1}(z^b - z^r)p^+(r)f(z)(z - \gamma_{b+1})}{z^b g(z) - f(z)} = q_0 + q_1 z + \sum_{i=b+2}^{b+m} \frac{l_i}{z - \gamma_i}, \quad (21)$$

where q_0, q_1, and l_i's are the constant coefficients of the partial fraction. Equating the coefficient of the like powers of z in both sides of Equation (21) after a little algebraic calculation, we obtain:

$$p^+(0) = -\frac{\sum_{i=b+2}^{b+m} \frac{-l_i}{\gamma_i} + q_0}{\gamma_{b+1}}, \tag{22}$$

$$p^+(1) = -\frac{\sum_{i=b+2}^{b+m} \frac{-l_i}{\gamma_i^2} + q_1}{\gamma_{b+1}}, \tag{23}$$

$$p^+(n) = -\frac{\sum_{i=b+2}^{b+m} \frac{-l_i}{\gamma_i^{n+1}} - p^+(n-1)}{\gamma_{b+1}}, \qquad n = 2, 3, \ldots, N-1, \tag{24}$$

where:

$$l_i = \frac{\sum_{r=0}^{b-1} (\gamma_i^b - \gamma_i^r) p^+(r) f(\gamma_i)(\gamma_i - \gamma_{b+1})}{\left[\frac{d}{dz}[z^b g(z) - f(z)]\right]_{z=\gamma_i}}, \quad i = b+2, b+3, \ldots, b+m.$$

Let the coefficient of z^0 and z from the Taylors' series expansion of the right hand side of Equation (20) be c_0 and c_1. Now, $q_0 = c_0 - \sum_{i=b+2}^{b+m} \frac{-l_i}{\gamma_i}$ and $q_1 = c_1 - \sum_{i=b+2}^{b+m} \frac{-l_i}{\gamma_i^2}$. The last probability $p^+(N)$ can be found using the normalization condition (19) and is given by: $p^+(N) = 1 - \sum_{j=0}^{N-1} p^+(j)$.

Thus, one can obtain post-departure epoch probabilities successfully for all possible values of offered load or traffic intensity ρ for this queueing model under consideration.

Remark 2. *It may be remarked here that all the probabilities $p^+(i)$ $(0 \leq i \leq N)$ can also be computed by solving the system of linear equations $\mathbf{p}^+\mathbf{P} = \mathbf{p}^+$ (see [17], p. 98) using the GTH algorithm or the RG-factorization technique (extensions of the matrix-analytic methods, see Neuts [30] and Li [31]), where $\mathbf{p}^+ = [p^+(0),\ p^+(1),\ \cdots,\ p^+(N)]$ and $\mathbf{P}$ is the transition probability matrix of the Markov chain at the embedded post-departure epochs of a batch. In that case, the computational complexities of the GTH algorithm and the RG-factorization technique can be computed as $O((N+1)^4)$, (see [32]) and $O(N^3)$ (see Li et al. [33]), respectively. On the other hand, in the case of the roots' method, which is adopted in this paper, the main goal is to calculate the $m+b$ roots of the associated characteristic Equation (8) with the probability generating function given by Equation (7). Without a loss of generality, we assume that the degree of the characteristic Equation (8) is $m+b$. It should be mentioned here that we can numerically locate the roots of a polynomial using Newton's method. The root extraction using Newton's method and multiplication of two numbers has equal complexity, see [34]. That is, if we want to calculate roots of the characteristic equation up to d_1 digits accuracy, then the computational complexity of per root extraction will be at most $O(d_1^2)$. The next part is to evaluate b unknowns by solving b simultaneous linear equations. There are several methods used to solve simultaneous equations, e.g., Cramer's rule, Gaussian elimination, LU decomposition, etc. One of the most efficient methods to determine the b unknowns is LU decomposition with complexity $O(b^3)$. The last part is a partial fraction decomposition which has computational complexity at most $O(m^2)$ (see Xin [35]), where m is the degree of the denominator polynomial of the corresponding rational function. From our computational experience, we have seen that when the buffer capacity N becomes large and since $b \leq N$, both the GTH algorithm and the RG-factorization technique take much longer to compute the post-departure epoch probabilities $\{p^+(r)\}_{r=0}^N$ than the roots' method used in this paper. Further, the roots' method also gives explicit closed-form analytic results. In Section 5, we have presented a comparison of computation time for above discussed three methods.*

Lemma 1. *The expression for the probability that the server is busy (ρ') is given by:*

$$\rho' = \frac{E(S)}{T} = \frac{1}{\mu T}, \tag{25}$$

where T is mean inter-departure time of service batches. An expression for T may be obtained as follows:

$$T = E(S) + \sum_{i=0}^{a-1} p^+(i) \cdot \left(\frac{a-i}{\lambda}\right). \tag{26}$$

Proof. The server's state can be either busy or idle at any instant. As T is the mean inter-departure time of service batches, $1/T$ corresponds to the departure rate. Since T is the mean inter-departure time, it can be derived by considering two cases, (1) when the server is busy and (2) when the server is idle. Thus:

$$T = E(S)\left(1 - \sum_{i=0}^{a-1} p^+(i)\right) + \sum_{i=0}^{a-1} p^+(i) \cdot \left(\frac{a-i}{\lambda} + E(S)\right), \tag{27}$$

which after simplification gives (26). The probability that the server is busy is given by Equation (25), which is the proportion of time the server is busy divided by the mean inter-departure time. □

Probability Distribution for Queue Length at a Random Epoch

We develop relation between a distribution and the probability that the number of customers in the queue at the embedded Markov points (post-departure epochs) using the argument of the Markov renewal theory and the semi-Markov process. In terms of notations, the random or arbitrary epoch probabilities are defined as follows:

$$p_0(n) = \lim_{t\to\infty} P\{N_q(t) = n,\ \xi(t) = 0\}, \quad 0 \le n \le a-1, \tag{28}$$

$$p_1(j) = \lim_{t\to\infty} P\{N_q(t) = j,\ \xi(t) = 1\}, \quad 0 \le j \le N, \tag{29}$$

where $N_q(t)$ is the number of customers in the queue at time t and $\xi(t)$ is the server's state at time t. If at time epoch t the server is busy, then $\xi(t) = 1$, and if at time epoch t the server is idle with i ($(0 \le i < a)$ customers waiting for service, then $\xi(t) = 0$. Therefore, $p_1(n)$ $(0 \le n \le N)$ denotes the probability of n customers in the queue (excluding the size of a batch in service) when the server is busy and $p_0(n)$ $(0 \le n \le a-1)$ denotes the probability of n customers in the queue when the server is idle, where '1' stands for busy state of the server; and '0' stands for idle state of the server.

Theorem 1. *The random-epoch (or arbitrary-epoch) probabilities* $\{p_0(n)\ (0 \le n \le a-1)$ *and* $p_1(j)\ (0 \le j \le N)\}$ *and post-departure-epoch probabilities* $\{p^+(n)\ (0 \le n \le N)\}$ *are related by:*

$$p_0(n) = p_0(n-1) + \frac{1}{\lambda T} p^+(n), \quad 0 \le n \le a-1, \tag{30}$$

$$p_1(j) = \frac{1}{\mu T}\left[\sum_{i=0}^{b} p^+(i)\widehat{k}_j + \sum_{r=1}^{\min\{j,\, N-b\}} p^+(b+r)\widehat{k}_{j-r}\right], \ 0 \le j \le N-1, \tag{31}$$

$$p_1(N) = \frac{1}{\mu T}\left[\sum_{i=0}^{b} p^+(i)\widehat{k}^c_N + \sum_{r=1}^{N-b} p^+(b+r)\widehat{k}^c_{N-r}\right], \tag{32}$$

where $p_0(-1) = 0$ *and* $\widehat{k}^c_j = \sum_{i=j}^{\infty} \widehat{k}_i\ (j \ge 0)$.

Proof. Using rate-in, rate-out, we can consider the state $\{N_q(t) = n,\ \xi(t) = 0\}$ in the limiting case of $t \to \infty$. This can change by arrivals or service completions. Thus, we have:

$$\lambda p_0(n) = \lambda p_0(n-1) + \frac{1}{T} p^+(n), \quad 0 \le n \le a-1. \tag{33}$$

The left-hand side of Equation (33) is the rate of leaving out of the state $\{n,\ 0\}$ by an arrival, while the first term of the right-hand side is the entering rate to the state $\{n,\ 0\}$ after an arrival and the second term is the entering rate to the state $\{n,\ 0\}$ after a departure. After a little simplification of (33), we get (30). Using rate-in, rate-out, we now consider the state $\{N_q(t) = n,\ \xi(t) = 1\}$ in the limiting case when $t \to \infty$. This can change by service completions. As $\widehat{k}_j$ is the probability of number of customers that arrive following an embedded Markov point until stationary elapsed service time, then we have the following result:

$$\mu \lim_{t\to\infty} P\{N_q(t) = j,\ \xi(t) = 1\} = \mu p_1(j) = \frac{1}{T}\sum_{i=0}^{b} p^+(i)\widehat{k}_j + \frac{1}{T}\sum_{r=1}^{j} p^+(b+r)\widehat{k}_{j-r}, \quad 0 \le j \le N-1. \tag{34}$$

On the left-hand side of (34), we are leaving the state $\{j,\ 1\}$ and on the right-hand side of (34), we are entering into the corresponding state $\{j,\ 1\}$. After simplifying (34), we obtain (31). An exactly similar argument as used in Equation (31) derives Equation (32). □

One may note that the results derived in Equations (30) and (31) are actually the same as those presented by Chaudhry and Gupta [17]; see Appendix A for this derivation. The rate-in and rate-out principle used to derive Equations (30) and (31) are much simpler than the supplementary variable method used in [17]. Moreover, one can check the correctness of the numerical results obtained by both the methods. Hereafter, let us define the pgf of $\{p_1(j)\}_0^N$ by $P_{qb}(z) = \sum_{j=0}^{N} p_1(j) z^j$, which is the pgf of the random epoch probabilities $\{p_1(j)\}_0^N$ when the server is busy. Using (31) and (32), one can check that $P_{qb}(1) = \sum_{j=0}^{N} p_1(j) = \frac{1}{\mu T}$, which is equal to ρ' as derived in Lemma 2. One can check the fact that $P_{qb}(1) = 1 - \sum_{i=0}^{a-1} p_0(i)$ using Equations (30) and (25) in the following way:

$$1 - \sum_{i=0}^{a-1} p_0(i) = \frac{1}{\mu T}, \tag{35}$$

which is, in fact, true as $\frac{1}{\mu T}$ represents ρ' which denotes the probability that the server is busy, i.e.:

$$\sum_{n=0}^{N} p_1(n) = 1 - \sum_{i=0}^{a-1} p_0(i) = \rho'.$$

Moreover, it is interesting to note from Equation (35) that:

$$1/T = [\mu(1 - \sum_{i=0}^{a-1} p_0(i))] = \Big[\frac{1}{\mu} + \sum_{i=0}^{a-1} p^+(i).\Big(\frac{a-i}{\lambda}\Big)\Big]^{-1}. \tag{36}$$

Finally, one may note that using the fact that Poisson arrivals see time average (PASTA), the pre-arrival epoch probabilities are equal to the random epoch probabilities $\{p_0(n)\ (0 \le n \le a-1)$ and $p_1(j)\ (0 \le j \le N)\}$. Therefore, the probability of loss or blocking of a random arrival customer is given by $P_{loss} = p_1(N)$.

3. Waiting-Time (In Queue) Distribution for a Random Customer

Let us denote $V_q^{(n)}$ as the waiting time in the queue of a random customer who belongs to those customers served in the n-th batch. Further, let $W_q(t) = \lim_{n\to\infty} P(V_q^{(n)} \le t)$ and in the limiting case as $n \to \infty$, we denote $V_q^{(n)}$ by V_q. Let us define $W_q^*(s) = \int_0^\infty e^{-st} dW_q(t)$ with $\Re(s) \geq 0$. Now, following Chaudhry and Templeton [36], one can observe that during the busy or idle state of the server, the total probability of arrivals during the waiting time (in queue) of a random customer is equal to the total probability of possible number of customers waiting in the queue. Thus, considering the random nature of arrivals and busy/idle state of the server, we obtain:

$$\sum_{i=0}^{a-1} p_0(i) z^i + P_{qb}(z) = W_q^*(\lambda' - \lambda' z), \tag{37}$$

where λ' $(= \lambda(1 - P_{loss}))$ which is the customer's actual arrival rate due to finite-buffer capacity. We note that differentiating (37) with respect to z and setting $z = 1$, we obtain Little's law as follows:

$$L_q = \sum_{i=0}^{a-1} i p_0(i) + \sum_{n=0}^{N} n p_1(n) = \lambda' E(V_q). \tag{38}$$

From Equation (37) using $s = \lambda' - \lambda' z$, after a little algebraic manipulation, we obtain:

$$W_q^*(s) = \sum_{i=0}^{a-1} p_0(i) \Big(1 - \frac{s}{\lambda'}\Big)^i + \sum_{j=0}^{N} p_1(j) \Big(1 - \frac{s}{\lambda'}\Big)^j. \tag{39}$$

Now, using Padé approximation of the right-hand side of Equation (39), one may approximate $W_q^*(s)$ by the rational function $\frac{\widehat{P}(s)}{\widehat{Q}(s)}$. We assume the degree of the numerator $\widehat{P}(s)$ as r_0 and that of the denominator $\widehat{Q}(s)$ as r_d $(> r_0)$. One can check the correctness of Padé approximation using the fact that $\frac{\widehat{P}(0)}{\widehat{Q}(0)} = \sum_{n=0}^{N} p_1(n) + \sum_{i=0}^{a-1} p_0(i) = 1$ and $-\frac{d}{ds}\{\frac{\widehat{P}(s)}{\widehat{Q}(s)}\}|_{s=0} = \frac{\sum_{i=0}^{a-1} i p_0(i) + \sum_{n=0}^{N} n p_1(n)}{\lambda'} = \frac{L_q}{\lambda'}$. Further, one can improve the Padé approximation using those two criteria. Since $W_q^*(s)$ is convergent or analytic in $\Re(s) \geq 0$, the r_d zeros of the denominator $\widehat{Q}(s)$ may be assumed as $s_1, s_2, \ldots, s_{r_d}$ with $\Re(s_i) < 0$. Using partial fractions, the right-hand side of Equation (39) may be written in the following form:

$$W_q^*(s) \simeq \frac{\widehat{P}(s)}{\widehat{Q}(s)} = \sum_{j=1}^{r_d} \frac{k_j}{s - s_j}. \tag{40}$$

Now, Equation (40) leads to the following equation:

$$w_q(t) = \sum_{j=1}^{r_d} k_j e^{s_j t}, \quad t > 0, \;\text{ with } \Re(s_j) < 0, \tag{41}$$

$$\text{where } \; k_j = \frac{\widehat{P}(s_j)}{\Big[\frac{d}{ds}[\widehat{Q}(s)]\Big]_{s=s_j}},$$

where $j = 1, 2, \ldots, r_d$.

Remark 3. *Here, we demonstrate how to practically approximate $W_q^*(s)$ by the rational function using the Padé approximation through a numerical example. For this purpose, we consider PH-type service-time distribution with the representation $(\boldsymbol{\beta}, \boldsymbol{T})$, where $\boldsymbol{\beta} = (0.2, 0.8)$ and $\boldsymbol{T} = \begin{pmatrix} -2 & 0 \\ 0 & -4 \end{pmatrix}$, so that $\mu = 3.333333$. The other parameters are taken as $a = 2, b = 5, \lambda = 10$*

with $\rho = \frac{\lambda}{b \cdot \mu} = 0.6$. After computing $p_0(i)$'s and $p_1(j)$'s, the effective arrival rate of customers can be obtained as $\lambda' = \lambda(1 - P_{loss}) = 8.928835$. From Equation (39), we get:

$$\begin{aligned} W_q^*(s) =& 0.217096 + 0.234453(1 - 0.111997s) + 0.108621(1 - 0.111997s)^2 \\ &+ 0.090139(1 - 0.111997s)^3 + 0.075175(1 - 0.111997s)^4 + 0.062991(1 - 0.111997s)^5 \\ &+ 0.045634(1 - 0.111997s)^6 + 0.031130(1 - 0.111997s)^7 + 0.018954(1 - 0.111997s)^8 \\ &+ 0.008689(1 - 0.111997s)^9 + 0.107117(1 - 0.111997s)^{10} \end{aligned}$$

We have used the 'pade' command in MAPLE 2015 and approximated $W_q^*(s)$ as $\frac{\widehat{P}(s)}{\widehat{Q}(s)}$ where:

$$\begin{aligned} \widehat{P}(s) &= 1 + 0.928374s + 0.498682s^2 + 0.147040s^3 + 0.127709s^4 \\ \widehat{Q}(s) &= 1 + 0.443441s + 0.991992s^2 + 0.140031s^3 + 0.125392s^4 + 0.572475s^5. \end{aligned}$$

Moreover, the correctness of the approximation can be checked by using the criteria given after the Equation (39):

$$\frac{\widehat{P}(0)}{\widehat{Q}(0)} = 1.000000, \text{ and } -\frac{d}{ds}\left\{\frac{\widehat{P}(s)}{\widehat{Q}(s)}\right\}\Big|_{s=0} = 0.350604 = \frac{L_q}{\lambda'}.$$

All of the roots of $\widehat{Q}(s)$ have real parts less than zero, which makes $W_q^*(s)$ convergent in $\Re(s) > 0$. Then we proceed further after writing $\frac{\widehat{P}(s)}{\widehat{Q}(s)}$ in terms of partial fractions.

Remark 4. *One may note here that inversion of the LST presented in Equation (39) may be done using several other methods of inverting the LSTs (or transforms) reported in the literature (see the excellent reviews on the numerical transform inversion by Abate and Whitt [37] and Abate et al. [38]). Abate and Whitt [37] use the Fourier-series method to invert the generating functions and the LSTs numerically. The Fourier-series method involves numerically integrating a standard inversion integral employing the trapezoidal rule. The greatest difficulty in this case is approximately calculating the infinite series obtained from the inversion integral. However, the method described above is free of such an approximation of numerical integration.*

Moments: Moments of order k (≥ 1) for V_q can be obtained using (41) as follows:

$$E[V_q^k] = \int_0^\infty t^k \cdot w_q(t)dt = \int_0^\infty t^k \cdot \sum_{j=1}^{r_d} k_j e^{s_j t} dt, \quad \text{where } \Re(s_j) < 0, \quad k \geq 1, \tag{42}$$

where $\Re(s_j) < 0$. Alternatively, one can get the k-th moment of the queueing-time distribution by differentiating (39) k times with respect to s and then setting $s = 0$ in the following way:

$$\begin{aligned} E[V_q^k] =& (-1)^k \left[\frac{d^k}{ds^k}\left\{W_q^*(s)\right\}\right]_{s=0} = (-1)^k \left[\frac{d^k}{ds^k}\left\{\sum_{i=0}^{a-1} p_0(i)\left(1 - \frac{s}{\lambda'}\right)^i + \right.\right. \\ & \left.\left.\sum_{j=0}^{N} p_1(j)\left(1 - \frac{s}{\lambda'}\right)^j\right\}\right]_{s=0}, \quad k \geq 1, \end{aligned} \tag{43}$$

which may work as a check on correctness of the moments obtained via (42).

Sojourn-time distribution: Sojourn-time distribution of a random customer is its queueing time, plus its batch service time. Therefore, the LST of sojourn time of random customer $W^*(s)$ is given by:

$$W^*(s) = W_q^*(s) \times B^*(s). \tag{44}$$

Using (41), we can invert $W^*(s)$. It gives the pdf $w(t)$ of the sojourn time distribution of a random customer and is given by:

$$w(t) = \int_0^t \sum_{j=1}^{r_d} k_j e^{s_j u} \cdot b(t-u)du, \quad t \geq 0, \tag{45}$$

where $\Re(s_j) < 0$. Now, using the probability density function $w(t)$, one can get a moment of order k for the random variable V (sojourn time in the system for a random customer) as $E[V^k] = \int_0^\infty t^k w(t)dt$ which may be described as follows:

$$E[V^k] = \int_0^\infty t^k \int_0^t \sum_{j=1}^{r_d} k_j e^{s_j u} \cdot b(t-u)dudt, \quad \text{where } \Re(s_j) < 0, \quad k \geq 1. \tag{46}$$

The cumulative distribution function (CDF) of the queueing and sojourn times may be obtained by the following results:

$$W_q(t) = \int_0^t w_q(u)du = \int_0^t \sum_{j=1}^{r_d} k_j e^{s_j u} dt, \quad t > 0, \quad \text{where } \Re(s_j) < 0, \tag{47}$$

$$W(t) = \int_0^t \int_0^v \sum_{j=1}^{r_d} k_j e^{s_j u} \cdot b(v-u)dudv, \quad t \geq 0, \quad \text{where } \Re(s_j) < 0. \tag{48}$$

4. Applications of the Queueing Model in the Performance Evaluation of Blockchain Systems

In a blockchain system, client transactions must wait for servers (e.g., miner, validator or orderer) to provide service (e.g., mining, validating or ordering) and finally get confirmed/served. One can find a detailed literature survey and a description of the performance evaluation of blockchain systems by Fan et al. [39]. Different consensus processes of blockchain systems can be modelled as different types of queueing systems. In a queueing system, the most important part is the quantitative evaluation of its performance measures, such as what is the expected number of transactions in the system, what is the transaction throughput of the system, and what is the average waiting time in the queue or in the system (i.e., sojourn time). In the blockchain called Bitcoin [40], the ledger is maintained and updated by the mining process. In the mining process, a bunch of nodes, called miners, compete to solve challenging puzzle-like problems, which consume much computation power. Transactions issued by users are grouped into a container called a block. The mining competition winner who first finds the algorithmic puzzle answer specialized for the block has the right to add the new block to the blockchain and gets incentives accordingly. Kawase and Kasahara [41] first built a modified $M/G^B/1$ queue (see Chaudhry and templeton [36] for the performance analysis of the $M/G^B/1$ queueing model) with batch service to model the Bitcoin mining process. Li et al. [42] introduced a new blockchain queueing model by decomposing the mining process into two exponential service stages: block generation and blockchain-building processes. The sum of both stages' times is regarded as the transaction confirmation time, which a general service-time distribution may represent. The distribution of the key performance indicators, such as the the transaction waiting time, may be defined by the random amount of time a transaction has to wait before it gets confirmation; see Geissler et al. [43] for details. Geyer et al. [44] modelled the solo ordering process of Hyperledger Fabric as an $M/M^B/1$ queueing system. To solve this model, the authors borrowed the results from the waiting time analysis of a general bulk service queueing model $M/M^{(a,b)}/1$ (see Medhi [7]). They considered the batch size range as either $a = b = B$ or $a = 1$ and $b = B$. Following the above discussions, we propose the queue length and waiting time analysis of a more general bulk-service $M/G^{(a,b)}/1/N$ queue to analyse the performance of a blockchain queueing system which includes the possibility that the buffer capacity $N \to \infty$. See the following Figure 1.

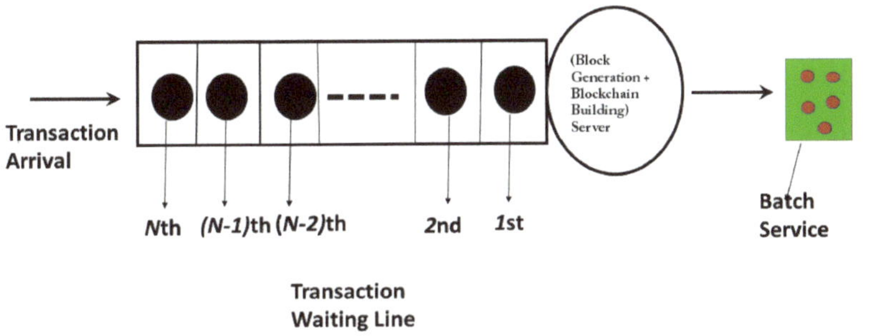

Figure 1. $M/G^{(a,b)}/1/N$ blockchain queueing model.

5. Numerical Results

In this section, the validity of the results obtained in the previous sections has been tested through numerical experiments. Several numerical experiments have been carried out for different model parameters in this connection. Out of them, a few numerical examples are presented in this section. The experiments have been carried out using MAPLE 2015 for Windows 10 (64-bit) OS environment. The results have been presented in this paper up to six decimal places due to a lack of space. However, upon specific requests, the authors can provide high-precision numerical calculations exhibited through MAPLE coding.

Example 1. *Here we will examine the case of traffic intensity (ρ) greater than 1. We construct a PH type distribution as the service-time distribution with the representation $(\boldsymbol{\beta_3}, \boldsymbol{T_3})$ to serve that purpose, where $\boldsymbol{\beta_3} = (0.2, 0.4, 0.1, 0.3)$, $\boldsymbol{T_3} = \begin{pmatrix} -3 & 1 & 0 & 1 \\ 1 & -5 & 1 & 0 \\ 0 & 2 & -4 & 2 \\ 1 & 0 & 1 & -4 \end{pmatrix}$ with $\mu = 1.619870$. The other parameters are chosen as $\lambda = 20, N = 20, a = 3, b = 10$ and $\rho = 1.234667$. Proceeding in a similar fashion mentioned in the above sections, we get the denominator of Equation (7) as:*

$$80000z^{14} - 384000z^{13} + 689800z^{12} - 549610z^{11} + 163885z^{10} + 8000z^3 - 29140z^2 + 35364z - 14299. \tag{49}$$

In this case, Equation (8) has eleven roots inside and on the unit disk and three roots outside the unit disk; $\gamma_1 = 1.00000$, $\gamma_2 = -0.745122$, $\gamma_3 = -0.615642 - 0.425348i$, $\gamma_4 = -0.615642 + 0.425348i$, $\gamma_5 = -0.267744 - 0.709653i$, $\gamma_6 = -0.267744 + 0.709653i$, $\gamma_7 = 0.190352 - 0.754816i$, $\gamma_8 = 0.190352 + 0.754816i$, $\gamma_9 = 0.620330 - 0.531710i$, $\gamma_{10} = 0.620330 + 0.531710i$, $\gamma_{11} = 0.967206$, $\gamma_{12} = 1.168465$, $\gamma_{13} = 1.223421$ and $\gamma_{14} = 1.331440$. The values of $p^+(n)$ and $p_1(n)$ are given in Table 1, from which the loss probability of a random arrival customer can be obtained as $P_{loss} = p_1(20) = 0.313256$.

Using the procedure discussed in the paper the CDF of queueing-time distribution $W_q(t)$ for the model is obtained as follows:

$$\begin{aligned} W_q(t) =& 0.999999 - 3.32859e^{-2.99754t} + 2.32864e^{-2.07443t}\cos(2.46909t) \\ &- 1.31795e^{-2.07443t}\sin(2.46909t), \quad t > 0. \end{aligned}$$

Moreover, using the method discussed in the paper the CDF of sojourn-time distribution $W(t)$ for the model is obtained as follows:

$$\begin{aligned} W(t) =& 0.999999 + 0.150422e^{-6.66337t} + 0.0602525e^{-4.47452t} \\ & - 0.0268614e^{-3.36865t} - 4.26332e^{-1.49347t} + 2.33519e^{-2.99754t} \\ & + 0.744376e^{-2.07443t}\cos(2.46909t) + 1.36014e^{-2.07443t}\sin(2.46909 * t), \quad t \geq 0. \end{aligned} \quad (50)$$

Table 1. Computation of post-departure epoch probabilities and random epoch probabilities (when the server is busy) for Example 1 with the parameters $\lambda = 20$, $\mu = 1.619870, a = 3, b = 10, N = 20$ and $\rho = 1.234667$.

n	$p^+(n)$	$p_1(n)$	n	$p^+(n)$	$p_1(n)$
0	0.028418	0.023753	11	0.034051	0.051154
1	0.027992	0.024237	12	0.035143	0.048347
2	0.027893	0.024816	13	0.036286	0.045447
3	0.028048	0.025475	14	0.037480	0.042453
4	0.028402	0.026200	15	0.038723	0.039359
5	0.028914	0.026983	16	0.040014	0.036162
6	0.029554	0.027819	17	0.041355	0.032858
7	0.030299	0.028702	18	0.042743	0.029443
8	0.031132	0.029630	19	0.044184	0.025912
9	0.032041	0.030600	20	0.324318	0.313256
10	0.033016	0.053875	-	-	-

$p_0(0) = 0.002270$, $p_0(1) = 0.004507$, $p_0(2) = 0.006736$.

Example 2. *Let us take the service time as Pareto power tail distribution (see Ramsay [45], and Roughan et al. [46]) having the distribution function $B(t) = 1 - (1+t)^{\alpha-1}$, where $\alpha = 1.5$, so that $\mu = 0.5$. To find the Laplace–Stieltjes transformation of the service-time distribution using Padé approximation in the following way:*

$$B^*(\lambda - \lambda z) = \lim_{N\to\infty} \sum_{n=0}^{N} z^n \int_0^\infty \frac{(\lambda t)^n}{n!} e^{-\lambda t} b(t)dt = \lim_{N\to\infty} \sum_{n=0}^{N} z^n c_n, |z| \leqslant 1, \quad (51)$$

where $\lim_{N\to\infty} \sum_{n=0}^{N} c_n = 1 \uparrow$, and $B^*(\lambda - \lambda z)$ has a probabilistic interpretation. Now the Padé approximation is applied to the function $B^*(\lambda - \lambda z)$. After using the Padé approximation of the function, we get a function $B^*(s)$ by replacing $\lambda - \lambda z = s$. For more details, see Avram et al. [47]. Using the above procedure $B^*(s)$ can be obtained as follows:

$$B^*(s) \approxeq \frac{0.00681 + 0.00608s + 0.00088s^2}{0.00681 + 0.01969s + 0.00552s^2 + 0.00059s^3}. \quad (52)$$

Moreover, let us take other parameters as $a = 4$, $b = 11$ and $\lambda = 9$ so that $\rho = \frac{\lambda}{b \cdot \mu} = 1.636364$. Using the method described in the preceding sections, we obtained the CDF of the queueing time and sojourn time of a random customer in the $M/Pareto^{(4,11)}/1/20$ queue. Now we take the characteristic Equation (8) as follows:

$$0.42684z^{14} - 1.72726z^{13} + 2.35122z^{12} - 1.05762z^{11} + 0.07113z^2 - 0.19698z + 0.13265 = 0 \quad (53)$$

In this case, Equation (8) has twelve roots inside and on the unit disk and two roots outside the unit disk; $\gamma_1 = 0.99999, \gamma_2 = -0.76442 - 0.21674i, \gamma_3 = -0.76442 + 0.21674i, \gamma_4 = -0.54018 - 0.58975i, \gamma_5 = -0.54018 + 0.58975i, \gamma_6 = -0.15091 - 0.79690i, \gamma_7 = -0.15091 + 0.79690i, \gamma_8 = 0.30129 - 0.77441i, \gamma_9 = 0.30129 + 0.77441i, \gamma_{10} = 0.698677 - 0.50934i, \gamma_{11} = 0.698677 + 0.50934i, \gamma_{12} = 0.95335, \gamma_{13} = 1.50214 - 0.34752i$ and $\gamma_{14} = 1.50214 + 0.34752i$. The values of $p^+(n)$ and $p_1(n)$ are given in the Table 2, from which the loss probability of a random *arrival customer can be obtained as* $P_{loss} = p_1(20) = 0.48373$. *Using the procedure discussed in the paper the CDF of queueing-time distribution* $W_q(t)$ *is obtained as follows:*

$$\begin{aligned} W_q(t) =& 0.999987 - 0.999986e^{-0.534383t}cos(0.375759t) \\ & - 2.25306e^{-0.534383t}sin(0.375759t), \quad t > 0. \end{aligned} \tag{54}$$

Moreover, using the method discussed in the paper the CDF of the sojourn-time distribution $W(t)$ *is obtained as follows:*

$$\begin{aligned} W(t) =& 0.99998 - 0.01082e^{-4.51669t}cos(3.12265t) + 0.01657e^{-4.51669t}sin(3.12265t) \\ & - 2.54227e^{-0.38563t} + 1.55311e^{-0.53438t}cos(0.37576t) \\ & - 0.66818e^{-0.534383t}sin(0.375759t), \quad t \geq 0. \end{aligned} \tag{55}$$

Table 2. Computation of post-departure epoch probabilities and arbitrary epoch probabilities (when the server is busy) for Example 1 with the parameters $\lambda = 9$, $\mu = 0.5, a = 4, b = 11, N = 20$ and $\rho = 1.636364$.

n	$p^+(n)$	$p_1(n)$	n	$p^+(n)$	$p_1(n)$
0	0.039995	0.015206	11	0.027209	0.037136
1	0.033106	0.014958	12	0.028567	0.035579
2	0.028170	0.015058	13	0.029995	0.033943
3	0.024985	0.015412	14	0.031490	0.032226
4	0.023199	0.015950	15	0.033052	0.030424
5	0.022451	0.016617	16	0.034684	0.028533
6	0.022435	0.017378	17	0.036392	0.026549
7	0.022917	0.018210	18	0.038180	0.024467
8	0.023728	0.019100	19	0.040052	0.022283
9	0.024757	0.040034	20	0.408697	0.483735
10	0.025931	0.038620	-	-	-

$p_0(0) = 0.002180$, $p_0(1) = 0.003985$, $p_0(2) = 0.005521$, $p_0(3) = 0.006884$.

Example 3. *We consider matrix-exponential (ME) service time with the representation* $b(x) = \boldsymbol{\alpha} e^{x\boldsymbol{T}}\boldsymbol{S}$, *where* $\boldsymbol{\alpha} = (4\pi^2 + 1, 0, 0)$, $\boldsymbol{T} = \begin{pmatrix} 0 & 1 & 0 \\ 0 & 0 & 1 \\ -4\pi^2 - 1 & -4\pi^2 - 3 & -3 \end{pmatrix}$, *and* $\boldsymbol{S} = (0,0,1)^t$ *with* $\mu = 0.952917$. *One may note that we have picked the service-time distribution from the paper of Akar [48] where he uses the state space method to analyse waiting-time distribution of ME/ME/1. The other parameters are taken as* $\lambda = 17.152511$, $N = 20$, $a = 4$, $b = 11$ *so that* $\rho = 1.636364$. *The characteristic Equation (8) is obtained as follows:*

$$5046.416z^{14} - 16021.87z^{13} + 17633.11z^{12} - 6698.132z^{11} + 40.47841 = 0, \tag{56}$$

which gives one roots $\gamma_1 = 1.000000$, *eleven roots inside the unit disk, and two roots out side the unit circle. We write those roots as* $\gamma_2 = -0.544626 - 0.145851i$, $\gamma_3 = -0.544626 + 0.145851i$, $\gamma_4 = -0.404969 - 0.405222i$, $\gamma_5 = -0.404969 + 0.405222i$, $\gamma_6 = -0.151554 - 0.573269i$, $\gamma_7 = -0.151554 + 0.573269i$, $\gamma_8 = 0.170621 - 0.606064i$, $\gamma_9 = 0.170621 + 0.606064i$, $\gamma_{10} = 0.515050 - 0.475234i$, $\gamma_{11} = 0.515050 + 0.475234i$, $\gamma_{12} = 0.903618$, $\gamma_{13} = 1.051120 - 0.371486i$, $\gamma_{14} = 1.051120 + 0.371486i$. *The value of* $p^+(n)$ *and* $p_1(n)$ *are given in Table 3, from which the loss probability of a random arrival customer can be obtained as* $P_{loss} = p_1(20) = 0.425405$.

Table 3. Computation of post-departure epoch probabilities and arbitrary epoch probabilities (when the server is busy) for Example 3 with the parameters $\lambda = 17.1525$, $\mu = 0.95291$, $a = 4$, $b = 11$, $N = 20$ and $\rho = 1.636364$.

n	$p^+(n)$	$p_1(n)$	n	$p^+(n)$	$p_1(n)$
0	0.001159	0.010570	11	0.029512	0.044812
1	0.003243	0.012151	12	0.031757	0.043051
2	0.005972	0.013721	13	0.034298	0.041149
3	0.009064	0.015284	14	0.037253	0.039083
4	0.012264	0.016862	15	0.040715	0.036826
5	0.015375	0.018491	16	0.044761	0.034344
6	0.018268	0.020220	17	0.049444	0.031603
7	0.020885	0.022100	18	0.054801	0.028564
8	0.023235	0.024187	19	0.060862	0.025189
9	0.025383	0.047969	20	0.454308	0.425405
10	0.027432	0.046448	-	-	-

$p_0(0) = 0.000064$, $p_0(1) = 0.000244$, $p_0(2) = 0.000575$, $p_0(3) = 0.001077$.

Using the procedure discussed in the paper the CDF of queueing-time distribution $W_q(t)$ is obtained as follows:

$$\begin{aligned} W_q(t) =& 0.999999 - 0.9999994e^{-1.185375t}cos(0.8279833t) \\ & - 2.372224e^{-1.185375t}sin(0.8279833t), \quad t > 0. \end{aligned} \tag{57}$$

Moreover, using the method discussed in the paper the CDF of the sojourn-time distribution $W(t)$ is obtained as follows:

$$\begin{aligned} W(t) =& 0.999992 - 0.127843e^{-0.440013t}cos(0.30778t) - 3.03358e^{-0.440013t}sin(0.30778t) \\ & - 0.00186616e^{-t}cos(6.28318t) + 0.000838452e^{-t}sin(6.28318t) - 0.870283e^{-t}, \quad t \geq 0. \end{aligned} \tag{58}$$

We have conducted an experiment to compare the results between Examples 2 and 3. We have plotted a graph depicting N versus P_{loss}, see Figure 2. However, here, we have considered a very high arrival rate ($\lambda = 17.15251$) for the matrix-exponential service-time distribution and $\lambda = 9$ for Pareto service-time distribution. In addition, ρ is the same for both the cases and greater then one. Here, we increase the value of N and as a result the loss probability slowly decreases in matrix-exponential case, but in the case of the Pareto service-time distribution the loss probability is higher than that of the matrix-exponential, service-time distribution.

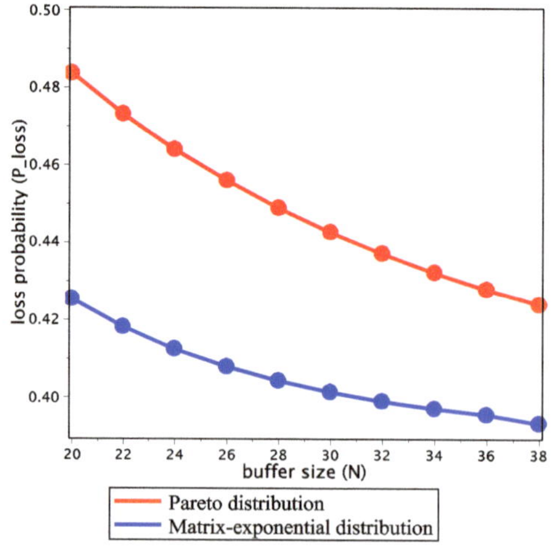

Figure 2. Effect of N on loss probabilities.

In Figure 3, we have plotted a graph of N versus mean sojourn time, when the distribution follows matrix-exponential, as well as the Pareto service-time distribution. Moreover, here we consider the same traffic intensity ($\rho = 1.636364$) for both the service-time distributions. One can observe from Figure 3 that the mean sojourn time for the Pareto service-time distribution is higher than that of the matrix-exponential, service-time distribution and both of them rise linearly upward.

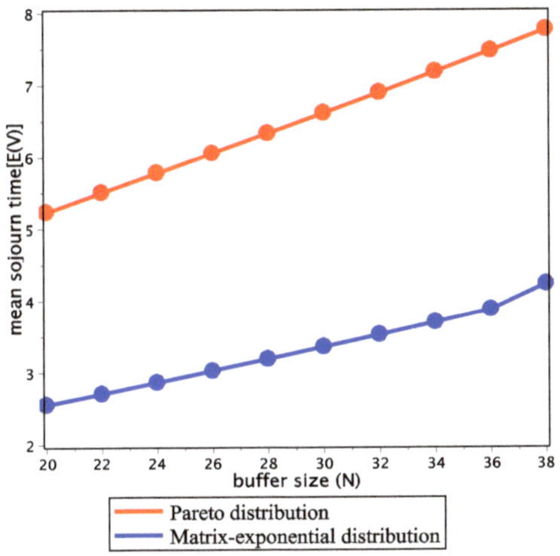

Figure 3. Effect of N on mean sojourn time.

Example 4. *Let us take the service time distribution as PH-type having the representation $(\boldsymbol{\beta_1}, \boldsymbol{T_1})$, where $\boldsymbol{\beta_1} = (0.25, 0.75)$ and $\boldsymbol{T_1} = \begin{pmatrix} -1 & 0 \\ 0 & -2 \end{pmatrix}$, so that $\mu = 1.6$. Moreover, let us take other parameters as $a = b = 5, \lambda = 6$ with $\rho = \frac{\lambda}{b \cdot \mu} = 0.75$. Using the method described in the preceding sections, we obtained the CDF of the sojourn time of a random customer for the $M/PH_2/1/20$ queue. To present a numerical comparison, we have used the CDF values of the sojourn time obtained in Example 4 of Yu and Tang [49]. In our example, if we take buffer space high (say, $N = 300$), our model behaves almost as an infinite-buffer, single-server queue, as the traffic intensity ($\rho = 0.75$) is less than one. From Table 4 and Figure 4, it can be observed that while the buffer size (N) is large, the values of the sojourn time CDF almost match the results of the infinite-buffer queue as obtained in [49]. Substituting $k = 1$ into Equation (46), we get the value of the expectation of the sojourn time in the system for a random customer, $E[V] = 1.626046$ when $N = 20$. If we take $N = 300$, the expectation can be obtained as $E[V] = 2.153832$. We also calculated $E[V] = 2.153820$ using the sojourn time CDF obtained from [49]. The denominator $(28.0z^5 - 45.0z^6 + 18.0z^7 - 6.25 + 5.25z)$ of Equation (7) when equated to zero gives one root equal to 1, four roots inside the unit disk and two roots outside the unit circle. $\gamma_6 = 1.080640, \gamma_7 = 1.265100$. Using these γ_i's, $p^{+*}(z)$ can be expressed by applying the partial-fraction method, as follows:*

$$p^{+*}(z) = D\Big(\frac{\widehat{l}_6}{z - \gamma_6} + \frac{\widehat{l}_7}{z - \gamma_7}\Big), \tag{59}$$

where $D = 1.000000$, $\widehat{l}_6 = -0.066828$, $\widehat{l}_7 = -0.045404$.

Table 4. Comparison among the CDF of sojourn times from Example 4 with the parameters $\lambda = 6$, $\mu = 1.6, a = 5, b = 5, N = 20/300$ and $\rho = 0.75$.

t	$W(t)$ from Present Paper with $N = 20$	$W(t)$ from Present Paper with $N = 300$	$W(t)$ from Yu and Tang [49] with $N \to \infty$
2	0.689965	0.618004	0.610358
4	0.975235	0.858918	0.856842
6	0.996731	0.946573	0.945808
8	0.999626	0.979706	0.979418
10	0.999957	0.992289	0.992180
12	0.999994	0.997070	0.997029
14	0.999999	0.998887	0.998871
16	0.999999	0.999577	0.999571

For Table 4, a mathematical expression of $W(t)$ for $N = 300$ is as follows:

$$\begin{aligned} W(t) =&(-0.058873e^{-0.78404t}\sin(3.74003t) - 0.0000129996e^{1.79961t} - 0.973426e^{1.51618t} \\ &+ 0.194208e^{-0.78404t}\cos(3.74003t) - 0.423864e^{0.356990t} + 0.999999e^{2t} \\ &+ 0.203411 - 0.000312514e^{t})e^{-2t}, \quad t \geq 0. \end{aligned}$$

Almost similar expressions have been obtained for all other cases, and hence, those are sometimes not presented in the paper.

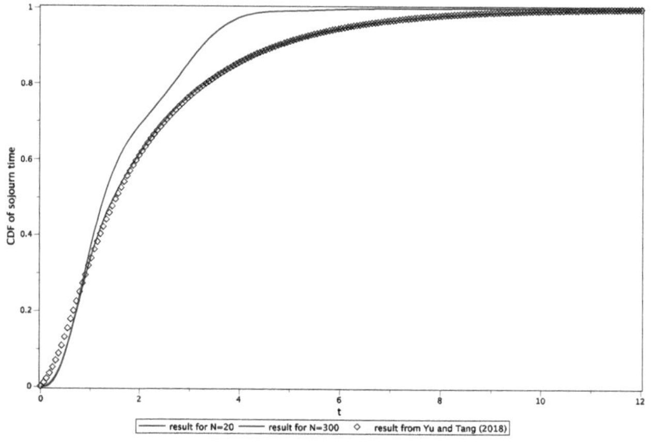

Figure 4. Comparison among the CDF of the sojourn times for different values of N from Example 4 of Yu and Tang [49].

Example 5. *In this example, we consider phase-type (PH) service-time distribution with the representation $(\boldsymbol{\beta_2}, \mathbf{T_2})$, where $\boldsymbol{\beta_2} = (0.1, 0.6, 0.3)$, $\mathbf{T_2} = \begin{pmatrix} -2 & 2 & 0 \\ 3 & -5 & 0 \\ 0 & 0 & -4 \end{pmatrix}$ with $\mu = 1$. The other parameters are taken as $\lambda = 10, N = 20, a = 3, b = 10$, and $\rho = 1$. We have from the denominator of (7) as $250z^{13} - 1025z^{12} + 1380z^{11} - 609z^{10} + 60z^2 - 160z + 104$, when equated to zero gives two roots equal to 1, nine roots inside the unit disk, and two roots outside the unit circle. $\gamma_{12} = 1.395684, \gamma_{13} = 1.636619$. Using these γ_i's, $p^{+*}(z)$ can be expressed by applying the partial-fraction method, as follows:*

$$p^{+*}(z) = D\left(\frac{\widehat{l}_{11}}{z-1} + \frac{\widehat{l}_{12}}{z-\gamma_{12}} + \frac{\widehat{l}_{13}}{z-\gamma_{13}}\right), \tag{60}$$

where the values of D, $\widehat{l}_{11}$, $\widehat{l}_{12}$, *and* $\widehat{l}_{13}$ *can be obtained from Equation (14) and partial fraction method as:*

$$D = -0.386247,\ \widehat{l}_{11} = 0.131752,\ \widehat{l}_{12} = -0.048451,\ \widehat{l}_{13} = -0.035298.$$

The obtained results are given in Table 5. From the table, the loss probability of a random arrival customer can be observed as $P_{loss} = p_1(20) = 0.221596$. *The CDF of queueing-time distribution* $W_q(t)$ *for the model is obtained using the approach discussed in the paper as follows:*

$$\begin{aligned} W_q(t) =& 0.999998 - 1.92626e^{-2.12386t}\cos(0.744301t) - 7.89988e^{-2.12386t}\sin(0.744301t) \\ &+ 0.926266e^{-1.1913t}\cos(2.05986t) + 1.26064e^{-1.1913t}\sin(2.05986t), \quad t > 0. \end{aligned}$$

Moreover, using similar methodology conferred in the paper, the CDF of sojourn-time distribution $W(t)$ *for the model is obtained as follows:*

$$\begin{aligned} W(t) =& [-5.0501e^{1.87614t}\sin(0.744301t) + 0.861564e^{2.8087t}\sin(2.05986t) \\ &+ 1.68092e^{1.87614t}\cos(0.744301t) - 0.183397e^{2.8087t}\cos(2.05986t) \\ &- 0.0876947e^{-2.37228t} - 1.44189e^{3.37228t} \\ &+ 0.999999e^{4t} - 0.967939]e^{-4t}, \quad t \geq 0. \end{aligned}$$

Table 5. Computation of post-departure epoch probabilities and random epoch probabilities (when the server is busy) for Example 5 with the parameters $\lambda = 10$, $\mu = 1$, $a = 3$, $b = 10$, $N = 20$ and $\rho = 1$.

n	$p^+(n)$	$p_1(n)$	n	$p^+(n)$	$p_1(n)$
0	0.029150	0.045410	11	0.050509	0.044813
1	0.036192	0.046814	12	0.050621	0.039850
2	0.040896	0.047768	13	0.050699	0.034880
3	0.044057	0.048419	14	0.050755	0.029904
4	0.046194	0.048866	15	0.050794	0.024925
5	0.047648	0.049174	16	0.050821	0.019943
6	0.048641	0.049388	17	0.050841	0.014959
7	0.049324	0.049537	18	0.050855	0.009973
8	0.049796	0.049640	19	0.050864	0.004987
9	0.050123	0.049713	20	0.050871	0.221596
10	0.050350	0.049764	-	-	-

$p_0(0) = 0.002858$, $p_0(1) = 0.006406$, $p_0(2) = 0.010415$.

Putting $k = 1$ *and* 2 *in Equation (42), we obtain the following:*

$$E[V_q] = 1.315271, E[V_q^2] = 2.318872, \text{and}$$

$$Var(V_q) = 0.588934.$$

Similarly, a moment of order n can be obtained for the sojourn time in the system for a random customer by substituting $k = 1$ *and* $k = 2$ *in Equation (46):*

$$E[V] = 2.315271, E[V^2] = 7.874414, \text{and}$$

$$Var(V) = 2.513934.$$

We have plotted loss probability (P_{loss}) *and mean sojourn time* ($E[V]$) *against the buffer size N by varying it from 10 to 45 in Figure 5, from which it is apparent that as the system capacity rises,* P_{loss} *decreases and the mean sojourn time increases linearly apparently.*

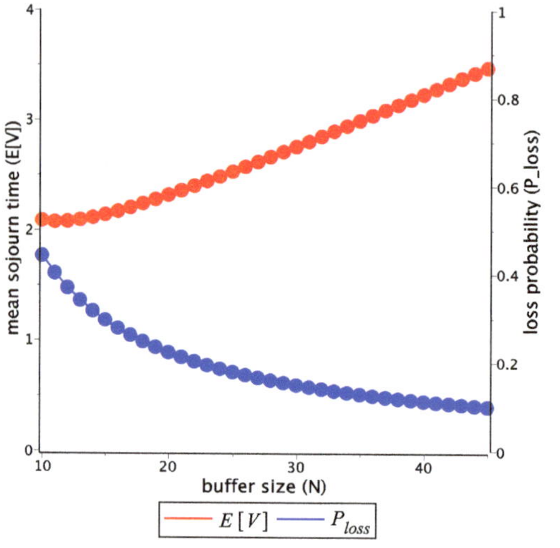

Figure 5. Buffer size (N) versus mean sojourn time ($E[V]$) and N versus loss probability (P_{loss}).

Example 6. *Medhi [7] deals with the waiting-time distribution for the $M/M^{(a,b)}/1$ queueing model under steady-state. In this example, we have analysed this model and compared it with the results of [7]. Here, the service time is assumed to be exponentially distributed with a mean of $\frac{1}{\mu}$ and $B^*(s) = \frac{\mu}{\mu+s}$. Let us take the parameters as $a = 3, b = 6, N = 200, \lambda = 6.7$, and $\mu = 1.7$ with $\rho = \frac{\lambda}{b\mu} = 0.656863$. Following a similar procedure, as described in the examples mentioned earlier, we have obtained the following results and compared our results with that of [7] in Table 6. Moreover, we have calculated the CDF of the waiting time in the queue and compared it with the results of [7]. Using the procedure discussed in the paper, $W_q(t)$ is obtained as follows:*

$$\begin{aligned} W_q(t) =& 0.999999 - 0.0177191e^{-5.20699t} - 0.125600e^{-5.10225t}\cos(10.5612t) \\ & - 0.257258e^{-5.10225t}\sin(10.5612t) - 0.856680e^{-0.905312t} \\ & + 0.0000107034e^{-0.335552t}, \quad t > 0. \end{aligned}$$

Using [7], $W_q(t)$ is obtained as follows:

$$\begin{aligned} W_q(t) =& 1.000000 - 0.856711e^{-0.90531t} - 0.434749e^{-6.7t} + 0.360203e^{-7.60531t} - 1.45641e^{-6.7t}t \\ & + 1.28304te^{-7.60531t}, \quad t > 0. \end{aligned}$$

The results have been exhibited in Table 7 and Figure 6.

Table 6. Performance measures of different quantities with that of [7].

Quantities	Expression	Results Obtained from the Present Paper	Results Obtained from Medhi [7]
Expected waiting time in the queue	$E(V_q)$	0.974087	0.974087
Second order moment of V_q	$E(V_q^2)$	2.093599	2.105153
Expected sojourn time in the system	$E(V)$	1.562322	1.562322

Table 7. Comparison between numerical results for CDF of queueing time of a random customer from Example 1 with the parameters $\lambda = 6.7, \mu = 1.7, a = 3, b = 6, N = 200$ and $\rho = 0.656863$.

t	$W_q(t)$ from the Present Paper	$W_q(t)$ from Medhi [7]
2	0.859884	0.859882
4	0.977085	0.977084
6	0.996252	0.996252
8	0.999387	0.999387
10	0.999900	0.999900
12	0.999984	0.999984
14	0.999997	0.999997

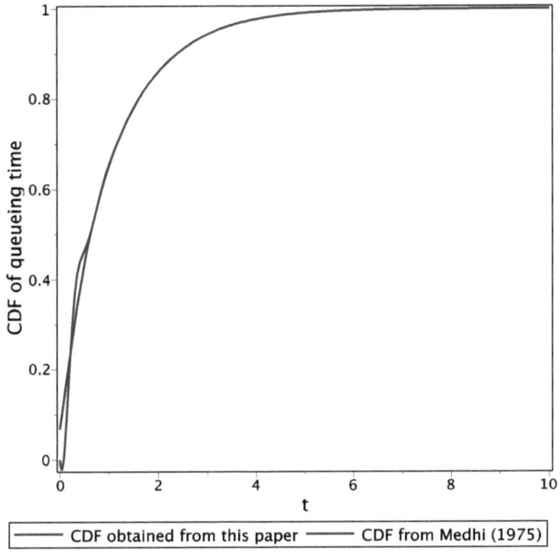

Figure 6. Comparison of CDF of queueing time obtained from this paper against that of Medhi [7].

Example 7. *In reference [49], the sojourn-time distribution for the $M/M^{[a]}/1$ queueing model under steady-state was numerically illustrated with the parameters $a = b = 4, \lambda = 3.0$, and $\mu = 1.0$, and $\rho = 0.75$. Reference [49] p. 1511, obtained the expression for $W(t)$ as follows:*

$$W(t) = 1.0 + 0.237540e^{-3t} + 0.339631te^{-3t} + 0.162036t^2e^{-3t} - 1.237543e^{-0.377695t}, \quad t \geq 0.$$

Using the procedure discussed in this paper $W(t)$ is obtained as follows:

$$\begin{aligned} W(t) =& 0.999999 + (0.141648e^{-0.23297t}\sin(7.12046t) + 0.183803e^{-6.82028t} \\ &+ 0.0214328e^{-0.23297t}\cos(7.12046t) - 1.20519e^{0.622305t} \\ &+ 0.0000105754e^{0.767445t} - 0.0000332082)e^{-t}, \quad t \geq 0. \end{aligned}$$

Using Medhi [7], $W(t)$ is obtained as follows:

$$\begin{aligned} W(t) =& 0.999999 - 0.0420232e^{-3.37770t}t^2 - 0.882464e^{-3.37770t}t - 1.20518e^{-0.3777t} \\ &- 0.621137e^{-3.37770} - 0.204900e^{-t} + 0.562490e^{-3.0t}t^2 \\ &+ 1.31248e^{-3.0t}t + 1.03123e^{-3.0t}, \quad t \geq 0. \end{aligned}$$

In this example, we have analysed this model and compared it with [7]'s results. Here, the service time is assumed to be exponentially distributed with a mean of $\frac{1}{\mu} = 1.0$ *and* $B^*(s) = \frac{1.0}{1.0+s}$. *Following a similar procedure, as described in the aforementioned examples, we have obtained the following results taking* $N = 300$ *and compared our results with that of [49] in Table 8. Also, we have compared them with the results of [7]. The results have been exhibited in Table 8. Further, it can be seen that* $E(V) = 3.147639$ *exactly matches with that of [7] while* $E(V) = 3.147645$ *is given in [49].*

Table 8. Comparison among the CDF of sojourn times from Example 7 with the parameters $\lambda = 3$, $\mu = 1.0,\ a = b = 4,\ N = 300$ and $\rho = 0.75$.

t	$W(t)$ from Present Paper with $N = 300$	$W(t)$ from Medhi [7]	$W(t)$ from Yu and Tang [49]
2	0.445537	0.415938	0.418568
4	0.733605	0.730294	0.726828
6	0.874932	0.874503	0.871656
8	0.941281	0.941208	0.939701
10	0.972411	0.972401	0.971670
12	0.987037	0.987036	0.986690
14	0.993910	0.993910	0.993746
16	0.997139	0.997139	0.997062
18	0.998656	0.998656	0.998620
20	0.999368	0.999368	0.999351
	$E(V) = 3.147639$	$E(V) = 3.147639$	$E(V) = 3.147645$

Computation Time Comparison

Finally, the computation time for determining the queue-length distribution at a post-departure epoch of a batch in an $M/PH^{(3,10)}/1/N$ queueing system using the roots' method, the GTH algorithm and the RG-factorization technique (extensions of the matrix-analytic method) have been compared. The computation time for these three techniques (which are already discussed in the previous remark) has been presented graphically in Figure 7. We used MAPLE 2015 in a PC with Intel(R) Core i7 processor @3.40 GHz with 8 GB RAM and the Windows 10 environment for this experiment. The PH representation is taken the same as used for Example 1. The mean arrival rate (λ) is taken as $\lambda = 20$. The buffer capacity (N) varies from 11 to 25, and for each case, the computation time (in seconds) has been noted down for computing the stationary queue-length distribution at a post-departure epoch of a batch. The computation times against each buffer capacity are presented in Figure 7. Moreover, from Figure 7, it may be observed that the computation time using the roots' method is lower than that of the RG-factorization technique. Further, from Figure 7, one can also see that the computation time for the roots' method and GTH algorithm are almost the same in the given range of the buffer capacity. One may observe that the computation time using the roots' method and GTH algorithm is almost parallel to the x-axis in the given range of N, see Figure 7.

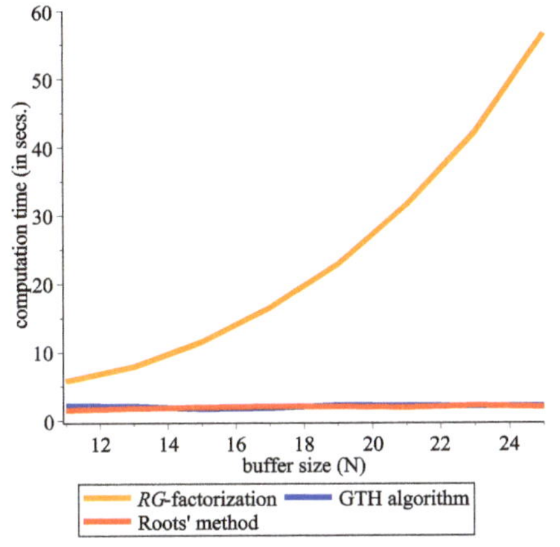

Figure 7. Computation time for different techniques with varying number of servers.

6. Conclusions

In this paper, the finite-buffer, bulk-service queue has been analysed, and the steady-state probability distributions at random- and post-departure epochs have been obtained. For this purpose, the roots of the characteristic equation are used. After that, the functional relation between the pgf of the queue-length distribution and the LST of the queueing-time distribution have been used to obtain the queueing time distribution when the server is busy. Moreover, we obtain the probability density functions of the sojourn time and the queueing-time distributions. The analysis of the stationary probabilities for the batch arrivals and batch service queues have problems in dealing with waiting-time distributions. Instead of the Poisson arrival process, one may consider a non-renewal (batch) arrival process (MAP or $BMAP$ arrival processes) to the same bulk-service queue studied in this paper, and in such a case, the stationary distributions may be achieved using a solution procedure similar to the present article. The analysis of stationary probability vectors for the corresponding batch size dependent bulk-service service queues, such as $BMAP/MSP_n^{(a,b)}/1/N\ (\infty)$ and $BMAP/G_n^{(a,b)}/1/N\ (\infty)$ are interesting problems and may attract researchers' attention in the near future.

Author Contributions: M.C.: Conceptualization, writing, review & Editing; A.D.B.: Methodology, software, writing—original draft, validation; S.B.: Investigation, data curation & visualization; V.G.: Supervision & project administration. All authors have read and agreed to the published version of the manuscript.

Funding: The first author received partial support from CDARP, Canada. The second author acknowledges the partial financial support from the Department of Science and Technology, New Delhi, India, under the research grant MTR/2021/000287.

Data Availability Statement: Any supporting data related to this research may be available upon request to the corresponding or any author of this article.

Conflicts of Interest: The authors declare no conflict of interest.

Appendix A

Remark A1. *It may be remarked here that the results derived in Equations (30) and (31) are actually the same as the corresponding expressions for these probabilities derived in Equations (2.18) and (2.19) presented by Chaudhry and Gupta [17], respectively. This fact can be verified as follows: Solving n linear equations given by (30) with n unknowns* $p_0(n)$ $(0 \le n \le a-1)$*, we obtain:*

$$p_0(n) = \frac{1}{\lambda T}\sum_{i=0}^{n} p^+(i) = \frac{\sum_{i=0}^{n} p^+(i)}{\rho b + \sum_{i=0}^{a-1}(a-i)p^+(i)}, \tag{A1}$$

where $\lambda T = \rho b + \sum_{i=0}^{a-1}(a-i)p^+(i)$ *using Equation (26). Equation (A1) is exactly the same as Equation (2.18) in [17]. Next, consider the following pgf as described before:*

$$\widehat{K}(z) \simeq \sum_{i=0}^{\infty} \widehat{k}_i z^i = \mu \frac{1 - B^*(\lambda - \lambda z)}{\lambda - \lambda z} = \frac{\mu}{\lambda}\Big(1 - \sum_{i=0}^{\infty} k_i z^i\Big)(1 + z + z^2 + \cdots), \tag{A2}$$

where $K(z) = B^*(\lambda - \lambda z) \simeq \sum_{i=0}^{\infty} k_i z^i$. *Now, comparing the coefficients of the like powers of z in both sides of Equation (A2), we obtain:*

$$\widehat{k}_n = \frac{\mu}{\lambda}(1 - \sum_{i=0}^{n} k_i), \quad n \geq 0. \tag{A3}$$

Substituting $\widehat{k}_n$ *from Equation (A3) on the right-hand side of Equation (31), using Equation (2) and after simplification, we get:*

$$p_1(n) = \frac{\sum_{i=n+1}^{\min\{b+n,N\}} p^+(i)}{\rho b + \sum_{i=0}^{a-1}(a-i)p^+(i)}, \quad 0 \leq n \leq N-1, \tag{A4}$$

which is exactly the same expression as given by Equation (2.19) in [17].

References

1. Chaudhry, M.L.; Templeton, J.G.C. *First Course in Bulk Queues*; John Wiley & Sons Inc.: Hoboken, NJ, USA, 1983.
2. Dshalalow, J.H. On modulated random measures. *Int. J. Stoch. Anal.* **1991**, *4*, 305–312. [CrossRef]
3. Dshalalow, J.H. A single-server queue with random accumulation level. *Int. J. Stoch. Anal.* **1991**, *4*, 203–210. [CrossRef]
4. Dshalalow, J.H. *Frontiers in Queueing: Models and Applications in Science and Engineering*; CRC Press: Boca Raton, FL, USA, 1997; Volume 7.
5. Neuts, M.F. A general class of bulk queues with Poisson input. *Ann. Math. Stat.* **1967**, *38*, 759–770. [CrossRef]
6. Borthakur, A. A Poisson queue with a general bulk service rule. *J. Assam. Sci. Soc.* **1971**, *14*, 162–167.
7. Medhi, J. Waiting time distribution in a Poisson queue with a general bulk service rule. *Manag. Sci.* **1975**, *21*, 777–782. [CrossRef]
8. Easton, G.; Chaudhry, M. The queueing system $E_k/M^{(a,b)}/1$ and its numerical analysis. *Comput. Oper. Res.* **1982**, *9*, 197–205. [CrossRef]
9. Medhi, J. *Recent Developments in Bulk Queueing Models*; Wiley Eastern Limited: Bridgewater, NJ, USA, 1984.
10. Powell, W.B. Stochastic Delays in Transportation Terminals: New Results in the Theory and Application of Bulk Queues. Ph.D. Thesis, Massachusetts Institute of Technology, Cambridge, MA, USA, 1981.
11. Chaudhry, M.L.; Madill, B.; Briere, G. Computational analysis of steady-state probabilities of $M/G^{(a,b)}/1$ and related nonbulk queues. *Queueing Syst.* **1987**, *2*, 93–114. [CrossRef]
12. Briere, G.; Chaudhry, M. Computational analysis of single-server bulk-service queues, $M/G^Y/1$. *Adv. Appl. Probab.* **1989**, *21*, 207–225.
13. Finch, P. The effect of the size of the waiting room on a simple queue. *J. R. Stat. Soc. Ser. B (Methodol.)* **1958**, *20*, 182–186. [CrossRef]
14. Chaudhry, M.L.; Gupta, U.C.; Agarwal, M. On exact computational analysis of distributions of numbers in systems for $M/G/1/N+1$ and $GI/M/1/N+1$ queues using roots. *Comput. Oper. Res.* **1991**, *18*, 679–694. [CrossRef]
15. Chaudhry, M.; Goswami, V. The Queue $Geo/G/1/N+1$ Revisited. *Methodol. Comput. Appl. Probab.* **2019**, *21*, 155–168. [CrossRef]
16. Singh, V. Finite waiting space bulk service system. *J. Eng. Math.* **1971**, *5*, 241–248. [CrossRef]
17. Chaudhry, M.L.; Gupta, U.C. Modelling and analysis of $M/G^{(a,b)}/1/N$ queue—A simple alternative approach. *Queueing Syst.* **1999**, *31*, 95–100. [CrossRef]
18. Gold, H.; Tran-Gia, P. Performance analysis of a batch service queue arising out of manufacturing system modelling. *Queueing Syst.* **1993**, *14*, 413–426. [CrossRef]
19. Goswami, V.; Patra, S.S.; Mund, G.B. Performance analysis of cloud computing centers for bulk services. *Int. J. Cloud Appl. Comput. (IJCAC)* **2012**, *2*, 53–65. [CrossRef]
20. Banerjee, A.; Gupta, U.C.; Chakravarthy, S.R. Analysis of a finite-buffer bulk-service queue under Markovian arrival process with batch-size-dependent service. *Comput. Oper. Res.* **2015**, *60*, 138—149. [CrossRef]
21. Pradhan, S.; Gupta, U.C. Analysis of an infinite-buffer batch-size-dependent service queue with Markovian arrival process. *Ann. Oper. Res.* **2019**, *277*, 161–196. [CrossRef]

22. Keilson, J.; Servi, L. A distributional form of Little's law. *Oper. Res. Lett.* **1988**, *7*, 223–227. [CrossRef]
23. Botta, R.F.; Harris, C.M.; Marchal, W.G. Characterizations of generalized hyperexponential distribution functions. *Stoch. Model.* **1987**, *3*, 115–148. [CrossRef]
24. Neuts, M.F. *Matrix-Geometric Solutions in Stochastic Models: An Algorithmic Approach*; Courier Corporation: Chelmsford, MA, USA, 1981.
25. Kendall, D.G. Some recent work and further problems in the theory of queues. *Theory Probab. Its Appl.* **1964**, *9*, 1–13. [CrossRef]
26. Kleinrock, L. *Queueing Systems, Volume 1: Theory*; John Wiley & Sons. Inc.: New York, NY, USA, 1975; p. 417.
27. Chaudhry, M. *QROOT Software Package*; A & A Publications: Kingston, ON, Canada, 1992.
28. Chaudhry, M.L. Numerical issues in computing steady-state queueing-time distributions of single-server bulk-service queues: $M/G^b/1$ and $M/G^d/1$. *ORSA J. Comput.* **1992**, *4*, 300–310. [CrossRef]
29. Chaudhry, M.L.; Harris, C.M.; Marchal, W.G. Robustness of root finding in single-server queueing models. *ORSA J. Comput.* **1990**, *2*, 273–286. [CrossRef]
30. Neuts, M.F. *Structured Stochastic Matrices of $M/G/1$ Type and Their Applications*; CRC Press: Boca Raton, FL, USA, 2021.
31. Li, Q.L. *Constructive Computation in Stochastic Models with Applications: The RG-Factorizations*; Springer Science & Business Media: Berlin/Heidelberg, Germany, 2011.
32. Gelenbe, E.; Lent, R. *Computer and Information Sciences III*; Springer: Berlin/Heidelberg, Germany, 2013.
33. Li, Q.L.; Lian, Z.; Liu, L. An RG-factorization approach for a $BMAP/M/1$ generalized processor-sharing queue. *Stoch. Model.* **2005**, *21*, 507–530. [CrossRef]
34. Borwein, J.M.; Borwein, P.B. *Pi and the AGM: A Study in the Analytic Number Theory and Computational Complexity*; Wiley-Interscience: Hoboken, NJ, USA, 1987.
35. Xin, G. A fast algorithm for partial fraction decompositions. *arXiv* **2004**, arXiv:math/0408189.
36. Chaudhry, M.; Templeton, J. The queuing system $M/G^B/l$ and its ramifications. *Eur. J. Oper. Res.* **1981**, *6*, 56–60. [CrossRef]
37. Abate, J.; Whitt, W. Numerical inversion of probability generating functions. *Oper. Res. Lett.* **1992**, *12*, 245–251. [CrossRef]
38. Abate, J.; Choudhury, G.L.; Whitt, W. An introduction to numerical transform inversion and its application to probability models. In *Computational Probability*; Springer: Berlin/Heidelberg, Germany, 2000; pp. 257–323.
39. Fan, C.; Ghaemi, S.; Khazaei, H.; Musilek, P. Performance evaluation of blockchain systems: A systematic survey. *IEEE Access* **2020**, *8*, 126927–126950. [CrossRef]
40. Nakamoto, S.; Bitcoin, A. A Peer-to-Peer Electronic Cash System. 2008. Available online: https://bitcoin.org/bitcoin.pdf (accessed on 17 March 2022).
41. Kawase, Y.; Kasahara, S. Transaction-confirmation time for bitcoin: A queueing analytical approach to blockchain mechanism. In Proceedings of the Queueing Theory and Network Applications: 12th International Conference, QTNA 2017, Qinhuangdao, China, 21–23 August 2017; Springer: Berlin/Heidelberg, Germany, 2017; pp. 75–88.
42. Li, Q.L.; Ma, J.Y.; Chang, Y.X. Blockchain queue theory. In Proceedings of the Computational Data and Social Networks: 7th International Conference, CSoNet 2018, Shanghai, China, 18–20 December 2018; Springer: Berlin/Heidelberg, Germany, 2018; pp. 25–40.
43. Geissler, S.; Prantl, T.; Lange, S.; Wamser, F.; Hossfeld, T. Discrete-time analysis of the blockchain distributed ledger technology. In Proceedings of the 2019 31st International Teletraffic Congress (ITC 31), Budapest, Hungary, 27–29 August 2019; pp. 130–137.
44. Geyer, F.; Kinkelin, H.; Leppelsack, H.; Liebald, S.; Scholz, D.; Carle, G.; Schupke, D. Performance perspective on private distributed ledger technologies for industrial networks. In Proceedings of the 2019 International Conference on Networked Systems (NetSys), Munich, Germany, 18 March 2019; pp. 1–8.
45. Ramsay, C.M. Exact waiting time and queue size distributions for equilibrium $M/G/1$ queues with Pareto service. *Queueing Syst.* **2007**, *57*, 147–155. [CrossRef]
46. Roughan, M.; Veitch, D.; Rumsewicz, M. Computing queue-length distributions for power-law queues. In Proceedings of the IEEE INFOCOM'98, the Conference on Computer Communications, Seventeenth Annual Joint Conference of the IEEE Computer and Communications Societies, Gateway to the 21st Century (Cat. No. 98), San Francisco, CA, USA, 29 March–2 April 1998; Volume 1, pp. 356–363.
47. Avram, F.; Banik, A.D.; Horváth, A. Ruin probabilities by Padé's method: simple moments based mixed exponential approximations (Renyi, De Vylder, Cramér–Lundberg), and high precision approximations with both light and heavy tails. *Eur. Actuar. J.* **2019**, *9*, 273–299. [CrossRef]
48. Akar, N. Solving the $ME/ME/1$ queue with state–space methods and the matrix sign function. *Perform. Eval.* **2006**, *63*, 131–145. [CrossRef]
49. Yu, M.; Tang, Y. Analysis of the sojourn time distribution for $M/G^L/1$ queue with bulk-service of exactly size L. *Methodol. Comput. Appl. Probab.* **2018**, *20*, 1503–1514. [CrossRef]

mathematics

MDPI

Article

Large Deviations for Hawkes Processes with Randomized Baseline Intensity

Youngsoo Seol

Department of Mathematics, Dong-A University, Busan 49315, Republic of Korea; prosul76@dau.ac.kr

Abstract: The Hawkes process, which is generally defined for the continuous-time setting, can be described as a self-exciting simple point process with a clustering effect, whose jump rate depends on its entire history. Due to past events determining future developments of self-exciting point processes, the Hawkes model is generally not Markovian. In certain special circumstances, it can be Markovian with a generator of the model if the exciting function is an exponential function or the sum of exponential functions. In the case of non-Markovian processes, difficulties arise when the exciting function is not an exponential function or a sum of exponential functions. The intensity of the Hawkes process is given by the sum of a baseline intensity and other terms that depend on the entire history of the point process, as compared to a standard Poisson process. It is one of the main methods used for studying the dynamical properties of general point processes, and is highly important for credit risk studies. The baseline intensity, which is instrumental in the Hawkes model, is usually defined for deterministic cases. In this paper, we consider a linear Hawkes model where the baseline intensity is randomly defined, and investigate the asymptotic results of the large deviations principle for the newly defined model. The Hawkes processes with randomized baseline intensity, dealt with in this paper, have wide applications in insurance, finance, queue theory, and statistics.

Keywords: Hawkes process; self-exciting point processes; intensity; baseline intensity; large deviations principles

MSC: 60G55; 60F05; 60F10

Citation: Seol, Y. Large Deviations for Hawkes Processes with Randomized Baseline Intensity. *Mathematics* **2023**, *11*, 1826. https://doi.org/10.3390/math11081826

Academic Editors: Gurami Tsitsiashvili and Alexander Bochkov

Received: 22 February 2023
Revised: 29 March 2023
Accepted: 11 April 2023
Published: 12 April 2023

1. Introduction

The Hawkes process is a self-exciting point process with a clustering effect and a jump rate that depends on its entire history [1]. There are many uses and benefits to Hawkes processes for modeling simple point processes. The Hawkes process exhibits self-exciting properties and clustering effects. Compared with a standard Poisson process, the intensity process for a point process consists of a summation of the baseline intensity and other terms that depend on the process's past. Hawkes processes are typically used in the modeling of high-frequency trading to express temporal phenomena in a stochastic process that evolves in continuous time. In addition to capturing the self-exciting property and clustering effects, the Hawkes process is a natural generalization of the Poisson process. This process is a variable model that is amenable to statistical analysis. As a result, it has a wide range of applications in criminology [2], finance [3], seismology [4], DNA modeling [5], neuroscience [6], machine learning, and artificial intelligent [7,8]. A general description of the Hawkes process is given below.

Let N be a simple point process on $\mathbb{R}$ and let $\mathcal{F}_t^{-\infty} := \sigma(N(C), C \in \mathcal{B}(\mathbb{R}), C \subset (-\infty, t])$ be an increasing family of $\sigma-$algebras. Any nonnegative $\mathcal{F}_t^{-\infty}$-progressively measurable process λ_t having

$$E\left[N(a,b]|\mathcal{F}_a^{-\infty}\right] = E\left[\int_a^b \lambda_s ds|\mathcal{F}_a^{-\infty}\right] \tag{1}$$

a. s. for all intervals $(a, b]$ is called an $\mathcal{F}_t^{-\infty}$-intensity of N. We use the notation $N_t := N(0, t]$ to denote the number of points in the interval $(0, t]$.

Let N be a simple point process containing an $\mathcal{F}_t^{-\infty}$-intensity (general Hawkes process)

$$\lambda_t := \lambda\left(\int_{-\infty}^{t} h(t-s)N(ds)\right), \tag{2}$$

where $\lambda(\cdot) : \mathbb{R}^+ \to \mathbb{R}^+$ is locally integrable and left continuous, $h(\cdot) : \mathbb{R}^+ \to \mathbb{R}^+$ and assuming $\|h\|_{L^1} = \int_0^\infty h(t)dt < \infty$. Here, $\int_{-\infty}^{t} h(t-s)N(ds)$ stands for $\int_{(-\infty,t)} h(t-s)N(ds)$. It is assumed that $N(-\infty, 0] = 0$, in other words, the history of the Hawkes process is nil. Typically, $h(\cdot)$ and $\lambda(\cdot)$ are known as the exciting function and rate function, respectively. The Hawkes process can either be linear or nonlinear. It is linear only when $\lambda(\cdot)$ is linear, otherwise, it is nonlinear. The above-mentioned model is non-Markovian. This is because the future evolution of this model depends upon its history of events and their occurrences. It is only in a special case that the above model behaves as Markovian. The Hawkes process has several applications, such as in criminology [2], finance [3], seismology [4], DNA modeling [5], and neuroscience [6]. Since it has both self-exciting and clustering properties, it is considered appropriate for certain financial applications. The clustering and self-exciting properties of the Hawkes processes have been utilized in modeling correlated defaults and estimating credit derivatives in finance; see Errais et al. [3] and Dassios and Zhao [9].

The linear Hawkes process [1] is much more commonly used than the non-linear one. Almost all applications of the Hawkes process use the linear model, such as immigration-birth representation [10], central limit theorems, the law of large numbers, large deviation principles, and the Bartlett spectrum [1,11–13]. In contrast, the nonlinear Hawkes process finds less application due to the lack of immigration birth representation and computational tractability. The nonlinear model was first reported by Brémaud and Massoulié [14] and has since been studied extensively [15–19], for example, the central limit theorem in [17], the large deviation principles [16], and applications in financial mathematics [19,20]. Jaisson and Rosenbaum presented limit behavior theorems and scaling limits for the rough fractional diffusion of the nearly unstable Hawkes model [21,22]. Seol studied the inverse process of Hawkes processes and reported limit theorems for the arrival time τ_n, arrival time [23]. Due to the continuous progress in storage technology, data-driven models have gained significant attention recently. However, the Hawkes process is defined in continuous-time settings, whereas events, in reality, are often recorded in discrete time. More importantly, the data should be collected in fixed intervals to avoid showing aggregate results. For instance, continuous-time Hawkes processes can model events occurring at unequal time intervals, whereas equal interval discrete-time Hawkes processes are required to model events occurring at fixed intervals. Seol proposed a 0–1 discrete case Hawkes process starting from an empty history and used it to study central limit theorems, laws of large numbers, and invariance principles [24]. Recently, Wang [25,26] studied the limit behaviors of a discrete-time Hawkes process with random marks and proved the large and moderate deviations for a discrete-time Hawkes process with marks.

Seol also investigated moderate deviation principles for Hawkes processes with random marks [27], and limit laws for the process of the compensator in Hawkes processes [28]. In addition to large time limits, Gao and Zhu also conducted studies on asymptotic results [29–32]. Furthermore, several studies related to the extension and modification of the classical Hawkes process have been reported in the literature. For example, in one study, the baseline intensity was considered to be inhomogeneous in time [33]. In another report, a dynamic contagion model was used where the immigrants were assumed to arrive following the Cox process with shot noise intensity [9]. Next, Wheatley et al. considered the renewal Hawkes process, which is a generalization of the classical Hawkes process, where immigrants arrive according to a renewal process instead of a Poisson process [34]. In addition to the studies mentioned above, other variations and extensions of the Hawkes process have also been reported in the literature [9,35–38]. Recently, Seol introduced the inverse Markovian Hawkes process and studied the limit theorems for the same. The inverse Markovian Hawkes process has the features of several existing models of the self-exciting process [39]. Seol also reported an extended version of the inverse Markovian Hawkes model [40] and non-Markovian inverse model [41]. In the present paper, linear Hawkes processes are employed with randomized baseline intensity, which has important applications. We also studied the limit theorems for the randomized baseline intensity of the Hawkes process. It is worth mentioning that Gao and Zhu studied the functional central limit theorems for the stationary case of the Hawkes process with its application to infinite-server queues [33]. They also studied large baseline intensity asymptotics. However, there are two main differences between their paper [33] and the current paper. First, in the present work, empty history was used, but Gao and Zhu used the stationary version of the Hawkes process [33]. Second, in the current paper, we studied scalar limit theorems for fixed t, whereas Gao and Zhu [33] considered functional limit theorems. The present paper is organized as follows: Section 2 presents some auxiliary results to support our main results, Section 3 presents our main results, and Section 4 contains the proofs of the main theorems.

2. Preliminaries

A review of the main problems is given in this section, along with an introduction to the classical results. Throughout the paper, we will be using a set of assumptions.

Assumption 1. *(1)* $\lambda(z) = \nu + z$, *for some* $\nu > 0$,
(2) $\|h\|_{L^1} < 1$ *where* $\|h\|_{L^1} = \int_0^\infty h(t)dt < \infty$.

Assumption 1 denotes that λ is a linear and increasing function. This implies that the Hawkes process represents immigration-birth very well, which is the main tool in the current paper.

A linear Hawkes process can be represented by an immigration-birth process, which is used in proving the theorem. The following is a general description of the immigration-birth representation. An immigrant arrives following a standard homogeneous Poisson process at a constant intensity $\nu > 0$. Then, based on the Galton–Watson tree, each immigrant generates a number of children [10]. Let η be the number of an immigrant's children, and η has a Poisson distribution with parameter $\|h\|_{L^1}$. Conditioned on the number of children of an immigrant, the time when a child is born has a probability density function of

$$\frac{h(\cdot)}{\|h\|_{L^1}}. \tag{3}$$

The immigration-birth representation signifies that N_t denotes the total number of immigrants and their descendants up to time t.

Some reviews on the results of the Hawkes processes are shown below.

Limit Theorems for Hawkes Processes

The limit theorems of both linear and nonlinear Hawkes processes have been studied by many researchers.

Linear case model: Since $\lambda(\cdot)$ is linear, say $\lambda(z) = \nu + z$ for some $\nu > 0$, and $\|h\|_{L^1} < 1$, there is an excellent immigration-birth representation, and the limit theorems can be represented more explicitly. Daley and Vere-Jones [42] investigated the proof of the law of large numbers for the linear case of the Hawkes process. Functional central limit theorems for linear cases of multivariate Hawkes processes with certain assumptions were studied by Bacry et al. [11]. Bordenave and Torrisi [12] proved that if $0 < \|h\|_{L^1} < 1$ and $\int_0^\infty th(t)dt < \infty$, then $(\frac{N_t}{t} \in \cdot)$ satisfies the large deviation principles. Zhu reported the moderate deviation principle for the linear case in continuous-time Hawkes processes [18], as well as the limit theorems for linear Hawkes processes with random marks [37].

Nonlinear case model: Since $\lambda(\cdot)$ is nonlinear, the usual immigration-birth representations are no longer relevant. This makes the nonlinear model much harder to study. Brémaud and Massoulié [14] proved that there are unique stationary versions of nonlinear Hawkes processes, under certain conditions, and nonstationary versions converge to equilibrium. The central limit theorem obtained by Zhu [15,17] demonstrated the large deviation for a special case of the nonlinear model when $h(\cdot)$ is exponential or a sum of exponentials. Zhu [16] proved a process-level large deviation principle for nonlinear Hawkes processes with general $h(\cdot)$, namely the level-3 large deviation principle and, therefore, by the contradiction principle, the level-1 large deviation principle for $(\frac{N_t}{t} \in \cdot)$. Before we proceed, let us review some limit theorem results of the linear Hawkes process from the literature. Daley and Vere-Jones [42] proved the law of large numbers for the linear Hawkes process, as follows.

$$\frac{N_t}{t} \to \frac{\nu}{1 - \|h\|_{L^1}} \text{ as } t \to \infty \tag{4}$$

The functional central limit theorem for the linear multivariate case of the Hawkes process, under certain assumptions, was obtained by Bacry et al. [11]. They proved that

$$\frac{N_{\cdot t} - \cdot \mu t}{\sqrt{t}} \to \sigma B(\cdot), \text{ as } t \to \infty, \tag{5}$$

where $B(\cdot)$ is a standard Brownian motion and

$$\mu = \frac{\nu}{1 - \|h\|_{L^1}} \text{ and } \sigma^2 = \frac{\nu}{(1 - \|h\|_{L^1})^3} \tag{6}$$

The convergence used in the above theorem is weak convergence on the space $D[0,1]$, which consists of càdlàg functions on $[0,1]$, and is equipped with the Skorokhod topology. Bordenave and Torrisi [12] proved that if $0 < \|h\|_{L^1} < 1$ and $\int_0^\infty th(t)dt < \infty$, then $(\frac{N_t}{t} \in \cdot)$ satisfies the large deviation principle with the good rate function $I(\cdot)$, which means that for any closed set $C \subset \mathbb{R}$,

$$\limsup_{t \to \infty} \frac{1}{t} \log \mathbb{P}(N_t/t \in C) \leq - \inf_{x \in C} I(x),$$

and for any open set $G \subset \mathbb{R}$,

$$\liminf_{t \to \infty} \frac{1}{t} \log \mathbb{P}(N_t/t \in G) \geq - \inf_{x \in G} I(x), \tag{7}$$

where

$$I(x) = \begin{cases} x\theta_x + \nu - \frac{\nu x}{\nu + \|h\|_{L^1} x} & \text{if } x \in (0, \infty) \\ \nu & \text{if } x = 0 \\ +\infty & \text{if } x \in (-\infty, 0). \end{cases} \tag{8}$$

where $\theta = \theta_x$ is the unique solution in $(-\infty, \|h\|_{L^1} - 1 - \log \|h\|_{L^1})$, of

$$\mathbb{E}(e^{\theta S}) = \frac{x}{\nu + x\|h\|_{L^1}}, x > 0 \tag{9}$$

where S in the above equation denotes the total number of descendants of an immigrant, including the immigrant himself.

The rate function described above $I(x)$ can be represented as a more explicit form. Note that (see [43] for detail), for all $\theta \in (-\infty, \|h\|_{L^1} - 1 - \log \|h\|_{L^1})$, $\mathbb{E}(e^{\theta S})$ satisfies

$$\mathbb{E}(e^{\theta S}) = e^{\theta} e^{\|h\|_{L^1} (\mathbb{E}(e^{\theta S}) - 1)}, \tag{10}$$

which implies that $\theta_x = \log\left(\frac{x}{\nu+x\|h\|_{L^1}}\right) - \|h\|_{L^1}\left(\frac{x}{\nu+x\|h\|_{L^1}} - 1\right)$. Substituting into the formula, we have

$$I(x) = \begin{cases} x\log\left(\frac{x}{\nu+x\|h\|_{L^1}}\right) - x + \|h\|_{L^1}x + \nu & \text{if } x \in (0,\infty) \\ \nu & \text{if } x = 0 \\ +\infty & \text{if } x \in (-\infty, 0). \end{cases} \tag{11}$$

Zhu [18] proved that if $\|h\|_{L^1} < 1$ and $\sup_{t>0} t^{3/2}h(t) \leq C < \infty$, then for any Borel set A and time sequence $\sqrt{n} \ll c(n) \ll n$, there exists a moderate deviation principle

$$-\inf_{x\in A^\circ} J(x) \leq \liminf_{t\to\infty} \frac{t}{c(t)^2}\log\mathbb{P}\left(\frac{1}{c(t)}(N_t - \mu t) \in A\right)$$
$$\leq \limsup_{t\to\infty} \frac{t}{c(t)^2}\log\mathbb{P}\left(\frac{1}{c(t)}(N_t - \mu t) \in A\right) \leq -\inf_{x\in \bar{A}} J(x) \tag{12}$$

where $J(x) = \frac{x^2(1-\|h\|_{L^1})^3}{2\nu}$.

3. Statement of the Main Results

This section states the main results of this paper. First, the asymptotic results for the randomized baseline intensity will be presented. The results of the large deviation principle of the Hawkes process with the randomized baseline intensity are shown below.

A Hawkes process N_t has the following intensity:

$$\lambda_t = \nu + \int_0^{t-} h(t-s)dN_s, \tag{13}$$

where $h(\cdot)$ is the exciting function, $\nu > 0$ is the baseline intensity.

We consider a Hawkes process N_t with randomized baseline intensity, i.e., ν follows a distribution on $[\nu_{\min}, \nu_{\max}]$, with $0 < \nu_{\min} < \nu_{\max} < \infty$ and a probability density function $p(\nu) > 0$ for every $\nu \in [\nu_{\min}, \nu_{\max}]$. Conditional on ν, N_t has the process of intensity

$$\lambda_t = \nu + \int_0^{t-} h(t-s)dN_s. \tag{14}$$

For constant ν, we know that $\mathbb{P}(N_t/t \in \cdot)$ satisfies the large deviation principle with the function of the rate

$$I(x;\nu) = \begin{cases} x\log\left(\frac{x}{\nu+x\|h\|_{L^1}}\right) - x + x\|h\|_{L^1} + \nu & \text{if } x \in [0,\infty) \\ +\infty & \text{otherwise} \end{cases}. \tag{15}$$

In this paper, we consider a modified model of the Hawkes process with randomized baseline intensity, and investigate whether $\mathbb{P}(N_t/t \in \cdot)$ satisfies a large deviation principle with a rate function that depends on the randomized baseline intensity, as shown below.

Theorem 1. *Considering Assumption 1 is satisfied, $\mathbb{P}(N_t/t \in \cdot)$ satisfies the large deviation principle with rate function:*

$$I(x) = \begin{cases} I(x;\nu_{\max}) & \text{if } x > \frac{\nu_{\max}}{1-\|h\|_{L^1}}, \\ 0 & \text{if } x \in [\frac{\nu_{\min}}{1-\|h\|_{L^1}}, \frac{\nu_{\max}}{1-\|h\|_{L^1}}], \\ I(x;\nu_{\min}) & \text{if } 0 \leq x < \frac{\nu_{\min}}{1-\|h\|_{L^1}}, \end{cases} \tag{16}$$

and $I(x) = +\infty$ otherwise.

4. The Proofs of the Theorems

This section furnishes the proofs of the main theorems. The following lemma plays a key role in proving the main theorem.

Lemma 1. *For any* $x > \frac{\nu_{\max}}{1-\|h\|_{L^1}}$,

$$\lim_{\delta\to 0}\lim_{t\to\infty}\frac{1}{t}\log\mathbb{P}\left(\frac{N_t}{t}\in B_\delta(x)\right) = -I(x;\nu_{\max}). \tag{17}$$

Proof. For every $x > \frac{\nu_{\max}}{1-\|h\|_{L^1}}$,

$$\begin{aligned}\mathbb{P}(N_t \geq xt) &\geq \mathbb{P}(\nu \geq \nu_{\max}-\epsilon)\mathbb{P}(N_t \geq xt|\nu \geq \nu_{\max}-\epsilon)\\ &\geq \int_{\nu_{\max}-\epsilon}^{\nu_{\max}} p(\nu)d\nu\mathbb{P}(N_t \geq xt|\nu = \nu_{\max}-\epsilon).\end{aligned}$$

Thus,

$$\liminf_{t\to\infty}\frac{1}{t}\log\mathbb{P}(N_t\geq xt) \geq -I(x;\nu_{\max}-\epsilon). \tag{18}$$

On the other hand,

$$\mathbb{P}(N_t \geq tx) \leq \mathbb{P}(N_t \geq tx|\nu = \nu_{\max}), \tag{19}$$

and, thus,

$$\limsup_{t\to\infty}\frac{1}{t}\log\mathbb{P}(N_t\geq xt) \leq -I(x;\nu_{\max}). \tag{20}$$

Since (20) holds for any sufficiently small $\epsilon > 0$, we conclude that for $x > \frac{\nu_{\max}}{1-\|h\|_{L^1}}$,

$$\lim_{t\to\infty}\frac{1}{t}\log\mathbb{P}(N_t\geq xt) = -I(x;\nu_{\max}). \tag{21}$$

This implies that for any $x > \frac{\nu_{\max}}{1-\|h\|_{L^1}}$,

$$\lim_{\delta\to 0}\lim_{t\to\infty}\frac{1}{t}\log\mathbb{P}\left(\frac{N_t}{t}\in B_\delta(x)\right) = -I(x;\nu_{\max}). \tag{22}$$

This completes the proof of Theorem 1. □

Lemma 2. *For any* $x < \frac{\nu_{\min}}{1-\|h\|_{L^1}}$,

$$\lim_{\delta\to 0}\lim_{t\to\infty}\frac{1}{t}\log\mathbb{P}\left(\frac{N_t}{t}\in B_\delta(x)\right) = -I(x;\nu_{\min}). \tag{23}$$

Proof. For every $x < \frac{\nu_{\min}}{1-\|h\|_{L^1}}$,

$$\begin{aligned}\mathbb{P}(N_t \geq xt) &\leq \mathbb{P}(\nu \geq \nu_{\min}-\epsilon)\mathbb{P}(N_t \geq xt|\nu \geq \nu_{\min}-\epsilon)\\ &\leq \int_{\nu_{\min}-\epsilon}^{\nu_{\min}} p(\nu)d\nu\mathbb{P}(N_t \geq xt|\nu = \nu_{\min}-\epsilon).\end{aligned}$$

Thus,

$$\liminf_{t\to\infty}\frac{1}{t}\log\mathbb{P}(N_t\geq xt) \leq -I(x;\nu_{\min}-\epsilon). \tag{24}$$

On the other hand,

$$\mathbb{P}(N_t \geq tx) \geq \mathbb{P}(N_t \geq tx|\nu = \nu_{\min}), \tag{25}$$

and, thus,

$$\limsup_{t\to\infty}\frac{1}{t}\log\mathbb{P}(N_t\geq xt)\geq -I(x;\nu_{\min}). \tag{26}$$

Since (26) holds for any sufficiently small $\epsilon>0$, we conclude that for $x<\frac{\nu_{\min}}{1-\|h\|_{L^1}}$,

$$\lim_{t\to\infty}\frac{1}{t}\log\mathbb{P}(N_t\geq xt)=-I(x;\nu_{\min}). \tag{27}$$

This implies that for any $x<\frac{\nu_{\min}}{1-\|h\|_{L^1}}$,

$$\lim_{\delta\to 0}\lim_{t\to\infty}\frac{1}{t}\log\mathbb{P}\left(\frac{N_t}{t}\in B_\delta(x)\right)=-I(x;\nu_{\min}). \tag{28}$$

This completes the proof of Theorem 2.
Similarly, we can show that for any $x<\frac{\nu_{\min}}{1-\|h\|_{L^1}}$,

$$\lim_{\delta\to 0}\lim_{t\to\infty}\frac{1}{t}\log\mathbb{P}\left(\frac{N_t}{t}\in B_\delta(x)\right)=-I(x;\nu_{\min}). \tag{29}$$

This completes the proof of Theorem 2. □

Lemma 3. *For any $x\in[\frac{\nu_{\min}}{1-\|h\|_{L^1}},\frac{\nu_{\max}}{1-\|h\|_{L^1}}]$, we have*

$$\lim_{\delta\to 0}\lim_{t\to\infty}\frac{1}{t}\log\mathbb{P}\left(\frac{N_t}{t}\in B_\delta(x)\right)=0. \tag{30}$$

Proof. Finally, for any $x\in[\frac{\nu_{\min}}{1-\|h\|_{L^1}},\frac{\nu_{\max}}{1-\|h\|_{L^1}}]$, we have for any sufficiently small $\delta>0$,

$$\begin{aligned}\mathbb{P}\left(\frac{N_t}{t}\in B_\delta(x)\right)\geq{}&\mathbb{P}\left(\frac{N_t}{t}\in B_\delta(x)\middle|\nu\in B_{\delta(1-\|h\|_{L^1})}(x(1-\|h\|_{L^1}))\cap[\nu_{\min},\nu_{\max}]\right)\\&\cdot\mathbb{P}\left(\nu\in B_{\delta(1-\|h\|_{L^1})}(x(1-\|h\|_{L^1}))\cap[\nu_{\min},\nu_{\max}]\right).\end{aligned}$$

Conditional on $(x-\delta)(1-\|h\|_{L^1})\leq\nu\leq(x+\delta)(1-\|h\|_{L^1})$,

$$\limsup_{t\to\infty}\frac{N_t}{t}\leq x+\delta, \tag{31}$$

$\mathbb{P}(\cdot|(x-\delta)(1-\|h\|_{L^1})\leq\nu\leq(x+\delta)(1-\|h\|_{L^1}))$-a.s. and

$$\liminf_{t\to\infty}\frac{N_t}{t}\geq x-\delta, \tag{32}$$

$\mathbb{P}(\cdot|(x-\delta)(1-\|h\|_{L^1})\leq\nu\leq(x+\delta)(1-\|h\|_{L^1}))$-a.s.
This implies that

$$\lim_{\delta\to 0}\lim_{t\to\infty}\frac{1}{t}\log\mathbb{P}\left(\frac{N_t}{t}\in B_\delta(x)\right)=0. \tag{33}$$

This completes the proof of Theorem 3. □

Proof of Theorem 1. Finally, for any $K>0$,

$$\limsup_{t\to\infty}\frac{1}{t}\log\mathbb{P}(N_t\geq K)\leq -I(K;\nu_{\max}), \tag{34}$$

which goes to $-\infty$ as $K\to\infty$.

Hence, with Lemmas 1–3 and the fact that (34) is satisfied, we conclude that $\mathbb{P}(N_t/t \in \cdot)$ satisfies a large deviation principle with the rate function:

$$I(x) = \begin{cases} I(x;\nu_{\max}) & \text{if } x > \frac{\nu_{\max}}{1-\|h\|_{L^1}}, \\ 0 & \text{if } x \in [\frac{\nu_{\min}}{1-\|h\|_{L^1}}, \frac{\nu_{\max}}{1-\|h\|_{L^1}}], \\ I(x;\nu_{\min}) & \text{if } 0 \le x < \frac{\nu_{\min}}{1-\|h\|_{L^1}}, \end{cases} \tag{35}$$

and $I(x) = +\infty$ otherwise. This completes the proof of Theorem 1. □

5. Discussion

Hawkes processes are generally defined for continuous-time settings as self-exciting simple point processes with a clustering effect, and their jump rate depends on the entire history of the process. Due to the fact that past events determine the future developments of self-exciting point processes, the Hawkes model is generally non-Markovian. However, under special circumstances, it can be Markovian with a generator of the model if the exciting function is an exponential function or the sum of exponential functions. Difficulties arise in the case of non-Markovian processes when the exciting function is not an exponential function or a sum of exponential functions. Many disciplines use point processes to analyze data, including criminology [2], finance [3], seismology [4], DNA modeling [5], neuroscience [6], and many others. There are a number of point processes, including the Poisson method, whose increments have independent timing, but real-life data generally do not demonstrate independent timing. Despite its importance for applications, many key theoretical results in the area remain unknown, including self-excitation, clustering, and contagion. In addition to capturing both the self-exciting property and clustering effect, the Hawkes process is a natural generalization of the Poisson process. Statistical analysis is possible for this process since it is variable. The large deviation theory describes the probability of rare events that depend on a parameter, which is one of the important techniques used to deal with some probabilistic properties. Moreover, the large deviation result is helpful in studying the ruin probabilities of a risk process when the claim arrivals follow a Hawkes process. The intensity process is one of the main methods used for studying the dynamic properties of general point processes, and is particularly important for credit risk analysis. The intensity of this point process is given by the sum of a baseline intensity plus other terms that depend on the entire history of the point process, compared with a standard Poisson process. The baseline intensity is usually defined for deterministic cases and plays a crucial role in the Hawkes process. In this paper, we consider a linear Hawkes model when the baseline intensity is randomly defined. Compared with previous results in the literature [19], we investigate several asymptotic results to observe financial applications. In particular, we have proven asymptotic results of the large deviation principle for the newly defined model, which has wide applications in insurance, finance, queue theory, and statistics. Therefore, this study has contributed to a better understanding of the large-time behavior of self-exciting point processes and provided numerous theoretical asymptotic behaviors of the Hawkes model, such as the large deviation principles for the randomized baseline intensity model, which was newly introduced in this article. In the future, we plan to perform additional experiments to verify the reliability of the Hawkes process model proposed in this article and consider more complex models that can be newly defined by the intensity processes of Hawkes models.

6. Conclusions

The Hawkes process, introduced by Hawkes [1], is a self-exciting point process, which means that future arrivals are increased by each arrival for a certain period of time. This process is an important class of the stochastic process, which found wide application in diverse areas, such as criminology [2], finance [3], seismology [4], DNA modeling [5], and neuroscience [6]. Recently, the Hawkes process was applied in machine learning and artificial intelligence [7,8]. In general, the model described in the current paper is non-Markovian. In the non-Markovian model, a self-exciting system with future evolution is controlled by the timing of past events. On the other hand, the Markovian model is used in special cases. In other words, the Hawkes process depends on the entire history and has a long memory. In addition to its self-exciting properties, it has clustering properties, which makes it appealing to applications in the financial industry. Although several limit theorems for linear and nonlinear Markovian cases are well known, there has been relatively little study of non-Markovian and inverse Hawkes processes. In a recent paper, Seol [39] presented a new inverse Markovian Hawkes process, mentioning the differences between the general Hawkes process and the inverse Markovian Hawkes process. The Hawkes process assumes that if there have been many jumps in the past, there will be more jumps in the future as well. Inverse Hawkes processes, however, predict larger jumps in the future as more jumps occurred in the past. Hawkes processes exhibit self-excitation that is dependent on the intensity of the process, whereas inverse Hawkes processes exhibit self-excitation that is related to the size of the jumps. Therefore, while the Hawkes process represents self-excitation in terms of frequency, the inverse Hawkes process represents self-excitation in terms of severity. The extended version of the inverse Markovian Hawkes process was introduced by Seol [40] and he studied several theoretical results. One of the best tools used for studying the dynamical properties of a general point process is the intensity process. In particular, the intensity process is important for studying credit risk. In the current paper, a more general (but active) model of the Hawkes process with randomized baseline intensity was considered. The large deviations principle for Hawkes processes with randomized baseline intensity was obtained in Theorem 1. Our results are important for various financial applications and add to the existing literature. Additionally, the findings of this paper have a broader impact on society beyond finance. In this paper, researchers and the general public can gain a deeper understanding of how complex systems cluster and self-excite. By using these techniques, we will build on, enhance, and better understand topical applications of self-exciting point processes. The general public and academics will be motivated to use them in the future. The newly defined model in this paper can be extended to other asymptotic results, such as the law of large numbers, central limit theorems, and moderate deviations in the future. In addition, we will consider the renewal Hawkes process and dynamic contagion model with several asymptotic theorems as potential projects.

Funding: This research is supported by a Dong-A University research grant.

Data Availability Statement: No new data were created or analyzed in this study. Data sharing is not applicable to this article.

Conflicts of Interest: The author declares no competing interest.

References

1. Hawkes, A.G. Spectra of some self-exciting and mutually exciting point process. *Biometrika* **1971**, *58*, 83–90. [CrossRef]
2. Mohler, G.O.; Short, M.B.; Brantingham, P.J.; Schoenberg, F.P.; Tita, G.E. Self-exciting point process modelling of crime. *J. Am. Stat. Assoc.* **2011**, *106*, 100–108. [CrossRef]
3. Errais, E.; Giesecke, K.; Goldberg, L. Affine Point Processes and Portfolio Credit Risk. *SIAM J. Financ. Math.* **2010**, *1*, 642–665. [CrossRef]
4. Hawkes, A.G.; Adamopoulos, L. Cluster models for earthquakesregional comparisons. *Bull. Int. Statist. Inst.* **1973**, *45*, 454–461.
5. Reynaud-Bouret, P.; Schbath, S. Adaptive estimation for Hawkes processes; application to genome analysis. *Ann. Statist.* **2010**, *38*, 2781–2822. [CrossRef]
6. Johnson, D.H. Point process models of single-neuron discharges. *J. Comput. Neurosci.* **1996**, *3*, 275–299. [CrossRef]
7. Bacry, E.; Bompaire, M.; Gaïffas, S.; Muzy, J.F. Muzy Sparse and low-rank multivariate Hawkes processes. *J. Mach. Learn. Res.* **2020**, *21*, 1–32.
8. Chiang, W.; Liu, X.; Mohler, G. Hawkes process modeling of COVID-19 with mobility leading indicators and spatial covariates. *Int. J. Forecast.* **2022**, *38*, 505–520. [CrossRef]
9. Dassios, A.; Zhao, H. A Dynamic Contagion Process. *Adv. Appl. Probab.* **2011**, *43*, 814–846. [CrossRef]
10. Hawkes, A.G.; Oakes, D. A cluster process representation of self-exciting process. *J. Appl. Prob.* **1974**, *11*, 493–503. [CrossRef]

11. Bacry, E.; Delattre, S.; Hoffmann, M.; Muzy, J.F. Scaling limits for Hawkes processes and application to financial statistics. *Stoch. Process. Appl.* **2012**, *123*, 2475–2499. [CrossRef]
12. Bordenave, C.; Torrisi, G.L. Large deviations of Poisson cluster processes. *Stoch. Models* **2007**, *23*, 593–625. [CrossRef]
13. Kelbert, M.; Leonenko, N.; Belitsky, V. On the Bartlett spectrum of randomized Hawkes processes. *Math. Commun.* **2013**, *18*, 393–407.
14. Brémaud, P.; Massoulié, L. Stability of nonlinear Hawkes processes. *Ann. Probab.* **1996**, *24*, 1563–1588. [CrossRef]
15. Zhu, L. Large deviations for Markovian nonlinear Hawkes processes. *Ann. Appl. Probab.* **2015**, *25*, 548–581. [CrossRef]
16. Zhu, L. Process-level large deviations for nonlinear Hawkes point processes. *Ann. l'IHP Probab. Stat.* **2014**, *50*, 845–871. [CrossRef]
17. Zhu, L. Central limit theorem for nonlinear Hawkes processes. *J. Appl. Prob.* **2013**, *50*, 760–771. [CrossRef]
18. Zhu, L. Moderate deviations for Hawkes processes. *Stat. Probab. Lett.* **2013**, *83*, 885–890. [CrossRef]
19. Zhu, L. Ruin probabilities for risk processes with non-stationary arrivals and subexponential claims. *Insur. Math. Econom.* **2013**, *53*, 544–550. [CrossRef]
20. Duffie, D.; Filipović, D.; Schachermayer, W. Affine processes and applications in finance. *Ann. Appl. Probab.* **2003**, *13*, 984–1053. [CrossRef]
21. Jaisson, T.; Rosenbaum, M. Limit theorems for nearly unstable Hawkes processes. *Ann. Appl. Probab.* **2015**, *25*, 600–631. [CrossRef]
22. Jaisson, T.; Rosenbaum, M. Rough fractional diffusions as scaling limits of nearly unstable heavy tailed Hawkes processes. *Ann. Appl. Probab.* **2016**, *26*, 2860–2882. [CrossRef]
23. Seol, Y. Limit theorem for inverse process T_n of linear Hawkes process. *Acta Math. Sin.* **2017**, *33*, 51–60. [CrossRef]
24. Seol, Y. Limit theorems of discrete Hawkes Processes. *Stat. Probab. Lett.* **2015**, *99*, 223–229. [CrossRef]
25. Wang, H. Limit theorems for a discrete-time marked Hawkes process. *Stat. Probab. Lett.* **2022**, *184*, 109368. [CrossRef]
26. Wang, H. Large and moderate deviations for a discrete-time marked Hawkes process. *Commun. Stat. Theory Methods* **2022**. [CrossRef]
27. Seol, Y. Moderate deviations for Marked Hawkes Processes. *Acta Math. Sin.* **2017**, *33*, 1297–1304. [CrossRef]
28. Seol, Y. Limit theorems for the compensator of Hawkes Processes. *Stat. Probab. Lett.* **2017**, *127*, 165–172. [CrossRef]
29. Gao, F.; Zhu, L. Some asymptotic results for nonlinear Hawkes processes. *Stoch. Process. Appl.* **2018**, *128*, 4051–4077. [CrossRef]
30. Gao, X.; Zhu, L. Limit theorems for Markovian Hawkes processes with a large initial intensity. *Stoch. Process. Appl.* **2018**, *128*, 3807–3839. [CrossRef]
31. Gao, X.; Zhu, L. Large deviations and applications for Markovian Hawkes processes with a large initial intensity. *Bernoulli* **2018**, *24*, 2875–2905. [CrossRef]
32. Gao, X.; Zhu, L. Functional central limit theorem for stationary Hawkes processes and its application to infinite-serve queues. *Queueing Sys.* **2018**, *90*, 161–206. [CrossRef]
33. Gao, X.; Zhou, X.; Zhu, L. Tranform Analysis for Hawkes processes with applications. *Quant. Financ.* **2018**, *18*, 265–282. [CrossRef]
34. Wheatley, S.; Filimonov, V.; Sorrette, D. The Hawkes process with renewal immigration & its estimation with an EM algorithm. *Comput. Stat. Data Anal.* **2016**, *94*, 120–135.
35. Fierro, R.; Leiva, V.; Mller, J. The Hawkes process with different exciting functions and its asymptotic behavior. *J. Appl. Prob.* **2015**, *52*, 37–54. [CrossRef]
36. Karabash, D.; Zhu, L. Limit theorems for marked Hawkes processes with application to a risk model. *Stoch. Models* **2015**, *31*, 433–451. [CrossRef]
37. Mehrdad, B.; Zhu, L. On the Hawkes Process with Different Exciting Functions. *arXiv* **2015**, arXiv:1403.0994.
38. Zhu, L. Limit theorems for a Cox-Ingersoll-Ross process with Hawkes jumps. *J. Appl. Prob.* **2014**, *51*, 699–712. [CrossRef]
39. Seol, Y. Limit theorems for an inverse Markovian Hawkes Processes. *Stat. Probab. Lett.* **2019**, *155*, 108580. [CrossRef]
40. Seol, Y. Asymptotics for an extended inverse Markovian Hawkes Process. *J. Korean Math. Soc.* **2021**, *58*, 819–833.
41. Seol, Y. Non-Markovian inverse Hawkes Process. *Mathematics* **2022**, *10*, 1413. [CrossRef]
42. Daley, D.J.; Vere-Jones, D. *An Introduction to the theory of Point Processes*, 2nd ed.; Springer: Berlin/Heidelberg, Germany, 2003; Volume I.
43. Jagers, P. *Branching Processes with Biological Applications*; John Wiley: London, UK, 1975.

mathematics

Article

Asymptotic Relations in Applied Models of Inhomogeneous Poisson Point Flows

Gurami Tsitsiashvili [1,*] and Marina Osipova [1,2]

[1] Institute for Applied Mathematics, Far Eastern Branch of Russian Academy of Sciences, 690041 Vladivostok, Russia; mao1975@list.ru
[2] Institute for Applied Mathematics, Far Eastern Federal University, 690922 Vladivostok, Russia
* Correspondence: guram@iam.dvo.ru; Tel.: +7-914-693-2749

Abstract: A model of a particle flow forming a copy of some image and the distance between the copy and the image are estimated using a special probability metric. The ability of the flow of balls to cover the surface, when grinding the balls, was investigated using formulas of stochastic geometry. Reconstruction of characteristics of an inhomogeneous Poisson flow by inaccurate observations is analysed using the Poisson flow point colouring theorem. The dependence of the Poisson parameter of the distribution of the number of customers in a queuing system with an infinite number of servers and a deterministic service time on the peak load created by an inhomogeneous input Poisson flow is estimated. All these models consist of an inhomogeneous Poisson flow of points and marks glued to each point of the flow and are characterised by their mass, area, volume, observability (or non-observability), and service time. The presence of an asymptotic power–law relationship between model objective functions and parameters of mark crushing is established. These results may be applied in nanotechnology, powder metallurgy, ecology, and consumer services in the implementation of the "Smart City" program. The proposed approach is phenomenological in nature and is justified by the results of real observations and experiments.

Keywords: Poisson point flow; stochastic geometry; safety margin; linear regression estimate; plan of experiment; prediction algorithms; polynomial

MSC: 60K20

Citation: Tsitsiashvili, G.; Osipova, M. Asymptotic Relations in Applied Models of Inhomogeneous Poisson Point Flows. *Mathematics* **2023**, *11*, 1881. https://doi.org/10.3390/math11081881

Academic Editors: Jiancang Zhuang and Elvira Di Nardo

Received: 2 February 2023
Revised: 5 April 2023
Accepted: 13 April 2023
Published: 15 April 2023

1. Introduction

The Poisson flow of points is widely used in queuing theory, in reliability theory, in mathematical biology and ecology, in astronomy, and in many other applied disciplines [1–5]. In this paper, we consider four systems described by inhomogeneous-labelled Poisson flows [6,7], to the points of which special types of stamps are attached (glued). The efficiency indicators of the systems and the parameter, characterising the grinding of grades, and the corresponding increase in the intensity of the flow are introduced. The dependence of the efficiency indicator on this parameter is investigated, and its asymptotic behaviour is analysed when the grinding parameter tends to infinity. The fast convergence of the efficiency indicator to zero is proven, which helps, in particular, to draw important conclusions about the simulated systems.

In the first model, the mark of each Poisson flow point on a flat rectangle or three-dimensional parallelepiped is its mass. In this way, a model of the image formed by a printer/copier or a $3D$ printer is obtained. The accuracy of reproduction by the real density of the flow of points of ideal density is investigated when the mass grinding parameter tends to infinity in a specially selected probabilistic metric [8]. This model is closely related to the tasks of nanotechnology.

In the second model, the mark of a Poisson flow point on a rectangle can be a square/circle, the centre of which is glued to this point. The set of stamps forms the so-called Boolean random set [2,9–13]. This applied model can characterise a surface covered with sprayed powder particles. As an indicator of efficiency, the mathematical expectation of the area of the rectangle part that is not covered with stamps is taken. The rate of convergence of this efficiency indicator to zero is investigated when the grinding parameter of powder particles tends to infinity. This model is applicable to the assessment of the quality of coatings created by the powder metallurgy method.

In the third model, the mark of the Poisson flow point on the plane is its colour red or white. Points coloured red are considered as observable flow points, and points coloured white are considered as unobservable [1]. The probability of flow points being coloured red is an unknown parameter that can be interpreted as interfering/hidden [14–17]. When analysing the flow of observed points, the interfering parameter is eliminated by the transition from the number of flow points in a certain area to the ratio of the number of flow points in a certain area to the number of flow points on the entire plane. Convergence in the probability of the relative frequency of the number of observed points in a certain region is established when the mathematical expectation of the number of observed points on the plane tends to infinity. This model arose when comparing the number of tiger tracks in different years in conditions when some tracks in the snow were invisible [18,19]. As a result, data on the accounting of animal tracks in a given year become industrial statistics and require special methods for their processing.

In the fourth model, the mark of a Poisson flow point on a straight line is a segment of length a, the left end of which is glued to each flow point. In this case, the points that are the right ends of the segments characterise the moments of leaving the queuing system of customers after their service during a deterministic time. This paper considers the case when the intensity of the input Poisson flow of customers into a system with an infinite number of devices has a pronounced maximum. The possibility of smoothing this maximum by reducing the service time is being investigated. This problem arises when analysing consumer service systems and conveyor production systems [20–23]. The task of smoothing the peak load was also considered in the papers [24–26] in relation to PC software testing systems. However, in these papers, the assumption was made about a special relationship between the intensity of the input flow and the intensity of service.

Such a statement of the question is caused by numerous problems for specialists in the subject areas when processing and interpreting the data they receive. These include observations of the long-term operation of copiers, in which the quality of duplicated images drops sharply. Apparently, this is due to the sintering of copier powder particles during prolonged heating. The dependence on the particle size was also observed in experiments to protect ball-bearing rings from collisions with balls by including polymer particles in the ball-bearing oil. It turned out that with a decrease in the size of these particles, the uptime of the ball bearing increases markedly. Another task was the observation of geographers for the detection of animal tracks, depending on meteorological conditions. In particular, the possibility of detecting a trace is significantly affected by the thickness of the snow cover. In addition, if such a trace study is carried out over a large area, then the financial conditions for organising observations also come to the fore. Many data-processing problems arise in the "Smart City" program. Traffic jams significantly affect the quality of public services. Therefore, recently, service centres, for example, sports complexes, are switching from hard copies of user subscriptions to electronic cards. Such maintenance allows users not to associate the time of their arrival for service with its periodic start.

The study of such phenomena creates an impetus for the development of mathematical and statistical methods for the analysis of complex systems. Such problems arise from various practical applications, and the use of mathematical models in solving them turns out to be a rather complicated procedure. Therefore, it is more convenient to build not very accurate, but rather easily calculated simulation models, in which the dependence of performance indicators on model parameters is explicit.

The construction of such models requires the combination of several rather heterogeneous mathematical techniques. Here, it is very convenient to model an inhomogeneous

Poisson flow of points with stamps pasted on them, the parameters of which are weight, size, etc., characteristics that are related to the intensity of the flow. The construction of asymptotic relations of the dependence of the efficiency index of such a system on the parameters of the marks requires the use of inequalities commonly used for probabilistic metrics when analysing the convergence rate in the limit theorems of probability theory. Formulas of stochastic geometry can also serve as a convenient tool for constructing such relations, especially in the framework of a Boolean model based on the Poisson flow construction. Another convenient technique for working with point flows of an inhomogeneous Poisson flow is the method of interfering (hidden) parameter. In this case, we can use well-known methods for analysing statistical samples in the presence of an interfering parameter. All these circumstances require, when constructing and studying the systems under consideration, the connection of heterogeneous mathematical elements: the choice of an inhomogeneous Poisson flow of points, the choice of stamps glued to the flow points, the choice of the ratio between the intensity of the flow of points and the characteristic of the brand, the choice of the system efficiency indicator and the choice of a mathematical method for analysing the dependence of the efficiency indicator on the parameter of stamps.

The established effects of the convergence of the system efficiency indicator to zero with the aspiration to infinity of the parameter of grinding marks are new. Methods of the theory of probability metrics [8], methods of the theory of random sets [2], methods of the theory of Poisson flows [1], and integral formulas for the parameter of the Poisson distribution of the number of points of the Poisson flow in a certain segment were used to study them. A characteristic feature of these models is a special choice of the efficiency indicator, the grinding parameter of the stamps glued to the flow points, and the relationship between the efficiency indicator and the grinding parameter. The proposed approach is phenomenological in nature. It is initiated and justified by the results of real observations and experiments.

2. Proximity of the Poisson Flow of Points to the Intensity Function When Grinding the Masses of Points

This mathematical model is based on the notion of a Poisson flow of Π points on r-dimensional Euclidean space E^r. The number of points of the Poisson flow in the Lebesgue-measurable subset $A \subseteq E^r$ does not depend on the number of points in any other Lebesgue-measurable subset $B \subseteq E^r : \ A \bigcap B = \oslash$ and obeys the Poisson distribution with the parameter $\Lambda(A)$. Next, we will assume that there exists a piecewise continuous function $\lambda(x), \ x \in E^r$ such that for any Lebesgue-measurable set $A \subseteq E^r$, the parameter $\Lambda(A) = \int_A \lambda(x)dx$. The function $\lambda(x), \ x \in E^r$, is called the intensity of the Poisson flow Π.

Consider Poisson point flow Π on r-dimensional rectangle X with a continuous intensity function $\lambda(x)$ such that

$$\Lambda(X) = \int_X \lambda(x)dx < \infty. \tag{1}$$

If X' is a Lebesgue-measurable subset in the rectangle X, then a random number of points $n(X')$ of the Poisson flow Π in the set X' has a Poisson distribution with the parameter

$$\Lambda(X') = \int_{X'} \lambda(x)dx. \tag{2}$$

Now let us denote Π_m Poisson flow on the rectangle X with the intensity function $m\lambda(x)$. Let us put $n_m(X')$ the number of flow points Π_m on the set X' and define random variables $N_m(X') = \dfrac{n_m(X')}{m}$. Here, the random variable $N_m(X')$ characterises the total mass of the flow points Π_m in the set X', assuming that the mass of each point is $1/m$.

We introduce the distance ρ_m between the point flow Π_m, in which each point has a mass $1/m$, and the intensity function $\lambda(x)$. To do this, we denote $\mathcal{X}$ as the partition of the rectangle X by a finite number $n(\mathcal{X})$ disjoint and Lebesgue-measurable subsets $X'_1, \ldots, X'_{n(\mathcal{X})}$. Let $\mathcal{A}$ be the set of all possible such partitions of $\mathcal{X}$; then, the distance ρ_m is determined by the equality

$$\rho_m = \left(\sup_{\mathcal{X} \in \mathcal{A}} E \sum_{i=1}^{n(\mathcal{X})} (\Lambda(X'_i) - N_m(X'_i))^2 \right)^{1/2}. \tag{3}$$

It follows from the properties of the Poisson flow Π_m that random variables $N_m(X'_i)$, $i = 1, \ldots, n(\mathcal{X})$ are independent. This property is based on the following statements.

In probability, statistics and related fields, a Poisson point process is a type of random mathematical object that consists of points randomly located on a mathematical space with the essential feature that the points occur independently of one another [2]. Consider a collection of disjoint and bounded subregions of the underlying space. By definition, the number of points of a Poisson point process in each bounded subregion will be completely independent of all the others. This property is known under several names such as complete randomness, complete independence [27], or independent scattering [4,6], and it is common to all Poisson point processes. In other words, there is a lack of interaction between different regions and the points in general [28], which motivates the Poisson process, being sometimes called a purely or completely random process [27].

Therefore, the equalities are fulfilled $\Lambda(X'_i) = EN_m(X'_i)$,

$$E \sum_{i=1}^{n(\mathcal{X})} (\Lambda(X'_i) - N_m(X'_i))^2 = \sum_{i=1}^{n(\mathcal{X})} E(\Lambda(X'_i) - N_m(X'_i))^2 =$$

$$= \sum_{i=1}^{n(\mathcal{X})} Var N_m(X'_i) = \frac{1}{m^2} \sum_{i=1}^{n(\mathcal{X})} m\Lambda(X'_i) = \frac{\Lambda(X)}{m},$$

from which, due to Formula (3), we obtain

$$\rho_m = \sqrt{\frac{\Lambda(X)}{m}}. \tag{4}$$

Relation (4) shows how, when splitting the mass of points of the Poisson flow Π_m into m of identical parts, the distance between the flow Π_m and the function $\lambda(x)$ tends to zero at $m \to \infty$.

Remark 1. *This ratio can be used to study the quality of work of 2D and 3D printers and copiers depending on the mass of individual flow points, which can be interpreted as the mass of flow particles.*

Remark 2. *The results obtained in this section have not been previously used in the analysis of copiers and 2D and 3D printers. They have made it possible to analyse the operation of these devices during heating and sintering, as well as during the grinding of powder particles used in these devices.*

3. Covering Points of the Poisson Flow on the Plane with Cubes of Variable Size

Suppose there is Poisson point flow Π_m with constant intensity $m\overline{\lambda}$, given on the unit square X, whose one side coincides with the axis of the abscissa. Let a square with a side of length $m^{-1/3}$, rotated by a random angle φ (between the axis of the abscissa and the diagonal of the glued square) be glued to each point of the flow Π_m. The angle φ has a uniform distribution on the segment $[0, \pi]$. Thus, the set of squares glued to X forms a random set obeying the Boolean model [2].

Denote S_m as the complement of this random set to the unit square of X. According to Steiner's theorem [2], for the mathematical expectation α_m of the area of a random set S_m, the relation is fulfilled

$$\alpha_m = \exp(-\overline{\lambda} m^{1/3}) \to 0, \ m \to \infty. \tag{5}$$

Formally, we can assume that every square glued to a flow point Π_m on the plane is the side of a cube parallel to the sputtering plane with a volume $1/m$. Then, the mathematical expectation of the total volume of cubes is $\beta_m = \overline{\lambda}$. In the case, when this total volume $\beta_m = \overline{\lambda} m^{-\gamma}$, $0 \leq \gamma < 1/3$, it is not difficult to obtain the ratio

$$\alpha_m = \exp(-\overline{\lambda} m^{1/3-\gamma}) \to 0, \ m \to \infty. \tag{6}$$

The obtained relations show that, when the cubes are crushed while maintaining or decreasing their average total volume, the area of the random set S_m decreases to zero.

Remark 3. *Along with shredding cubes, the proposed approach can be applied to coating a flat surface with shredding balls. In this case, instead of squares, circles are considered, which are projections of balls with a diameter of $m^{-1/3}$ on the plane. Then, the relations are fulfilled*

$$\alpha_m = \exp(-\pi \overline{\lambda} m^{1/3}/4) \to 0, \ m \to \infty, \ \beta_m = \pi \overline{\lambda}/6. \tag{7}$$

Remark 4. *Such limiting ratios can be useful in analysing the quality of the coating obtained by powder metallurgy. Another possible application of these results may be to evaluate the coating of the ice surface with notches in order to reduce the possibility of people sliding on it.*

Remark 5. *The results of this section are based on the analysis of powder metallurgy processes. They show that the grinding of powder particles makes it possible to increase the protective properties of the particle layer on the metal surface. This one is being investigated analytically. The results are very difficult to repeat using the equations of gas dynamics and the processes of gluing solid particles to a metallic surface. The results of this section are the development of well-known results in the field of stochastic geometry, which continue to develop to date [11].*

4. Method of Eliminating the Interfering Parameter in Statistics of Poisson Points Flow

Let the Lebesgue-measurable and disjoint regions be distinguished on the plane G_k, $k = 1, \ldots, n$. Poisson point flow Π with continuous intensity $\lambda(x)$ is given, and the relations are fulfilled

$$\overline{\lambda}_k = \int_{G_k} \lambda(x)dx < \infty, \ k = 1, \ldots, r, \ \overline{\lambda} = \sum_{k=1}^{r} \overline{\lambda}_k.$$

Denote $\Lambda_k = \dfrac{\overline{\lambda}_k}{\overline{\lambda}}$ and consider the flow Π_m with the intensity function $m\lambda(x)$. Let each point of the flow Π_m be independent of other points and, from its coordinates with probability p, enter the flow $\overline{\Pi}_m$. Then, the flow $\overline{\Pi}_m$ due to the point colouring theorem of the Poisson flow [1] is Poisson with intensity $pm\lambda(x)$. Therefore, the number n_k of flow points $\overline{\Pi}_m$ in the subdomain G_k has a Poisson distribution with the parameter $pm\overline{\lambda}_k$, and the sum $n = \sum_{k=1}^{r} n_k$ has a Poisson distribution with the parameter $pm\overline{\lambda}$.

Theorem 1. *The convergence in probability of the random variable $N_k = \dfrac{n_k}{n}$ to the parameter Λ_k is valid when $m \to \infty$.*

Proof. Due to the properties of the Poisson distribution, the relations are fulfilled

$$En_k = Var\ n_k = pm\overline{\lambda}_k, \ k = 1, \ldots, r; \ En = Var\ n = pm\overline{\lambda}. \tag{8}$$

It follows from the equalities (8) that

$$Var\,\frac{n_k}{pm\overline{\lambda}_k} = \frac{1}{pm\overline{\lambda}_k},\ Var\frac{n}{pm\overline{\lambda}} = \frac{1}{pm\overline{\lambda}}.$$

From Chebyshev's inequality, we obtain that for any ε, $0 < \varepsilon < 1$, and for $m \to \infty$

$$P\left(1-\varepsilon \le \frac{n_k}{pm\overline{\lambda}_k} \le 1+\varepsilon\right) \ge 1 - \frac{1}{pm\overline{\lambda}_k\varepsilon^2} \to 1,\ k = 1,\dots,r,$$
$$P\left(1-\varepsilon \le \frac{n}{pm\overline{\lambda}} \le 1+\varepsilon\right) \ge 1 - \frac{1}{pm\overline{\lambda}\varepsilon^2} \to 1. \tag{9}$$

Using Formulas (8) and (9), it is not difficult for any ε, $0 < \varepsilon < \frac{1}{2}$, to obtain the inequality

$$P\left(1-2\varepsilon \le \frac{N_k}{\Lambda_k} \le 1+4\varepsilon\right) \ge P\left(\frac{1-\varepsilon}{1+\varepsilon} \le \frac{N_k}{\Lambda_k} \le \frac{1+\varepsilon}{1-\varepsilon}\right) \ge 1 - \frac{1}{pm\overline{\lambda}_k\varepsilon^2} - \frac{1}{pm\overline{\lambda}\varepsilon^2}. \tag{10}$$

From the inequality $\Lambda_k \le 1$, and from Formula (10), we obtain the relation

$$P(-2\varepsilon \le N_k - \Lambda_k \le 4\varepsilon) \ge 1 - \frac{1}{pm\lambda_k\varepsilon^2} - \frac{1}{pm\lambda\varepsilon^2} \to 1,\ m \to \infty. \tag{11}$$

This proves the statement of Theorem 1. □

Therefore, the relation N_k, for $m \to \infty$, is a consistent estimate of the parameter Λ_k, free of probability p, playing the role of an interfering parameter in this problem.

Remark 6. *Using Inequalities (11), we can assume that* $p = p(m) = m^{-\delta}$, $0 \le \delta < 1$, *and establish convergence by probability* N_k *to* Λ_k *at* $m \to \infty$.

Remark 7. *This problem arises when comparing accounts of the number of animal tracks in the snow when the interfering parameter takes on different values caused by different meteorological and economic characteristics in different years.*

Remark 8. *The results obtained in this section show that the analysis of traces of rare animals requires more careful processing of the results obtained. This becomes especially important when traces are investigated over a large area and data collection turns into an industrial statistics procedure.*

5. Peak Loads in the Queuing System $M|D|\infty$

Consider a queuing system with a nonstationary Poisson input flow of intensity $\lambda(t)$, $t \ge 0$, with a deterministic service time a and an infinite number of servers. Let the intensity of Poisson input flow $\lambda(t)$, $0 \le t \le T$, be a continuously differentiable function, and at the point t_*, $0 < t_* < T - a$ has a single extremum maximum.

Denote $n(t)$ as the number of customers in the queuing system at time t. It is obvious that the random variable $n(t)$ has a Poisson distribution with the parameter

$$\Lambda(t) = \int_{\max(0,t-a)}^{t} \lambda(u)du. \tag{12}$$

Designate $\lambda(t_*) = \lambda^*$, $\Lambda^* = \sup_{0\le t\le T} \Lambda(t)$.

Let us investigate the dependence of the efficiency indicator Λ^* on the parameters λ^*, a. First, focus on the analysis of the smoothing of the peak load determined by the intensity of the input stream $\lambda(t)$, with a decrease in the parameter a. It is obvious that the inequality is true

$$\Lambda^* \le a\lambda^*. \tag{13}$$

Thus, reducing the parameter a leads to a decrease in the value of Λ^*.

Let us now proceed to a more detailed study of the dependence Λ^* on the parameter a. From Formula (12), it follows that the equalities are fulfilled

$$\Lambda(t) = \int_0^t \lambda(u)du \ \ 0 \le t \le a, \ \Lambda(t) = \int_{t-a}^t \lambda(u)du, \ \ a \le t \le T. \tag{14}$$

Now calculate the derivative of twice piecewise continuously differentiable function $\Lambda(t)$ for $t \ne a$:

$$\dot{\Lambda}(t) = \lambda(t), \ \ 0 \le t < a \le T, \ \dot{\Lambda}(t) = \lambda(t) - \lambda(t-a), \ \ a < t \le T. \tag{15}$$

Theorem 2. *The function $\Lambda(t)$ has a single extremum-= maximum t^* on the segment $[0,T]$ and $a \le t^* \le t_* + a$.*

Proof. It follows from Formula (15) that the inequalities are satisfied

$$\dot{\Lambda}(t) \ge 0, \ \ 0 \le t < a; \ \dot{\Lambda}(t) \le 0, \ t_* + a \le t \le T. \tag{16}$$

(1) Assume that $0 \le t_* \le a, \ \lambda(0) \ge \lambda(a)$. From (16), we have that the function $\Lambda(t)$ is non-decreasing on $[0,a]$ and non-increasing on $[t_* + a, T]$. Calculate now

$$\sup_{a<t\le t_*+a} \dot{\Lambda}(t) = \sup_{a<t\le t_*+a} (\lambda(t) - \lambda(t-a)) \le \sup_{a<t\le t_*+a} \lambda(t) - \inf_{a<t\le t_*+a} \lambda(t-a) =$$

$$= \lambda(a) - \inf_{0<t\le t_*} \lambda(t) = \lambda(a) - \lambda(0) \le 0.$$

Consequently, the function $\Lambda(t)$ has single extremum maximum at the point $t^* = a$.
(2) Assume that $0 \le t_* \le a, \ \lambda(0) \le \lambda(a)$,; then, from (15) and (16), we have that $\dot{\Lambda}(a+0) = \lambda(a) - \lambda(0) \ge 0, \ \dot{\Lambda}(t_* + a) \le 0$. Now calculate

$$\sup_{a\le t\le t_*+a} \ddot{\Lambda}(t) = \sup_{a<t\le t_*+a} \dot{\lambda}(t) - \inf_{a<t\le t_*+a} \dot{\lambda}(t) \le 0. \tag{17}$$

Hence, we obtain the existence of a single root of the function $\dot{\Lambda}(t)$ and thus a single extremum maximum t^* for the function $\Lambda(t)$ and the inequality $a \le t^* \le t_* + a$.
(3) If $a \le t_*$, then from Formulas (15) and (16), we have

$$\dot{\Lambda}(t) \ge 0, \ 0 \le t < a; \ \dot{\Lambda}(t) \ge 0, \ a < t \le t_*; \dot{\Lambda}(t) \le 0, \ t_* + a \le t \le T.$$

Analogously with (17), we lead to

$$\dot{\Lambda}(t_*) \ge 0, \ \dot{\Lambda}(t_* + a) \le 0; \ \ \sup_{t_*\le t\le t_*+a} \ddot{\Lambda}(t) \le 0.$$

Consequently, we obtain the existence of a single extremum maximum t^* for the function $\Lambda(t)$ and the inequality $a \le t_* \le t^* \le t_* + a$, and thus Theorem 2 is proved. □

Let us now proceed to illustrate the results obtained in a numerical example. Suppose that for some $T > a > 0$, the function

$$\lambda(t) = \frac{\exp(-(t-b)^2/(2c))}{\sqrt{2\pi c}}, \ 0 < t < T - a, \ \lambda(t) = 0, \ t \le 0 \text{ or } T - a \le t. \tag{18}$$

Then, for $T = 10, \ b = 5, \ c = 1$, the function $\Lambda(t)$ looks like this.

Figure 1 shows how when the parameter a increases, the maximum Λ^* and the point t^* increase.

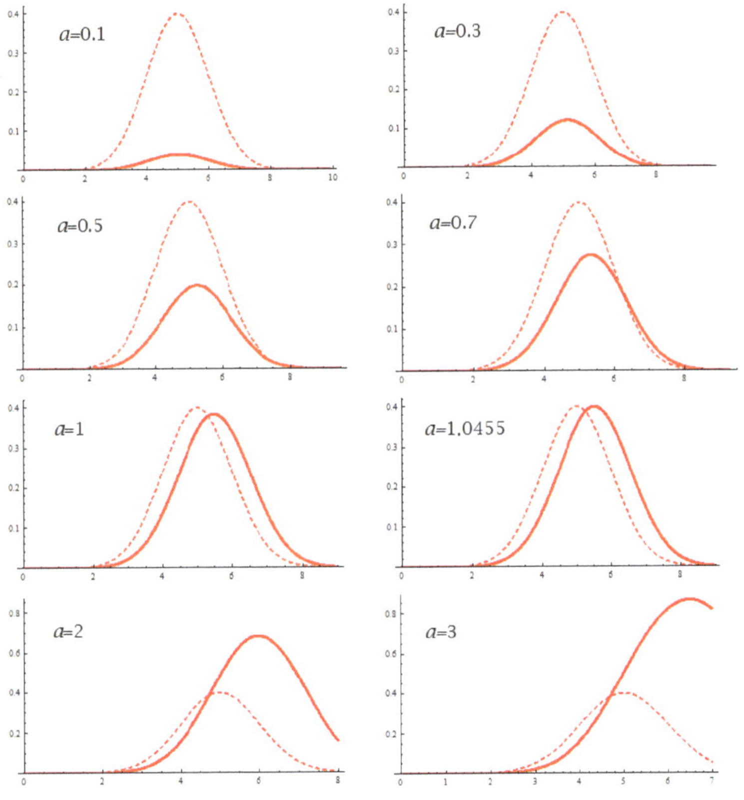

Figure 1. Graphs of functions $\lambda(t)$ (dotted line), $\Lambda(t)$ (solid line).

Remark 9. *Such a task arises when modelling sports complexes, cinemas, and production conveyor systems.*

Remark 10. *The results obtained in this section arose when comparing the previous one, based on hard copy subscriptions with periodic access of users to the service system, and the new one, using electronic cards and allowing users at any time that was convenient for them. The known methods of calculating queuing systems with variable input flow intensity [29,30] are quite complex, and their application for the* $M|D|\infty$ *system seems superfluous.*

6. Discussion

As possible extensions of the problems considered in this paper, there may be a transition from the Poisson flow model of points to more complex and, perhaps, more adequate models of point flows. The construction of asymptotic relations for the dependence of the flow intensity on the parameter of stamps is not always convenient. It uses very strict requirements for a small (large) parameter characterising the stamp glued to the point (mass, size, probability of detection, time of service of stamps). In the application plan, it may be sufficient to relax these requirements. Such a weakening may lead to a transition from the asymptotic analysis of these dependencies to numerical calculations. There may also be a situation when the presence of a small (large) parameter in the models may not be completely appropriate to the model under consideration. For example, in nanotechnology models, the aspiration to zero of the size or mass of particles contradicts physical conditions.

This circumstance requires greater caution when constructing a mathematical model of the analysed system and, as a consequence, greater selectivity when choosing a mathematical research method. The results obtained in this paper are applied to the analysis of new technical systems. These systems periodically arise in various applied fields and become a source of new problems at the junction of mathematical modelling and system analysis.

7. Conclusions

A model of the flow of particles forming a copy of some image was considered, and it is shown how the distance between the image formed by them and the original image decreases when the particles are crushed. The ability of flow particles to close the surface protected by them during their grinding was investigated. A model for restoring the characteristics of an inhomogeneous Poisson flow with inaccurate observation of flow points and with increasing flow intensity was analysed. The dependence of the Poisson parameter of the distribution of the customer number in a queuing system with an infinite number of servers and a deterministic service time on the peak load created by the Poisson input flow was estimated.

The obtained results indicate the presence of power–law upper bounds for various efficiency indicators, depending on the parameter characterising the grinding of stamps glued to the flow points. They were confirmed by observations of the studied systems. The established effects of the convergence of the system efficiency indicator to zero with the aspiration to infinity of the parameter of grinding marks were obtained. These results were established using methods of the theory of probability metrics, methods of the theory of random sets, methods of the theory of Poisson flows and their colouring points, and on integral formulas for the parameter of the Poisson distribution of the point numbers in the Poisson flow on a certain segment. A characteristic feature of these models is a special choice of the efficiency indicator, the grinding parameter of the stamps glued to the flow points, and the relationship between the efficiency indicator and the grinding parameter. The results presented in this paper may be applied in nanotechnology, in powder metallurgy, in ecology, and in consumer services during the implementation of the "Smart City" program. The proposed approach is phenomenological in nature and is justified by the results of real observations and experiments.

Author Contributions: Conceptualisation, methodology, formal analysis, G.T.; manifold numerical, M.O. All authors have read and agreed to the published version of the manuscript.

Funding: This research received no external funding. The research was carried out within the state assignment for IAM FEB RAS (N 075-01290-23-00).

Institutional Review Board Statement: Not applicable.

Informed Consent Statement: Not applicable.

Data Availability Statement: Data supporting the reported results were obtained by M.O.

Conflicts of Interest: The authors declare no conflict of interest.

References

1. Kingman, J.F.C. *Poisson Processes;* Oxford Science Publications: Oxford, UK, 1993.
2. Chiu, S.N.; Stoyan, D.; Kendall, W.S.; Mecke, J. *Stochastic Geometry and Its Applications;* John Wiley and Sons: New York, NY, USA, 2013.
3. Cox, D.R.; Isham, V. *Point Processes;* Chapman and Hall: London, UK, 1980.
4. Streit, R.L. *Poisson Point Processes: Imaging, Tracking, and Sensing;* Metron Inc.: Reston, VA, USA. 2010.
5. Tijms, H.C. *A First Course in Stochastic Models;* John Wiley and Sons: New York, NY, USA, 2003.
6. Moller, J; Waagepetersen, R.P. *Statistical Inference and Simulation for Spatial Point Processes;* Chapman and Hall: London, UK, 2003.
7. Haenggi, M. *Stochastic Geometry for Wireless Networks;* Cambridge University Press: Cambridge, UK, 2013.
8. Zolotarev, V. *Modern Theory of Summation of Random Variables;* VSP: Utrecht, The Netherlands, 1997.
9. Stoyan, D. On some qualitative properties of the Boolean model of stochastic geometry. *ZAMM-J. Appl. Math. Mech./Zeitschrift für Angewandte Mathematik und Mechanik* **1979**, *59*, 447–454. [CrossRef]
10. Stoyan, D. Stereological formulae for size distribution through marked point processes. *Probab. Math. Stat.* **1982**, *2*, 161–166.

11. Brandell, M.; Brands, D.; Maike, S.; Rheinbach, O.; Schroder, J.; Schwarz, A.; Stoyan, D. Effective hyperelastic material parameters from microstructures constructed using the planar Boolean model. *Comput. Mech.* **2022**, *69*, 1295–1321. [CrossRef]
12. Hmamouche, Y. *Applications of Stochastic Geometry in the Modelling and Analysis of Wireless Networks. Networking and Internet Architecture [cs.NI]*; Ecole Nationale Superieure Mines-Telecom Atlantique: Paris, France, 2020; ffNNT: 2020IMTA0212ff. fftel-03199210.
13. Hmamouche, Y.; Benjillali, M.; Saoudi, S.; Yanikomeroglu, H.; Di Renzo, M. New Trends in Stochastic Geometry for Wireless Networks: A Tutorial and Survey. *Proc. IEEE* **2021**, *109*, 1200–1252. [CrossRef]
14. Cox, D.R.; Hinkley, D.V. *Theoretical Statistics*; Chapman and Hall: London, UK, 1979.
15. Young, G.A.; Smith, R.L. *Essentials of Statistical Inference*; Cambridge University Press: Cambridge, UK, 2005.
16. Holevo, A.S. *Quantum Systems, Channels, Information*; De Gruyter: Berlin, Germany, 2013.
17. Holevo, A.S. Quantum noise as noncommutative stationary random process. *Int. J. Mod. Phys.* **2022**, *A37*, 2243011. [CrossRef]
18. Bocharnikov, V.N.; Krasnopeev, S.M. World Forest monitoring by network of indigenous peoples. In Proceedings of the 18th ESRI Annual User Conference, San Diego, CA, USA, 24–28 July 1998.
19. Murzin, A.A. Construction of the Amur tiger population dynamics model and forecast calculations. Its number for the period from 2015 to 2022. *Int. J. Appl. Fundam. Res.* **2018**, *11*, 333–340. (In Russian)
20. Boyle, D.E.; Yates, D.C.; Yeatman E.M. Urban Sensor Data Streams: London 2013. *IEEE Internet Comput.* **2013**, *17*, 12–20. [CrossRef]
21. Greenfield, A. *Against the Smart City*; Verso: London, UK, 2013.
22. Porshnev, S.V.; Karelin, I.A. Investigation of the features of a non-stationary single-channel queuing system in the context of the number of applications served. *Cloud Sci.* **2017**, *4*, 366–374. (In Russian)
23. Vasilyev, S.A.; Tsareva , G.O. Analysis of Queueing Systems with an Infinite Number of Servers and a Small Parameter. *RUDN J. MIPh* **2018**, *26*, 167-–175. [CrossRef]
24. Pham, H. *Software Reliability*; Springer: Singapore, 2000.
25. Yamada, S. *Software Reliability Modelling, Fundamentals and Applications*; Springer: Tokyo, Japan, 2014.
26. Minamino, Y.; Makita, Y.; Inoue, S.; Yamada, S. Efficiency evaluation of software faults correction based on queuing simulation. *Mathematics* **2022**, *10*, 1438. [CrossRef]
27. Daley, D.J.; Vere-Jones, D. *An Introduction to the Theory of Point Processes: Volume II: General Theory and Structure*; Springer: Berlin/Heidelberg, Germany, 2007.
28. Feller, W. *An Introduction to Probability Theory and Its Applications, Volume 2*; John Wiley and Sons: Hoboken, NJ, USA, 1971.
29. Ivnitsky, V. *Theory of Queuing Networks*; Fizmatlit: Moscow, Russia, 2004. (In Russian)
30. Katrakhov, V; Ryzhkov, D. *Introduction to the Functional-Analytical Method in the Dynamic Theory of Queuing*; DVGU Publishing House: Vladivostok, Vladivostok, 2004. (In Russian)

mathematics

Article

Event-Triggered Extended Dissipativity Fuzzy Filter Design for Nonlinear Markovian Switching Systems against Deception Attacks

Mourad Kchaou [1,]* and Mohamed Amin Regaieg [2]

1 Department of Electrical Engineering, College of Engineering, University of Ha'il, Hail 2440, Saudi Arabia
2 Lab-STA, LR11ES50, National School of Engineering of Sfax, University of Sfax, Sfax 3038, Tunisia; med-amine.regaieg@enis.tn
* Correspondence: m.kchaou@uoh.edu.sa

Abstract: This article is concerned with the adaptive-event-triggered filtering problem as it relates to a class of nonlinear discrete-time systems characterized by interval Type-2 fuzzy models. The system under investigation is susceptible to Markovian switching and deception attacks. It is proposed to implement an improved event-triggering mechanism to reduce the unnecessary signal transmissions on the communication channel and formulate the extended dissipativity specification to quantify the transient dynamics of filtering errors. By resorting to the linear matrix inequality approach and using the information on upper and lower membership functions, stochastic analysis establishes sufficient conditions for the existence of the desired filter, ensuring the mean-squared stability and extended dissipativity of the augmented filtering system. Further, an optimization-based algorithm (PSO) is proposed for computing filter gains at an optimal level of performance. The developed scheme was finally tested through experimental numerical illustrations based on a single-link robot arm and a lower limbs system.

Keywords: networked IT2 fuzzy systems; Markov jump process; adaptive-event-triggered scheme; extended dissipative filtering; deception attacks; PSO

MSC: 03E75; 62M05

Citation: Kchaou, M.; Regaieg, M.A. Event-Triggered Extended Dissipativity Fuzzy Filter Design for Nonlinear Markovian Switching Systems against Deception Attacks. *Mathematics* **2023**, *11*, 2064. https://doi.org/10.3390/math11092064

Academic Editors: Gurami Tsitsiashvili and Alexander Bochkov

Received: 4 March 2023
Revised: 2 April 2023
Accepted: 24 April 2023
Published: 26 April 2023

1. Introduction, Notations, and Outline

Throughout this section, we outline the literature review, the notations, and the study objectives.

1.1. Bibliographic Review

Dynamical systems are frequently subject to random changes because of component failures, changes in subsystem interconnections, or environmental changes. These processes motivate researchers to examine and characterize Markovian jump systems (MJSs). A Markovian jump system is a stochastic hybrid system that has finite modes of operation, where jumps between modes are controlled by a Markov process, Several recent research studies have focused on MJSs, resulting in significant publications about fault tolerance, target tracking, manufacturing, networked control, and multi-agent systems [1–6]. Despite many successes in this research, most practical systems are highly nonlinear, so linear MJSs impose marked limitations in real-world applications. In addition to overcoming this drawback, the T-S fuzzy models have recently been applied to deal with nonlinear complex systems due to their effective approximation of smoothly nonlinear models [7–11]. In general, the T-S fuzzy system encompasses Type-1 and Type-2 fuzzy systems. Type-1 models typically have crisp membership functions, while Type-2 models are used to deal with parameter uncertainties by incorporating upper and lower membership functions [12–15].

On the other hand, networked control systems (NCSs) are a type of control system in which many components of the system are connected via shared communication networks, such as actuators, controllers, and sensors [16,17]. Under resource-constrained scenarios, such as information interactions between agents with limited communication resources, it is difficult for physical systems to obtain energy supplies in a dynamic environment. Hence, in designing a control strategy, it is important to consider economy and cost, and the above-mentioned features will inevitably be subject to some constraints. Two types of techniques are generally applied to address this issue. The first one is the intermittent control proposed in [18], wherein the control resource is designed to conserve energy in conjunction with the communication connection condition. The second approach is the event-triggered control, in which the gain or resource is selected when there is a relatively large error state between the sensor and the actuator. This technique is also considered an adequate method of addressing the energy constraints in communications networks. In recent years, the ETS has received considerable attention with numerous achievements that relate to ET H_∞ control, ET fault detection, ET sliding-mode controller design, ET consensus control, and ET state estimation systems, as summarized in [19–27] and the references therein. Since communication networks are open and network resources are limited, studies on cyber security issues based on communication-saving regulations for NCSs are extremely important. In particular, deception attacks pose a significant threat to the normal operation of NCSs since they are continually introducing errors into the correct data [28–30]. It is generally adopted to evade detection mechanisms for which the error vectors are small or even offensive in nature, thus trying to destroy the working state of the system simply by accumulating errors. Due to the wide range of applications for parameter estimation, target tracking, as well as system monitoring, the filtering problem has attracted substantial research attention in the past few decades. Furthermore, there is no doubt that the Kalman filtering method is widely used in many filter design strategies for signal processing, control, and optimization [31–33]. In many practical cases, however, the Kalman filter does not always fulfil the requirement that the external noise is caused by white processes with known statistical properties. For input signals with non-statistical characteristics, the performance criteria can be used to compensate for the disadvantage of Kalman filtering. The filtering issue aims at developing filters that satisfy prescribed performance levels concerning the output error and disturbance input. It is noteworthy that several efficient filtering strategies have been proposed up to this point, including the H_∞ filtering [19,34–36], passive/dissipative filtering [37–39], and peak-to-peak filtering [40,41]. In this paper, we were first motivated by the extended dissipativity performance, which has been introduced to deal with the (Q, R, S)-dissipative and $L_2 - L_\infty$ filtering problem in a unified framework. Further, the quoted papers examined the extended dissipative performance, which can be converted into four different performances using different parameters [10,29,42].

As an attractive approach, the event-triggered method has been recently used to cope with the filtering problem for different classes of systems, and substantial research attention with some outstanding results can be found in [36,43,44]. To mention a few, a singular neural network with time-varying delays and Markovian jump parameters was investigated in [23] to deal with the event-triggered dissipative filtering problem. An event-triggered communication mechanism was employed by the authors of [24], for the design of piecewise fuzzy diagnostic observers for discrete-time TS fuzzy systems. As has been shown in previous studies, it has been found that the above-mentioned traditional event-triggered mechanisms can reduce communication resources and improve transmission effectiveness. However, these mechanisms are designed with constant thresholds that are not capable of adjusting themselves based on actual transmission conditions. This leads to the proposal of adaptive-event-triggered mechanisms (AETMs), in which the triggering threshold is dynamically adjusted during the operation of the system. There was an investigation in [28] of the design of H_∞ filters for discrete-time networked systems with an adaptive-event-triggered mechanism and hybrid cyber-attacks. In [45], a new dynamic-event-triggered protocol was developed to address the problem of filtering for affine systems presented by the Takagi–Sugeno fuzzy model. In [20], an observer-based finite-time H_∞ controller was designed to accommodate discrete-time-varying systems with adaptive-event-triggered mechanisms. Reference [35] addressed the problem of adaptive-event-triggered H_∞ filtering for discrete-time-delayed neural networks with missing measurements that occur randomly. It can be noted that the adaptive law proposed in [46], which has a lower bound, might result in conservative conclusions. Additionally, the AETMs proposed by [47,48] each suffer from a singular problem and may degenerate into traditional periodic-time-triggered mechanisms, which may limit their application in practice. As a result of the above considerations, the motivation appears to be the development of a new AETM that is capable of improving the existing ones. Moreover, our involvement with IT-2 fuzzy systems was motivated by the fact that, unlike Type-1 TS fuzzy systems, they do not share the same membership functions with fuzzy filters or controllers. By doing so, we can overcome the network delay problem affecting the fuzzy filter's membership functions when using the parallel distribution compensation (PDC) approach with the Type-1 T-S fuzzy systems.

1.2. Objective and Outline

We were inspired by the discussion above to pursue this study, which examines the event-triggered filter problem for a class of uncertain nonlinear systems that incorporate Markov jump switching:

(i) As an alternative to existing filtering schemes developed for Type-1 fuzzy systems [34,38,49], this study proposes a novel filter design for Markovian jump interval Type-2 fuzzy systems that simultaneously consider an event-triggered communication scheme and a deception attack.

(ii) This work introduces a novel adaptive-event-triggered communication scheme that improves the use of network resources as opposed to [50–52], which assumed the triggering parameters are constant. This mechanism was shown to alleviate network bandwidth and reduce system conservatism effectively by comparing it with other strategies [35,48,53,54].

(iii) As described in [30,55], an improved matrix decoupling approach, which can be implemented by selecting some constants, provides greater flexibility when designing filters. This study, in contrast to previous studies, used a meta-heuristic technique based on PSO to identify the parameters accurately.

We illustrate our scheme with experimental numerical results on a single-link robot arm and an isokinetic rehabilitation system and verify the effectiveness of the proposed scheme by conducting a feasibility study. The paper is organized as follows. Section 2 introduces preliminary results in which the model and assumptions are presented, as well as the problem under study is identified. In Section 3, the main conclusions of the article are presented and derived. An optimization algorithm is presented in Section 4 for determining filter gains and for obtaining optimal performance levels. To illustrate the potential applications of the proposed scheme, as well as to validate its effectiveness, Section 5 presents the computational framework for this work and experimental numerical illustrations of the proposed methodology on a single-link robot arm and an isokinetic rehabilitation system. In Section 6, we discuss some of the findings and limitations of our proposal, as well as suggestions for further investigation.

1.3. Notations

Table 1 presents the abbreviations, acronyms, and notations used in this article.

Table 1. Abbreviations, acronyms, and notations used in the present document.

Symbol	Definition
$\mathbb{R}$	set of real numbers
n	dimension of the Euclidean space
$\boldsymbol{X} \in \mathbb{R}^{n \times m}$	$n \times m$ real matrix
$\boldsymbol{X} > 0$	real symmetric positive definite matrix $\boldsymbol{X}$
$\\|\boldsymbol{X}\\|$	norm of the matrix $\boldsymbol{X}$
$\boldsymbol{X}^\top$	transpose of the matrix $\boldsymbol{X}$
$\text{sym}(\boldsymbol{X})$	$\boldsymbol{X} + \boldsymbol{X}^\top$
$*$	term that is induced by symmetry of a matrix
$\lambda_{\max}()$	the maximal eigenvalue of a matrix
$\mathbb{E}$	mathematical expectation
$\pi_{pq} = P(r_{k+1} = q \mid \bar{r}_k = p)$	transition probability from states p to q
r	number of `if-then` rules
$\bar{r}_k$	discrete-time Markov process
LMI	linear matrix inequalities
MJS	Markovian jump systems
NCS	networked control systems
ET	event-triggered
T-S	Takagi–Sugeno
IT-2	interval Type-2

2. System Description and Problem Formulation

2.1. Markovian Jump T-S-Fuzzy-Model-Based NCS

Consider a class of nonlinear discrete-time Markovian jump systems, which can be expressed by the following T-S fuzzy model on a complete probability space $(\Omega, \mathcal{F}, \mathcal{P})$.

$$\begin{aligned} &\mathbf{R}_i: \text{ If } \theta_1(\boldsymbol{x}(k)) \text{ is } \boldsymbol{M}_i^1 \text{ and If } \theta_2(\boldsymbol{x}(k)) \text{ is } \boldsymbol{M}_i^2 \cdots \text{If } \theta_s(\boldsymbol{x}(k)) \text{ is } \boldsymbol{M}_i^s, \text{ Then} \\ &\begin{cases} \boldsymbol{x}(k+1) = \boldsymbol{A}_i(\bar{r}_k)\boldsymbol{x}(k) + \boldsymbol{B}_{wi}(\bar{r}_k)\boldsymbol{w}(k) \\ \boldsymbol{y}(k) = \boldsymbol{C}_{2i}(\bar{r}_k)\boldsymbol{x}(k) \\ \boldsymbol{z}(k) = \boldsymbol{C}_{1i}(\bar{r}_k)\boldsymbol{x}(k) + \boldsymbol{D}_{1i}(\bar{r}_k)\boldsymbol{w}(k) \end{cases} \end{aligned} \tag{1}$$

where $\boldsymbol{\theta}(k) = [\theta_1(\boldsymbol{x}(k)), \theta_2(\boldsymbol{x}(k)), \cdots, \theta_s(\boldsymbol{x}(k))]$ are measurable premise variables of the system; $\boldsymbol{M}_i^\iota$, $\iota = 1, 2, \cdots, s$ are Type-2 fuzzy sets; $i \in \mathbb{S} \triangleq \{1, 2, \ldots, r\}$ is the number of rules. Vectors $\boldsymbol{x}(k) \in \mathbb{R}^n$, $\boldsymbol{y}(k) \in \mathbb{R}^{n_y}$, $\boldsymbol{w}(k) \in \mathbb{R}^m$, and $\boldsymbol{z}(k) \in \mathbb{R}^q$ define, respectively, the state, the output, the disturbance input, and the measured output. Matrices $\boldsymbol{A}_i(\bar{r}_k)$, $\boldsymbol{B}_{wi}(\bar{r}_k)$, $\boldsymbol{C}_{1i}(\bar{r}_k)$ $\boldsymbol{D}_{1i}(\bar{r}_k)$, and $\boldsymbol{C}_{2i}(\bar{r}_k)$ are known with appropriate dimensions. According to the i-th rule, the firing strength consists of the interval sets described as

$$\mathfrak{M}_i = \left[\underline{\mu}_i(\boldsymbol{x}(k)) \quad \bar{\mu}_i(\boldsymbol{x}(k))\right], \tag{2}$$

where

$$\underline{\mu}_i(\boldsymbol{x}(k)) = \prod_{\iota=1}^{s} \underline{\omega}_{\boldsymbol{M}_i^\iota(\boldsymbol{\theta}(k))} \geq 0, \quad \bar{\mu}_i(\boldsymbol{x}(k)) = \prod_{\iota=1}^{s} \bar{\omega}_{\boldsymbol{M}_i^\iota(\boldsymbol{\theta}(k))} \geq 0$$

$$\bar{\mu}_i(\boldsymbol{x}(k)) \geq \underline{\mu}_i(\boldsymbol{x}(k)) \geq 0, \quad \bar{\omega}_{\boldsymbol{M}_i^\iota(\boldsymbol{\theta}(k))} \geq \underline{\omega}_{\boldsymbol{M}_i^\iota(\boldsymbol{\theta}(k))} \geq 0 \tag{3}$$

As a result, the global fuzzy model can be inferred in the following manner:

$$\begin{cases} x(k+1) &= \sum_{i=1}^{r} \mu_i(x(k))\Big(A_i(\bar{r}_k)x(k) + B_{wi}(\bar{r}_k)w(k)\Big) \\ y(k) &= \sum_{i=1}^{r} \mu_i(x(k))\Big(C_{2i}(\bar{r}_k)x(k)\Big) \\ z(k) &= \sum_{i=1}^{r} \mu_i(x(k))\Big(C_{1i}(\bar{r}_k)x(k) + D_{1i}(\bar{r}_k)w(k)\Big) \end{cases} \tag{4}$$

The normalized membership function $\mu_i(x(k))$ satisfies $\sum_{i=1}^{r} \mu_i(x(k)) = 1$ and is defined as

$$\mu_i(x(k)) = \frac{\underline{\alpha}_i(x(k))\underline{\mu}_i(x(k)) + \bar{\alpha}_i(x(k))\bar{\mu}_i(x(k))}{\sum_{i=1}^{r} \underline{\alpha}_i(x(k))\underline{\mu}_i(x(k)) + \bar{\alpha}_i(x(k))\bar{\mu}_i(x(k))},$$

The weighting coefficients $\underline{\alpha}_i(x(k))$ and $\bar{\alpha}_i(x(k)$ satisfy

$$0 \le \underline{\alpha}_i(x(k)), \bar{\alpha}_i(x(k)) \le 1,\ \underline{\alpha}_i(x(k)) + \bar{\alpha}_i(x(k)) = 1 \tag{5}$$

and are related to the uncertain parameters of the model. These nonlinear functions may not be known, but exist and satisfy (5). Stochastic process $\{\bar{r}_k,\ k \ge 0\}$ is a Markov chain taking values in a finite set $\mathbb{N} = \{1, \cdots, N\}$ with transition probability $\Pi = \left[\pi_{pq}\right]$ defined as $\pi_{pq} = Pr(\bar{r}(k+1) = q | \bar{r}_k = p)$ satisfying $\pi_{pq} \ge 0$ and $\sum_{q=1}^{N} \pi_{pq} = 1$ for all $p, q \in \mathbb{N}$.

2.2. Event-Triggered Schemes

This paper assumed the measurement outputs are sent over communication channels and an event-triggered mechanism is implemented to conserve network resources and determine whether the latest sampled data packet can be transmitted to the filter system (see Figure 1). Event-triggered schemes are generally based on an event generator defined as a logic function that determines whether the sampled data should be transmitted or not. Accordingly, the event generator function is defined as follows:

$$U(k, e_y(k)) = e_y^\top(k)\Theta e_y(k) - \sigma(k)y^\top(k)\Theta y(k) \tag{6}$$

where $e_y(k) = y(k) - y(k_i)$ is the output error, $y(k_i)$ is the previously transmitted data at time k_i, and $\sigma(k)$ is the event-triggered threshold. The current data transmitted depend on the following constraint:

$$k_{i+1} = \inf\{k \in \mathbb{N} | k > k_i, U(k, e_y(k)) > 0\} \tag{7}$$

where $0 \le k_0 \le k_1 \le \cdots k_i \le \cdots$ are the sequence of ET instances. $\sigma(k)$ in (6) can be adaptively adjusted based on the following adaptive law defined by $\Theta > 0$ such that

$$\Delta\sigma(k) = \sigma(k+1) - \sigma(k) = -\varepsilon_0\sigma^2(k)e_y^\top(k)\Theta e_y(k) \tag{8}$$

where ε_0 is a positive constant. When $\varepsilon_0 = 0$ and $0 < \sigma(k) < 1$, it is clear that $\sigma(k)$ becomes a constant.

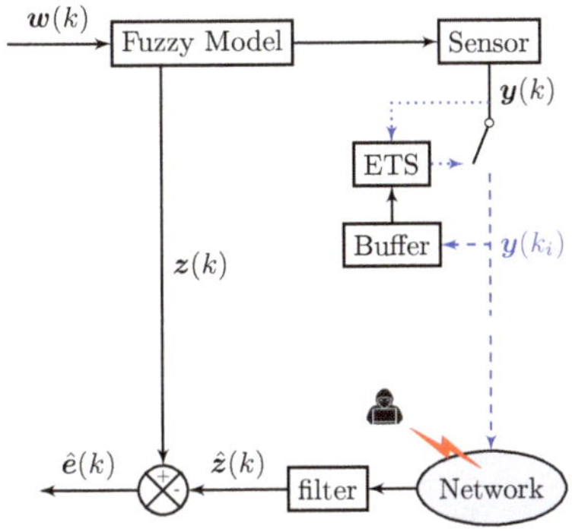

Figure 1. The framework of filtering system under deception attacks.

An important feature of networked control systems is network delay, which is caused by the limited bit rate of communication channels, the waiting period during which the packets are sent, or the propagation and processing of signals. As a result, $\tau(k) \in [0, \tau_M)$ was included in this paper, where τ_M is a positive integer. Based on the behavior of ZOH, $\bar{\boldsymbol{y}}(k)$ holds the value of $\boldsymbol{y}_{k_i}$ in the interval $[k_i + \tau_{k_i},\ k_{i+1} + \tau_{k_{i+1}})$. Thus, we have

$$\bar{\boldsymbol{y}}(k) = \boldsymbol{y}_{k_i}, \quad k \in [k_i + \tau_{k_i},\ k_{i+1} + \tau_{k_{i+1}}) \tag{9}$$

There are then two cases that need to be discussed:

Case 1 If $k_i + \tau_M + 1 > k_{i+1} + \tau_{k_{i+1}} - 1$, we define $\tau(k) = k - k_i, k \in [k_i + \tau_{k_i},\ k_{i+1} + \tau_{k_{i+1}})$. It is obvious that $\tau_{k_i} \leq \tau(k) \leq (k_{i+1} - k_i) + \tau_{k_{i+1}} - 1 \leq 1 + \tau_M$.

Case 2 If $k_i + 1 + \tau_M < k_{i+1} + \tau_{k_{i+1}}$, we define the intervals $[k_i + \tau_{k_i}, k_i + 1 + \tau_M)$, $[k_i + \tau_M + j,\ k_i + \tau_M + j + 1)$, $(j = 1, 2, \cdots)$. Due to $\tau_{k_i} \leq \tau_M$, there exists a positive integer j^* such that

$$k_i + \tau_M + j^* < k_{i+1} + \tau_{k_{i+1}} < k_i + \tau_M + j^* + 1$$

Hence, $\boldsymbol{y}(k_i)$ and $\boldsymbol{y}(k_i + j)$ with $j = 1, 2, \cdots, j^* - 1$ satisfy (7), and the time interval $[k_i + \tau_{k_i},\ k_{i+1} + \tau_{k_{i+1}})$ can be divided as

$$[k_i + \tau_{k_i},\ k_{i+1} + \tau_{k_{i+1}}) = \bigcup_{j=0}^{j^*} \boldsymbol{I}_j \tag{10}$$

where $\boldsymbol{I}_0 = [k_i + \tau_{k_i},\ k_i + \tau_M + 1)$, $\boldsymbol{I}_j = [k_i + \tau_M + j,\ k_i + \tau_M + j + 1)$ and $\boldsymbol{I}_{j^*} = [k_i + \tau_M + j^*,\ k_{i+1} + \tau_{k_{i+1}})$. Define

$$\tau(k) = \begin{cases} k - k_i, & k \in \boldsymbol{I}_0 \\ k - k_i - j, & k \in \boldsymbol{I}_j \\ k - k_i - j^*, & k \in \boldsymbol{I}_{j^*} \end{cases}$$

Obviously, we obtain

$$\begin{cases} \tau_{k_i} \leq \tau(k) \leq 1 + \tau_M, & k \in \boldsymbol{I}_0 \\ \tau_{k_i} \leq \tau_M \leq \tau(k) \leq 1 + \tau_M, & k \in \boldsymbol{I}_j \\ \tau_{k_i} \leq \tau_M \leq \tau(k) \leq 1 + \tau_M, & k \in \boldsymbol{I}_{j^*} \end{cases}$$

Therefore, we have $\tau_m \leq \tau_{k_i} \leq \tau(k) \leq \tau_M + 1, k \in [k_i + \tau_{k_i}, k_{i+1} + \tau_{k_{i+1}})$. In the first case, we define $e_y(k) = 0$; in the second case, $e_y(k)$ is defined as

$$e_y(k) = \begin{cases} 0, & k \in I_0 \\ y(k_i) - y(k_i + j), & k \in I_j \\ y(k_i) - y(k_i + j^*), & k \in I_{j^*} \end{cases}$$

Following the above discussion, the following relationship holds using (9):

$$\bar{y}(k) = y_{k_i} = y(k - \tau(k)) + e_y(k), \quad k \in [k_i + \tau_{k_i}, k_{i+1} + \tau_{k_{i+1}}) \tag{11}$$

Remark 1. *It is important to point out that, to mitigate the communication burden and enhance the utilization of network resources, the adaptive-event-triggered mechanism is widely used in the analysis and design of networked systems [26,27,30]. Based on the above-mentioned literature, we adopted the adaptive-event-triggered mechanism* (8) *to investigate the filtering problem of Markov IT-2 fuzzy systems subject to cyber-attack.*

In addition, according to (8), *the event-triggered interval is greater than zero [56], which indicates that the proposed event-triggered transmission mechanism does not incorporate Zeno's behavior.*

2.3. Deception Attacks

Furthermore, this paper discusses networked T-S fuzzy systems with deception attacks. The open networked environment leads to the possibility of data being modified through deception attacks in a random manner, where a stochastic process is adopted to describe this scenario.

During deception attacks, transmitted data may take the following form:

$$y_\zeta(k) = \bar{y}(k) + \zeta(k)(-\bar{y}(k)) + a(\bar{y}(k)) \tag{12}$$

Here, $\zeta(k)$ represents the Bernoulli random variable that meets the following conditions:

$$Pr\{\zeta(k) = 1\} = \bar{\zeta}, Pr\{\zeta(k) = 0\} = 1 - \bar{\zeta}$$

$a(\bar{y}(k))$ represents the embedded signal launched by the attacker. It was assumed that $a(\bar{y}(k))$ is a sufficiently smooth continuous nonlinear function satisfying the sector condition:

$$\|a(\bar{y}(k))\| \leq \|Gy(k)\| \tag{13}$$

where G is a known constant matrix.

2.4. Fuzzy Filter

To estimate the signal z_k in Model (4), the following fuzzy filter is defined:

$\mathbf{R}_j$: If $\vartheta_1(\hat{x}(k))$ is $\mathcal{N}_j^1$ and If $\vartheta_2(\hat{x}(k))$ is $\mathcal{N}_j^2 \cdots$ If $\vartheta_v(\hat{x}(k))$ is $\mathcal{N}_j^v$, Then

$$\begin{cases} \hat{x}(k+1) & = \hat{A}_j(\bar{r}_k)\hat{x}(k) + \hat{B}_j(\bar{r}_k)y_\zeta(k) \\ \hat{z}(k) & = \hat{C}_j(\bar{r}_k)\hat{x}(k) \end{cases}$$

where $\hat{x}(k)$ is the filter state, $\hat{z}(k)$ is the filter output, $\vartheta(k) = [\vartheta_1(\hat{x}(k)), \vartheta_2(\hat{x}(k)), \ldots, \vartheta_v(\hat{x}(k))]$ are the premise variables, and $\mathcal{N}_j^\varsigma$, $\varsigma = 1, 2, \cdots, v$ stands for the Type-2 fuzzy sets of the filter; $\hat{A}_j(\bar{r}_k)$, $\hat{B}_j(\bar{r}_k)$ and $\hat{C}_j(\bar{r}_k)$ are unknown matrices that should be designed.

$$\mathfrak{N}_j = \begin{bmatrix} \underline{\nu}_j(\hat{x}(k)) & \bar{\nu}_j(\hat{x}(k)) \end{bmatrix}, \tag{14}$$

where

$$\underline{\nu}_j(\hat{x}(k)) = \prod_{\varsigma=1}^{v} \underline{\omega}_{\mathcal{N}_j^{\varsigma}(\vartheta(k))} \geq 0, \quad \bar{\nu}_j(\hat{x}(k)) = \prod_{\varsigma=1}^{v} \bar{\omega}_{\mathcal{N}_j^{\varsigma}(\vartheta(k))} \geq 0$$

$$\underline{\nu}_j(\hat{x}(k)) \geq \bar{\nu}_j(\hat{x}(k)) \geq 0, \quad \bar{\omega}_{\mathcal{N}_j^{\varsigma}(\vartheta(k))} \geq \underline{\omega}_{\mathcal{N}_j^{\varsigma}(\vartheta(k))} \geq 0 \tag{15}$$

The global fuzzy filter is defined as follows:

$$\begin{cases} \hat{x}(k+1) &= \sum_{j=1}^{r} \nu_j(\hat{x}(k))(\hat{A}_j(\bar{r}_k)\hat{x}(k) + \hat{B}_j(\bar{r}_k)y_{\zeta}(k)) \\ \hat{z}(k) &= \sum_{j=1}^{r} \nu_j(\hat{x}(k))(\hat{C}_j(\bar{r}_k)\hat{x}(k)) \end{cases} \tag{16}$$

$$\nu_j(\hat{x}(k)) = \frac{\underline{\beta}_j(\hat{x}(k))\underline{\nu}_j(\hat{x}(k)) + \bar{\beta}_j(\hat{x}(k))\bar{\nu}_j(\hat{x}(k))}{\sum_{l=1}^{r}(\underline{\beta}_l(\hat{x}(k))\underline{\nu}_l(\hat{x}(k)) + \bar{\beta}_l(\hat{x}(k))\bar{\nu}_l(\hat{x}(k)))}, \ \nu_j(\hat{x}(k)) \geq 0, \ \sum_{j=1}^{r} \nu_j(\hat{x}(k)) = 1 \tag{17}$$

The two nonlinear functions $\underline{\beta}_j(\hat{x}(k))$ and $\bar{\beta}_j(\hat{x}(k))$ satisfy

$$0 \leq \underline{\beta}_j(\hat{x}(k)), \bar{\beta}_j(\hat{x}(k)) \leq 1, \ \underline{\beta}_j(\hat{x}(k)) + \bar{\beta}_j(\hat{x}(k)) = 1 \tag{18}$$

For simplicity, in the sequel, $\mu_i(x(k))$ and $\nu_j(\hat{x}(k))$ are denoted as μ_i and $\hat{\nu}_j$, respectively.

Let $\hat{e}(k) = z(k) - \hat{z}(k)$. Taking (4) and (16) together, we can represent the filtering error system as follows:

$$\begin{cases} \tilde{x}(k+1) &= \sum_{i=1}^{r}\sum_{j=1}^{r} \mu_i\hat{\nu}_j\Big(\tilde{A}_{ij}^p\tilde{x}(k) + (1-\bar{\zeta})\tilde{A}_{dij}^p\tilde{x}(k-\tau(k)) + (1-\bar{\zeta})\tilde{B}_{ej}^p e_y(k) \\ &\quad +\bar{\zeta}\tilde{B}_{ej}^p a(\bar{y}(k)) + \tilde{B}_{1i}^p w(k) + (\zeta(k)-\bar{\zeta})\tilde{B}_g g(\tilde{x}(k))\Big) \\ \hat{e}(k) &= \sum_{i=1}^{r}\sum_{j=1}^{r} \mu_i\hat{\nu}_j\Big(\tilde{C}_{ij}^p\tilde{x}(k) + \tilde{D}_i^p w(k)\Big) \end{cases} \tag{19}$$

where $\tilde{x}(k) = \left[x^\top(k),\ \hat{x}^\top(k)\right]^\top$ $\tilde{A}_{ij}^p = \begin{bmatrix} A_i^p & 0 \\ 0 & \hat{A}_j^p \end{bmatrix}$, $\tilde{A}_{dij}^p = \begin{bmatrix} 0 & 0 \\ \hat{B}_j^p C_{2i}^p & 0 \end{bmatrix}$, $\tilde{B}_{ej}^p = \begin{bmatrix} 0 \\ \hat{B}_j^p \end{bmatrix}$, $\tilde{B}_{1i}^p = \begin{bmatrix} B_{1i}^p \\ 0 \end{bmatrix}$, $\tilde{B}_g = \begin{bmatrix} 0 \\ I \end{bmatrix}$, $\tilde{C}_{ij}^p = \begin{bmatrix} C_{1i}^p & -\hat{C}_j^p \end{bmatrix}$, $\tilde{D}_i^p = D_{1i}^p$, $g(\tilde{x}(k)) = \hat{B}_j^p(a(\bar{y}(k)) - C_{2i}^p x(k-\tau(k)) - e_y(k))$.

2.5. Problem Formulation

Based on the above description, the objective of this article is to design the parameters of Filter (16) such that System (19) is asymptotically mean-squared stable with extended dissipative performance, i.e.:

1. System (19) is mean-squared stable;
2. Under zero initial conditions, the following criterion holds for $w(k) \neq 0$:

$$\mathbb{E}\left\{\sum_{k=0}^{K_f} J(k)\right\} \geq \sup_{0\leq k\leq K_f} \mathbb{E}\{\hat{e}^\top(k)\Omega_4\hat{e}(k)\} \tag{20}$$

where $J(k) = \hat{e}^\top(k)\Omega_1\hat{e}(k) + 2\hat{e}^\top(k)\Omega_2 w(k) + w^\top(k)\Omega_3 w(k)$, and matrices Ω_1, Ω_2, Ω_3 and Ω_4 satisfy the following assumption:

Assumption 1 ([57]). *For known real matrices $\Omega_1 = \Omega_1^\top \leq 0$, Ω_2, $\Omega_3 = \Omega_3^\top \geq 0$, and $\Omega_4 = \Omega_4^\top = (\Omega_4^+)^\top \Omega_4^+ \geq 0$, it was assumed that:*

(i) $\| D_i^p \| \| \Omega_4 \| = 0$;
(ii) $(\| \Omega_1 \| + \| \Omega_2 \|) \| \Omega_4 \| = 0$.

Remark 2. *From (20), the extended dissipativity criterion includes H_∞, strict dissipativity, $l_2 - l_\infty$, or passivity performances as special cases according to the different values of the matrices:*

- *If $\Omega_1 = -I$, $\Omega_2 = 0$, $\Omega_3 = \gamma^2 I$, and $\Omega_4 = 0$, Inequality (20) reduces to an H_∞ performance requirement.*
- *If $\Omega_3 = R - \gamma I$ $(\gamma > 0)$ and $\Omega_4 = 0$, Inequality (20) corresponds to a strict dissipativity.*
- *If $\Omega_1 = 0$, $\Omega_2 = 0$, $\Omega_3 = \gamma^2 I$, and $\Omega_4 = I$, Inequality (20) reduces to an $l_2 - l_\infty$ performance.*
- *If $\hat{e}(k)$ and $w(k)$ have the same dimension and $\Omega_1 = 0$, $\Omega_2 = I$, $\Omega_3 = \gamma I$, and $\Omega_4 = 0$, the passivity will be obtained.*

Next, we present some preliminaries that are crucial for the derivation of our major conclusions.

Definition 1 ([58]). *The filtering error system (19) is said to be stochastically mean-squared stable if the following condition holds:*

$$\mathbb{E}\{ \sum_{k=0}^{\infty} \|\tilde{x}(k)\|^2 | (\tilde{x}(k_0), \bar{r}_0)\} < \infty$$

for any initial condition $(\tilde{x}(k_0), \bar{r}_0)$.

Lemma 1 ([59]). *For given positive integers τ_m and τ_M, a positive matrix Z, symmetric matrices, Y_{11} and Y_{22}, and matrices Y_{12}, T_1, and T_2, such that*

$$\begin{bmatrix} Y_{11} & Y_{12} & T_1 \\ * & Y_{22} & T_2 \\ * & * & Z \end{bmatrix} \geq 0 \tag{21}$$

the following inequality holds:

$$-\sum_{s=k-\tau_M}^{k-\tau_m-1} \eta^\top(s) Z \eta(s) \leq \psi^\top(k)\Big(\tau_r(Y_{11} + \frac{1}{3}Y_{22}) + \text{sym}\Big\{T_1 e_{12} + T_2 e_{123}\Big\}\Big)\psi(k)$$

where $\tau_r = \tau_M - \tau_m \geq 1$, $\eta(k) = x(k+1) - x(k)$, and $\psi(k) = [x^\top(k-\tau_m) \;\; x^\top(k-\tau_M) \;\; \frac{1}{\tau_r+1}\sum_{s=k-\tau_M}^{k-\tau_m-1} x^\top(s)]^\top$, $e_{12} = [I \;\; -I \;\; 0]$, and $e_{123} = [I \;\; I \;\; -2I]$.

3. Main Results

3.1. Extended Dissipative Analysis

This section concerns the derivation of sufficient conditions in such a way that the filtering error system (19) is stochastically mean-squared stable with extended dissipativity performance.

Theorem 1. *Consider the Markovian jump IT-2 fuzzy system (4) and Filter (16). For given matrices $\Theta > 0$, Ω_1, Ω_2, Ω_3, and Ω_4 satisfying Assumption 1 and positive scalars ϱ_s, if matrices P^p, Q_v, $v = 1,2,3$, Z_s, X_{1s}, Y_{1s}, Z_{1s}, T_{1s}, T_{2s}, T_{3s}, F_s^p, and $s = 1,2$ and scalars $\tau > 0$ and $\sigma_0 > 1$ exist such that the following conditions hold:*

$$\begin{bmatrix} \boldsymbol{X}_{11} & \boldsymbol{X}_{12} & \boldsymbol{T}_{11} \\ * & \boldsymbol{X}_{22} & \boldsymbol{T}_{12} \\ * & * & \boldsymbol{Z}_1 \end{bmatrix} > 0 \begin{bmatrix} \boldsymbol{Y}_{11} & \boldsymbol{Y}_{12} & \boldsymbol{T}_{21} \\ * & \boldsymbol{Y}_{22} & \boldsymbol{T}_{22} \\ * & * & \boldsymbol{Z}_2 \end{bmatrix} > 0 \begin{bmatrix} \boldsymbol{Z}_{11} & \boldsymbol{Z}_{12} & \boldsymbol{T}_{13} \\ * & \boldsymbol{Z}_{22} & \boldsymbol{T}_{23} \\ * & * & \boldsymbol{Z}_2 \end{bmatrix} > 0 \tag{22}$$

$$\begin{bmatrix} -\boldsymbol{P}^p & (\tilde{\boldsymbol{\mathcal{C}}}_{ij}^p)^\top (\boldsymbol{\Omega}_4^+)^\top \\ * & -\boldsymbol{I} \end{bmatrix} < 0 \tag{23}$$

$$\begin{cases} \tilde{\boldsymbol{\Phi}}_{ii}^p < 0 \\ \tilde{\boldsymbol{\Phi}}_{ij}^p + \varrho_1 \tilde{\boldsymbol{\Phi}}_{ji}^p < 0 \\ \tilde{\boldsymbol{\Phi}}_{ij}^p + \varrho_2 \tilde{\boldsymbol{\Phi}}_{ji}^p < 0, \ 1 \le i < j \le r \end{cases} \tag{24}$$

then, System (19) is stochastically mean-squared stable and extended dissipative. The remaining matrices are defined as

$$\tilde{\boldsymbol{\Phi}}_{ij}^p = \begin{bmatrix} \tilde{\boldsymbol{\Psi}}_{ij}^p + \hat{\boldsymbol{\Psi}}^p & (\tilde{\boldsymbol{\mathcal{B}}}_{ij}^p)^\top - (\tilde{\boldsymbol{\mathcal{C}}}_{ij}^p)^\top \boldsymbol{\Omega}_2 & (\tilde{\boldsymbol{\mathcal{C}}}_{ij}^p)^\top \sqrt{-\boldsymbol{\Omega}_1} \\ * & -sym\{(\tilde{\boldsymbol{D}}_i^p)^\top \boldsymbol{\Omega}_2\} - \boldsymbol{\Omega}_3 & (\tilde{\boldsymbol{D}}_i^p)^\top \sqrt{-\boldsymbol{\Omega}_1} \\ * & * & -\boldsymbol{I} \end{bmatrix}$$

$$\tilde{\boldsymbol{\Psi}}_{ij}^p = \begin{bmatrix} \tilde{\boldsymbol{\Psi}}_{11ij}^p & \boldsymbol{0} & (1-\bar{\zeta})\boldsymbol{F}_1^p \tilde{\boldsymbol{A}}_{dij}^p & \boldsymbol{0} & \boldsymbol{0} & \boldsymbol{0} & \boldsymbol{0} & \tilde{\boldsymbol{\Psi}}_{18ij}^p & (1-\bar{\zeta})\boldsymbol{F}_1^p \tilde{\boldsymbol{B}}_{ej}^p & \bar{\zeta}\boldsymbol{F}_1^p \tilde{\boldsymbol{B}}_{ej}^p \\ * & -\boldsymbol{Q}_1 & \boldsymbol{0} & \boldsymbol{0} & \boldsymbol{0} & \boldsymbol{0} & \boldsymbol{0} & \boldsymbol{0} & \boldsymbol{0} & \boldsymbol{0} \\ * & * & \tilde{\boldsymbol{\Psi}}_{33}^p & \boldsymbol{0} & \boldsymbol{0} & \boldsymbol{0} & \boldsymbol{0} & (1-\bar{\zeta})(\boldsymbol{F}_2^p \tilde{\boldsymbol{A}}_{dij}^p)^\top & (\boldsymbol{C}_{2i}^p)^\top \boldsymbol{\Theta} & \boldsymbol{0} \\ * & * & * & -\boldsymbol{Q}_2 & \boldsymbol{0} & \boldsymbol{0} & \boldsymbol{0} & \boldsymbol{0} & \boldsymbol{0} & \boldsymbol{0} \\ * & * & * & * & \boldsymbol{0} & \boldsymbol{0} & \boldsymbol{0} & \boldsymbol{0} & \boldsymbol{0} & \boldsymbol{0} \\ * & * & * & * & * & \boldsymbol{0} & \boldsymbol{0} & \boldsymbol{0} & \boldsymbol{0} & \boldsymbol{0} \\ * & * & * & * & * & * & \boldsymbol{0} & \boldsymbol{0} & \boldsymbol{0} & \boldsymbol{0} \\ * & * & * & * & * & * & * & \tilde{\boldsymbol{\Psi}}_{88}^p & (1-\bar{\zeta})\boldsymbol{F}_2^p \tilde{\boldsymbol{B}}_{ej}^p & \bar{\zeta}\boldsymbol{F}_1^p \tilde{\boldsymbol{B}}_{ej}^p \\ * & * & * & * & * & * & * & * & (1-\sigma_0) & \boldsymbol{0} \\ * & * & * & * & * & * & * & * & * & -\tau \boldsymbol{I} \end{bmatrix}$$

$$\tilde{\boldsymbol{\mathcal{B}}}_{ij}^p = \begin{bmatrix} (\boldsymbol{F}_1^p \tilde{\boldsymbol{B}}_{1i}^p)^\top & \boldsymbol{0} & \boldsymbol{0} & \boldsymbol{0} & \boldsymbol{0} & \boldsymbol{0} & \boldsymbol{0} & (\boldsymbol{F}_2^p \tilde{\boldsymbol{B}}_{1i}^p)^\top & \boldsymbol{0} & \boldsymbol{0} \end{bmatrix}$$

$$\tilde{\boldsymbol{\mathcal{C}}}_{ij}^p = \begin{bmatrix} \tilde{\boldsymbol{C}}_{ij}^p & \boldsymbol{0} & \boldsymbol{0} & \boldsymbol{0} & \boldsymbol{0} & \boldsymbol{0} & \boldsymbol{0} & \boldsymbol{0} & \boldsymbol{0} & \boldsymbol{0} & \boldsymbol{0} \end{bmatrix}, \ \boldsymbol{X}^p = \sum_{q=1}^{N} \pi_{pq} \boldsymbol{P}^q$$

$$\begin{aligned} \tilde{\boldsymbol{\Psi}}_{11ij}^p &= sym(\boldsymbol{F}_1^p(\tilde{\boldsymbol{A}}_{ij}^p - \boldsymbol{I})) + (\boldsymbol{X}^p - \boldsymbol{P}^p) + \boldsymbol{Q}_1 + \boldsymbol{Q}_2 + (\tau_r + 1)\boldsymbol{Q}_3 \\ \tilde{\boldsymbol{\Psi}}_{33}^p &= -\boldsymbol{Q}_3 + (\boldsymbol{C}_{2i}^p)^\top \boldsymbol{\Theta} \boldsymbol{C}_{2i}^p + (\boldsymbol{C}_{2i}^p)^\top \boldsymbol{G}^\top \boldsymbol{G} \boldsymbol{C}_{2i}^p, \ \tilde{\boldsymbol{\Psi}}_{18ij}^p = -\boldsymbol{F}_1^p + \boldsymbol{X}^p + (\boldsymbol{F}_2^p(\tilde{\boldsymbol{A}}_{ij}^p - \boldsymbol{I}))^\top \\ \tilde{\boldsymbol{\Psi}}_{88}^p &= -\mathrm{sym}(\boldsymbol{F}_2^p) + \boldsymbol{X}^p + \tau_m \boldsymbol{Z}_1 + \tau_r \boldsymbol{Z}_2 \\ \hat{\boldsymbol{\Psi}}^p &= \boldsymbol{\Pi}_1^\top \left(\tau_m(\boldsymbol{X}_{11} + \frac{1}{3}\boldsymbol{X}_{22}) + \mathrm{sym}\left\{ \boldsymbol{T}_{11} \boldsymbol{e}_{12} + \boldsymbol{T}_{12} \boldsymbol{e}_{123} \right\} \right) \boldsymbol{\Pi}_1 \\ &\quad + \boldsymbol{\Pi}_2^\top \left(d_r(\boldsymbol{Y}_{11} + \frac{1}{3}\boldsymbol{Y}_{22}) + \mathrm{sym}\left\{ \boldsymbol{T}_{21} \boldsymbol{e}_{12} + \boldsymbol{T}_{22} \boldsymbol{e}_{123} \right\} \right) \boldsymbol{\Pi}_2 \\ &\quad + \boldsymbol{\Pi}_3^\top \left(d_r(\boldsymbol{Z}_{11} + \frac{1}{3}\boldsymbol{Z}_{22}) + \mathrm{sym}\left\{ \boldsymbol{T}_{31} \boldsymbol{e}_{12} + \boldsymbol{T}_{32} \boldsymbol{e}_{123} \right\} \right) \boldsymbol{\Pi}_3 \end{aligned}$$

$$\boldsymbol{\Pi}_1 = \mathrm{col}\left\{ \boldsymbol{e}_1 \ \boldsymbol{e}_2 \ \boldsymbol{e}_5 \right\}, \ \boldsymbol{\Pi}_2 = \mathrm{col}\left\{ \boldsymbol{e}_3 \ \boldsymbol{e}_4 \ \boldsymbol{e}_6 \right\}, \ \Pi_3 = \mathrm{col}\left\{ \boldsymbol{e}_2 \ \boldsymbol{e}_3 \ \boldsymbol{e}_7 \right\}$$

$$\boldsymbol{e}_i = \Big[\underbrace{\boldsymbol{0}, \cdots, \boldsymbol{0}}_{l-1} \quad \boldsymbol{I}_{\lambda(l)} \quad \underbrace{\boldsymbol{0}, \cdots, \boldsymbol{0}}_{11-l} \Big] \in \mathbb{R}^{\lambda(l) \times (16n + n_y + m)}, \ l = 1, 2, \cdots, 11$$

$$\lambda(l) = \begin{cases} 2n & l = 1, 2, \cdots 8 \\ n_y & l = 9, 10 \\ m & l = 11 \end{cases}$$

Proof. The stochastic stability is demonstrated for the filtering error system (19) with $w(k) = 0$ by considering a Lyapunov functional defined as $V(k) = V_1(k) + V_2(k) + V_3(k) + V_4(k)$:

$$V_1(k) = \tilde{x}^\top(k)P(\bar{r}_k)\tilde{x}(k) \tag{25}$$

$$V_2(k) = \sum_{s=k-\tau_m}^{k-1} \tilde{x}^\top(s)Q_1\tilde{x}(s) + \sum_{s=k-\tau_M}^{k-1} \tilde{x}^\top(s)Q_2\tilde{x}(s) + \sum_{\theta=-\tau_M}^{-\tau_m}\sum_{s=k+\theta}^{k-1} \tilde{x}^\top(s)Q_3\tilde{x}(s) \tag{26}$$

$$V_3(k) = \sum_{\theta=-\tau_m}^{-1}\sum_{s=k+\theta}^{k-1} \eta^\top(s)Z_1\eta(s) + \sum_{\theta=-\tau_M}^{-\tau_m-1}\sum_{s=k+\theta}^{k-1} \eta^\top(s)Z_2\eta(s) \tag{27}$$

$$V_4(k) = \frac{1}{2\varepsilon_0}\left(\frac{1}{\sigma(k)} - \sigma_0\right)^2 \tag{28}$$

where $\eta(k) = \tilde{x}(k+1) - \tilde{x}(k)$.

Let $\Delta V(k)$ be the forward difference of $V(k)$. Along the trajectories of System (19), we can demonstrate the following:

$$\begin{aligned}\mathbb{E}\{\Delta V_1(k)\} &= \mathbb{E}\Big\{V_1(k+1) - V_1(k)|\tilde{x}(k), \bar{r}_k = p\Big\} \\ &= \mathbb{E}\Big\{\tilde{x}^\top(k+1)X^p\tilde{x}(k+1)\Big\} - \tilde{x}^\top(k)P^p\tilde{x}(k) \\ &= \mathbb{E}\Big\{\eta^\top(k)X^p\eta(k)\Big\} + \tilde{x}^\top(k)(X^p - P^p)\tilde{x}(k) + 2\tilde{x}^\top(k)X^p\eta(k)\end{aligned} \tag{29}$$

$$\mathbb{E}\{\Delta V_2(k)\} \le \xi^\top(k)\Big(e_1^\top(Q_1 + Q_2 + (\tau_r+1)Q_3)e_1 - e_2^\top Q_1 e_2 - e_4^\top Q_2 e_4 - e_3^\top Q_3 e_3\Big)\xi(k) \tag{30}$$

$$\begin{aligned}\mathbb{E}\{\Delta V_3(k)\} &= \eta^\top(k)\Big(\tau_m Z_1 + \tau_r Z_2\Big)\eta(k) - \sum_{s=k-\tau_m}^{k-1}\eta^\top(s)Z_1\eta(s) \\ &\quad - \sum_{s=k-\tau_M}^{k-\tau(k)-1}\eta^\top(s)Z_2\eta(s) - \sum_{s=k-\tau(k)}^{k-\tau_m-1}\eta^\top(s)Z_2\eta(s)\end{aligned} \tag{31}$$

$$\mathbb{E}\{\Delta V_4(k)\} = -\frac{1}{\varepsilon_0}\left(\frac{1}{\sigma(k)} - \sigma_0\right)\frac{\Delta\sigma(k)}{\sigma^2(k)} = \left(\frac{1}{\sigma(k)} - \sigma_0\right)e_y^\top(k)\Theta e_y(k) \tag{32}$$

By Lemma 1, we can derive

$$-\sum_{s=k-\tau_m}^{k-1}\eta^\top(s)Z_1\eta(s) \le \xi^\top(k)\Pi_1^\top\left(\tau_m(X_{11} + \frac{1}{3}X_{22}) + \mathrm{sym}\Big\{T_{11}e_{12} + T_{12}e_{123}\Big\}\right)\Pi_1\xi(k) \tag{33}$$

$$-\sum_{s=k-\tau_M}^{k-\tau(k)-1}\eta^\top(s)Z_2\eta(s) \le \xi^\top(k)\Pi_2^\top\left(d_r(Y_{11} + \frac{1}{3}Y_{22}) + \mathrm{sym}\Big\{T_{21}e_{12} + T_{22}e_{123}\Big\}\right)\Pi_2\xi(k) \tag{34}$$

$$-\sum_{s=k-\tau(k)}^{k-\tau_m-1}\eta^\top(s)Z_2\eta(s) \le \xi^\top(k)\Pi_3^\top\left(d_r(Z_{11} + \frac{1}{3}Z_{22}) + \mathrm{sym}\Big\{T_{31}e_{12} + T_{32}e_{123}\Big\}\right)\Pi_3\xi(k) \tag{35}$$

where

$$\begin{aligned}\xi(k) = \mathrm{col}\Big\{&\tilde{x}(k), \tilde{x}(k-\tau_m), \tilde{x}(k-\tau(k)), \tilde{x}(k-\tau_M), \frac{1}{\tau_m+1}\sum_{s=k-\tau_m}^{k-1}\tilde{x}(s), \\ &\frac{1}{\tau_M - \tau(k) + 1}\sum_{s=k-\tau_M}^{k-\tau(k)-1}\tilde{x}(s), \frac{1}{\tau(k) - \tau_m + 1}\sum_{s=k-\tau(k)}^{k-\tau_m-1}\tilde{x}(s), \eta(k),\ e_y(k),\ a(\bar{y}(k))\Big\}.\end{aligned}$$

From (19), the following null equation holds:

$$
\begin{aligned}
&\mathbb{E}\Big\{\boldsymbol{\xi}^{\top}(k)\mathfrak{F}^{p}\sum_{i=1}^{r}\sum_{j=1}^{r}\mu_i\hat{\vartheta}_j\Big((\tilde{\boldsymbol{A}}_{ij}^{p}-\boldsymbol{I})\tilde{\boldsymbol{x}}(k)+(1-\bar{\zeta})\tilde{\boldsymbol{A}}_{dij}^{p}\tilde{\boldsymbol{x}}(k-\tau(k))+(1-\bar{\zeta})\tilde{\boldsymbol{B}}_{ej}^{p}\boldsymbol{e}_y(k)+\bar{\zeta}\tilde{\boldsymbol{B}}_{ej}^{p}\boldsymbol{a}(k)\\
&+(\zeta(k)-\bar{\zeta})\tilde{\boldsymbol{B}}_{g}\boldsymbol{g}(\tilde{\boldsymbol{x}}(k))\Big)\Big)\Big\}=\\
&\mathbb{E}\Big\{\boldsymbol{\xi}^{\top}(k)\mathfrak{F}^{p}\sum_{i=1}^{r}\sum_{j=1}^{r}\mu_i\hat{\vartheta}_j\Big[(\tilde{\boldsymbol{A}}_{ij}^{p}-\boldsymbol{I})\quad \boldsymbol{0}\quad (1-\bar{\zeta})\tilde{\boldsymbol{A}}_{dij}^{p}\quad \boldsymbol{0}\quad \boldsymbol{0}\quad \boldsymbol{0}\quad \boldsymbol{0}\quad -\boldsymbol{I}\quad (1-\bar{\zeta})\tilde{\boldsymbol{B}}_{ej}^{p}\quad \bar{\zeta}\tilde{\boldsymbol{B}}_{ej}^{p}\Big]\boldsymbol{\xi}(k)\Big\}=\boldsymbol{0}
\end{aligned}
\tag{36}
$$

where $\mathfrak{F}^{p}=\begin{bmatrix}(\boldsymbol{F}_1^{p})^{\top} & \boldsymbol{0} & \boldsymbol{0} & \boldsymbol{0} & \boldsymbol{0} & \boldsymbol{0} & \boldsymbol{0} & (\boldsymbol{F}_2^{p})^{\top} & \boldsymbol{0} & \boldsymbol{0}\end{bmatrix}^{\top}$.

In view of (6) and considering (32), it can be established that

$$
\begin{aligned}
(\frac{1}{\sigma(k)}-\sigma_0)\boldsymbol{e}_y^{\top}(k)\boldsymbol{\Theta}\boldsymbol{e}_y(k)&\le \boldsymbol{y}^{\top}(k_i)\boldsymbol{\Theta}\boldsymbol{y}(k_i)-\sigma_0\boldsymbol{e}_y^{\top}(k)\boldsymbol{\Theta}\boldsymbol{e}_y(k)\\
&\le\begin{bmatrix}\tilde{\boldsymbol{x}}(k-\tau(k))\\ \boldsymbol{e}_y(k)\end{bmatrix}^{\top}\begin{bmatrix}(\boldsymbol{C}_{2i}^{p})^{\top}\boldsymbol{\Theta}\boldsymbol{C}_{2i}^{p} & (\boldsymbol{C}_{2i}^{p})^{\top}\boldsymbol{\Theta}\\ * & (1-\sigma_0)\boldsymbol{\Theta}\end{bmatrix}\begin{bmatrix}\tilde{\boldsymbol{x}}(k-\tau(k))\\ \boldsymbol{e}_y(k)\end{bmatrix}
\end{aligned}
\tag{37}
$$

Moreover, from (13), we have

$$
\begin{aligned}
&-\tau\boldsymbol{a}^{\top}(\bar{\boldsymbol{y}}(k))\boldsymbol{a}(\bar{\boldsymbol{y}}(k))+\tau\boldsymbol{y}^{\top}(k)\boldsymbol{G}^{\top}\boldsymbol{G}\boldsymbol{y}(k)\\
&=\begin{bmatrix}\tilde{\boldsymbol{x}}(k-d(k))\\ \boldsymbol{a}(\bar{\boldsymbol{y}}(k))\end{bmatrix}^{\top}\begin{bmatrix}\tau(\tilde{\boldsymbol{C}}_{2i}^{p})^{\top}\boldsymbol{G}^{\top}\boldsymbol{G}\tilde{\boldsymbol{C}}_{2i}^{p} & \boldsymbol{0}\\ * & -\tau\boldsymbol{I}\end{bmatrix}\begin{bmatrix}\tilde{\boldsymbol{x}}(k-d(k))\\ \boldsymbol{a}(\bar{\boldsymbol{y}}(k))\end{bmatrix}\ge\boldsymbol{0}
\end{aligned}
\tag{38}
$$

Combining (29)–(37), one obtains

$$
\mathbb{E}\{\Delta \boldsymbol{V}(k)\}\le\sum_{i=1}^{r}\sum_{j=1}^{r}\mu_i\hat{\vartheta}_j\Big(\boldsymbol{\xi}^{\top}(k)\tilde{\boldsymbol{\Lambda}}_{ij}^{p}\boldsymbol{\xi}(k)\Big)
\tag{39}
$$

where $\tilde{\boldsymbol{\Lambda}}_{ij}^{p}=\boldsymbol{\Psi}_{ij}^{p}+\hat{\boldsymbol{\Psi}}^{p}$.

Assume that the membership functions are subjected to the following asynchronous constraints as described in [36]:

$$
\begin{aligned}
\hat{\vartheta}_j&=\rho_j(k)\mu_j\\
|\mu_j-\hat{\vartheta}_j|&\le\delta_j
\end{aligned}
\tag{40}
$$

where ρ_j, δ_j ($j\in\mathbb{S}$) are some positive constants. In light of the asynchronous constraints (40), it is evident that

$$
0<\rho^1\le 1-\frac{\delta_j}{\underline{\mu}_j}\le 1-\frac{\delta_j}{\mu_j}\le\rho_j(k)\le 1+\frac{\delta_j}{\mu_j}\le 1+\frac{\delta_j}{\underline{\mu}_j}\le\rho^2
\tag{41}
$$

where ρ^1 and ρ^2 are the lower and upper values of $\rho_j(k)$. Thus, it is easy to obtain

$$
\varrho_1=\frac{\rho^1}{\rho^2}\le\frac{\min\rho_i(k)}{\max\rho_j(k)}\le\frac{\rho_i(k)}{\rho_j(k)}\le\frac{\max\rho_i(k)}{\min\rho_j(k)}\le\frac{\rho^2}{\rho^1}=\varrho_2
\tag{42}
$$

As a result of the relation (40), this yields

$$
\sum_{i=1}^{r}\sum_{j=1}^{r}\mu_i\hat{\vartheta}_j\tilde{\boldsymbol{\Lambda}}_{ij}^{p}=\sum_{i=1}^{r}\sum_{j=1}^{r}\rho_j(k)\mu_i\mu_j\tilde{\boldsymbol{\Lambda}}_{ij}^{p}=\sum_{i=1}^{r}\rho_i(k)\mu_i^2\tilde{\boldsymbol{\Lambda}}_{ii}^{p}+\sum_{i=1}^{r}\sum_{j>i}^{r}\rho_j(k)\mu_i\mu_j\Big\{\tilde{\boldsymbol{\Lambda}}_{ij}^{p}+\frac{\rho_i(k)}{\rho_j(k)}\tilde{\boldsymbol{\Lambda}}_{ji}^{p}\Big\}
\tag{43}
$$

Furthermore, assuming the relationship (43) is valid, there exists $\epsilon_1 > 0$ and $\epsilon_2 > 0$ satisfying $\epsilon_1 + \epsilon_2 = 1$ such that

$$\frac{\rho_i(k)}{\rho_j(k)} = \epsilon_1 \varrho_1 + \epsilon_2 \varrho_2 \tag{44}$$

Thus, it can be obtained from the previous equation that

$$\tilde{\boldsymbol{\Lambda}}^p_{ij} + \frac{\rho_i(k)}{\rho_j(k)} \tilde{\boldsymbol{\Lambda}}^p_{ji} = \sum_{l=1}^{2} \epsilon_l (\tilde{\boldsymbol{\Lambda}}^p_{ij} + \varrho_l \tilde{\boldsymbol{\Lambda}}^p_{ji}) \tag{45}$$

Moreover, according to (24), (43), and (45), it can be deduced that $\tilde{\boldsymbol{\Lambda}}^p_{ij} < 0$ and $\sum_{i=1}^{r} \sum_{j=1}^{r} \mu_i \hat{\vartheta}_j \tilde{\boldsymbol{\Lambda}}^p_{ij} < 0$. Accordingly, (39) provides the following:

$$\mathbb{E}\{\Delta \boldsymbol{V}(k)\} \leq \lambda \boldsymbol{\xi}^\top(k) \boldsymbol{\xi}(k) \tag{46}$$

where $\lambda < 0$ is the largest eigenvalue of $\tilde{\boldsymbol{\Lambda}}^p_{ij}$. Thus, from (46) results

$$\mathbb{E}\Big\{ \sum_0^\infty \tilde{\boldsymbol{x}}^\top(k) \tilde{\boldsymbol{x}}(k) \Big\} \leq \mathbb{E}\Big\{ \sum_0^\infty \boldsymbol{\xi}^\top(k) \boldsymbol{\xi}(k) \Big\} \leq \frac{1}{\lambda} \mathbb{E}\{ \sum_0^\infty \Delta \boldsymbol{V}(k) \} \leq -\frac{1}{\lambda} \boldsymbol{V}(0) < \infty. \tag{47}$$

According to Definition 1, System (19) is stochastically mean-squared stable.

Now, the following index is suggested to examine the extended dissipativity of (19):

$$J = \mathbb{E}\{ \sum_{k=0}^{k_f} J(k) \} \tag{48}$$

where $J(k) = \hat{\boldsymbol{e}}^\top(k) \boldsymbol{\Omega}_1 \hat{\boldsymbol{e}}(k) + 2\hat{\boldsymbol{e}}^\top(k) \boldsymbol{\Omega}_2 \boldsymbol{w}(k) + \boldsymbol{w}^\top(k) \boldsymbol{\Omega}_3 \boldsymbol{w}(k)$.

According to a similar procedure outlined above, by denoting $\boldsymbol{\psi}_0(k) = \mathrm{col}\Big\{ \xi(k),\ \boldsymbol{w}(k) \Big\}$, one can conclude that

$$\begin{aligned}
\mathbb{E}\{\Delta \boldsymbol{V}(k)\} - J(k) &= \mathbb{E}\{\Delta \boldsymbol{V}(k)\} - \boldsymbol{\psi}_0^\top(k) \Big\{ ([\tilde{\boldsymbol{C}}^p_{ij}\, \tilde{\boldsymbol{D}}^p_i])^\top \boldsymbol{\Omega}_1 [\tilde{\boldsymbol{C}}^p_{ij}\, \tilde{\boldsymbol{D}}^p_i] + \boldsymbol{e}_1^\top (\tilde{\boldsymbol{C}}^p_{ij})^\top \boldsymbol{\Omega}_2 \boldsymbol{e}_{11} \\
&\quad + \boldsymbol{e}_{11}^\top (sym\{(\tilde{\boldsymbol{D}}^p_i)^\top \boldsymbol{\Omega}_2\} + \boldsymbol{\Omega}_3) \boldsymbol{e}_{11} \boldsymbol{\psi}_0(k) \Big\} \\
&\leq \sum_{i=1}^{r} \sum_{j=1}^{r} \mu_i \hat{\vartheta}_j \boldsymbol{\psi}_0{}^\top(k) \Bigg(\begin{bmatrix} \tilde{\boldsymbol{\Lambda}}^p_{ij} & (\tilde{\boldsymbol{\mathcal{B}}}^p_{ij})^\top - (\tilde{\boldsymbol{C}}^p_{ij})^\top \boldsymbol{\Omega}_2 \\ * & -sym\{(\tilde{\boldsymbol{D}}^p_i)^\top \boldsymbol{\Omega}_2\} - \boldsymbol{\Omega}_3 \end{bmatrix} \\
&\quad - ([\tilde{\boldsymbol{C}}^p_{ij}\, \tilde{\boldsymbol{D}}^p_i])^\top \boldsymbol{\Omega}_1 [\tilde{\boldsymbol{C}}^p_{ij}\, \tilde{\boldsymbol{D}}^p_i] \Bigg) \boldsymbol{\psi}_0(k)
\end{aligned} \tag{49}$$

In light of the above proof, it can be concluded that

$$\sum_{i=1}^{r} \sum_{j=1}^{r} \mu_i \hat{\vartheta}_j \tilde{\boldsymbol{\Phi}}^p_{ij} < 0 \tag{50}$$

By applying the Schur complement property to $\tilde{\boldsymbol{\Phi}}^p_{ij} < 0$, it is obvious to derive from (49) that

$$\mathbb{E}\{\Delta \boldsymbol{V}(k)\} - J(k) < 0 \tag{51}$$

Under Assumption 1, we validated that the condition (20) holds:

(i) When $\mathbf{\Omega}_4 = \mathbf{0}$, by summating (51) from 0 to k_f, the following condition holds under zero initial conditions:

$$\mathbb{E}\{\sum_{k=0}^{k_f} J(k)\} > \mathbb{E}\{V(k_f+1) - V(0)\} = \mathbb{E}\{V(k_f+1)\} > 0 \tag{52}$$

According to Assumption 1 with $\mathbf{\Omega}_4 = 0$, Condition (20) holds.

(ii) When $\mathbf{\Omega}_4 \neq 0$, it follows from Assumption 1 that: $\tilde{\boldsymbol{D}}_i^p = \mathbf{0}$, $\mathbf{\Omega}_1 = \mathbf{0}$, $\mathbf{\Omega}_2 = \mathbf{0}$, and $\mathbf{\Omega}_3 > 0$. In this case, the following condition holds:

$$\mathbb{E}\{\sum_{k=0}^{k_f-1} J(k)\} > \mathbb{E}\{V(k_f)\} > \mathbb{E}\{\tilde{\boldsymbol{x}}^\top(k_f)\boldsymbol{P}^p\tilde{\boldsymbol{x}}(k_f)\} > 0 \tag{53}$$

From (23), we have

$$(\tilde{\boldsymbol{C}}_{ij}^p)^\top \mathbf{\Omega}_4 \tilde{\boldsymbol{C}}_{ij}^p < \boldsymbol{P}^p \tag{54}$$

In the case of $k = k_f$, we obtain

$$\mathbb{E}\{\sum_{k=0}^{k_f-1} J(k)\} > \mathbb{E}\{\tilde{\boldsymbol{x}}^\top(k_f)\boldsymbol{P}^p\tilde{\boldsymbol{x}}(k_f)\} > \mathbb{E}\{\hat{e}(k_f)^\top \mathbf{\Omega}_4 \hat{e}(k_f)\} \tag{55}$$

Due to $\mathbf{\Omega}_1 = \mathbf{0}$, $\mathbf{\Omega}_2 = \mathbf{0}$, and $\mathbf{\Omega}_3 > 0$, we obtain

$$\mathbb{E}\{\sum_{k=0}^{k_f} J(k)\} > \sup_{0 \le k \le k_f} \mathbb{E}\{\hat{e}(k)^\top \mathbf{\Omega}_4 \hat{e}(k)\} \tag{56}$$

Hence, we can conclude, according to (20), that the filtering error system (19) is extended dissipative. □

3.2. Filter Design

Theorem 2. *System (19) is stochastically stable and extended dissipative for given matrices $\mathbf{\Theta} > 0$, $\mathbf{\Omega}_1$, $\mathbf{\Omega}_2$, $\mathbf{\Omega}_3$, and $\mathbf{\Omega}_4$ satisfying Assumption 1, positive scalars ϱ_s, and tuning parameters a_s, b_s, $s = 1,2$, if the matrices $\boldsymbol{P}^p$, $\boldsymbol{Q}_v$, $v = 1,2,3$, $\boldsymbol{Z}_s$, $\boldsymbol{X}_{1s}$, $\boldsymbol{Y}_{1s}$, $\boldsymbol{Z}_{1s}$, $\boldsymbol{T}_{1s}$, $\boldsymbol{T}_{2s}$, $\boldsymbol{T}_{3s}$, $\boldsymbol{F}_{1s}^p$, $\boldsymbol{F}_{2s}^p$, $\boldsymbol{U}^p$, $\hat{\boldsymbol{\mathcal{A}}}_j^p$, $\hat{\boldsymbol{\mathcal{B}}}_j^p$, and $\hat{\boldsymbol{\mathcal{C}}}_j^p$ and positive scalar $\sigma_0 > 1$ exist such that (22), (23), and the following conditions are satisfied:*

$$\begin{cases} \tilde{\mathbf{Y}}_{ii}^p < 0 \\ \tilde{\mathbf{Y}}_{ij}^p + \varrho_1 \tilde{\mathbf{Y}}_{ji}^p < 0 \\ \tilde{\mathbf{Y}}_{ij}^p + \varrho_2 \tilde{\mathbf{Y}}_{ji}^p < 0, \ 1 \le i < j \le r, \quad p \in \mathbf{N} \end{cases} \tag{57}$$

where

$$
\begin{aligned}
\tilde{\mathbf{Y}}_{ij}^{p} &= \begin{bmatrix} \tilde{\boldsymbol{\Gamma}}_{ij}^{p} + \hat{\boldsymbol{\Psi}}^{p} & (\tilde{\boldsymbol{\mathcal{B}}}_{ij}^{p})^{\top} - (\tilde{\boldsymbol{\mathcal{C}}}_{ij}^{p})^{\top}\boldsymbol{\Omega}_2 & (\tilde{\boldsymbol{\mathcal{C}}}_{ij}^{p})^{\top}\sqrt{-\boldsymbol{\Omega}_1} \\ * & -sym\{(\tilde{\boldsymbol{D}}_{i}^{p})^{\top}\boldsymbol{\Omega}_2\} - \boldsymbol{\Omega}_3 & (\tilde{\boldsymbol{D}}_{i}^{p})^{\top}\sqrt{-\boldsymbol{\Omega}_1} \\ * & * & -\boldsymbol{I} \end{bmatrix} \\
\tilde{\boldsymbol{\Gamma}}_{ij}^{p} &= \begin{bmatrix}
\tilde{\boldsymbol{\Gamma}}_{11ij}^{p} & \mathbf{0} & (1-\bar{\zeta})\mathbb{A}_{d1ij}^{p} & \mathbf{0} & \mathbf{0} & \mathbf{0} & \mathbf{0} & \tilde{\boldsymbol{\Gamma}}_{18ij}^{p} & (1-\bar{\zeta})\mathbb{B}_{e1j}^{p} & \bar{\zeta}\mathbb{B}_{e1j}^{p} \\
* & -\boldsymbol{Q}_1 & \mathbf{0} & \mathbf{0} & \mathbf{0} & \mathbf{0} & \mathbf{0} & \mathbf{0} & \mathbf{0} & \mathbf{0} \\
* & * & \tilde{\boldsymbol{\Psi}}_{33ij}^{p} & \mathbf{0} & \mathbf{0} & \mathbf{0} & \mathbf{0} & \bar{\zeta}(\mathbb{A}_{d2ij}^{p})^{\top} & (\boldsymbol{C}_{2i}^{p})^{\top}\boldsymbol{\Theta} & \mathbf{0} \\
* & * & * & -\boldsymbol{Q}_2 & \mathbf{0} & \mathbf{0} & \mathbf{0} & \mathbf{0} & \mathbf{0} & \mathbf{0} \\
* & * & * & * & \mathbf{0} & \mathbf{0} & \mathbf{0} & \mathbf{0} & \mathbf{0} & \mathbf{0} \\
* & * & * & * & * & \mathbf{0} & \mathbf{0} & \mathbf{0} & \mathbf{0} & \mathbf{0} \\
* & * & * & * & * & * & \mathbf{0} & \mathbf{0} & \mathbf{0} & \mathbf{0} \\
* & * & * & * & * & * & * & \tilde{\boldsymbol{\Psi}}_{88}^{p} & (1-\bar{\zeta})\mathbb{B}_{e2j}^{p} & (1-\bar{\zeta})\mathbb{B}_{e2j}^{p} \\
* & * & * & * & * & * & * & * & (1-\sigma_0)\boldsymbol{I} & \mathbf{0} \\
* & * & * & * & * & * & * & * & * & -\tau\boldsymbol{I}
\end{bmatrix} \\
\tilde{\boldsymbol{\Gamma}}_{11ij}^{p} &= sym(\mathbb{A}_{1ij}^{p}) + (\boldsymbol{X}^{p} - \boldsymbol{P}^{p}) + \boldsymbol{Q}_1 + \boldsymbol{Q}_2 + (\tau_r + 1)\boldsymbol{Q}_3 \\
\tilde{\boldsymbol{\Gamma}}_{18ij}^{p} &= -\boldsymbol{F}_1^{p} + \boldsymbol{X}^{p} + (\mathbb{A}_{2ij}^{p})^{\top} \\
\boldsymbol{F}_s^{p} &= \begin{bmatrix} \boldsymbol{F}_{1s}^{p} & a_s\boldsymbol{U}^{p} \\ \boldsymbol{F}_{2s}^{p} & b_s\boldsymbol{U}^{p} \end{bmatrix} \mathbb{A}_{sij}^{p} = \begin{bmatrix} \boldsymbol{F}_{1s}^{p}\boldsymbol{A}_i^{p} & a_s\hat{\boldsymbol{\mathcal{A}}}_j^{p} \\ \boldsymbol{F}_{2s}^{p}\boldsymbol{A}_i^{p} & b_s\hat{\boldsymbol{\mathcal{A}}}_j^{p} \end{bmatrix} \mathbb{A}_{dsj}^{p} = \begin{bmatrix} a_s\hat{\boldsymbol{\mathcal{B}}}_j^{p}\boldsymbol{C}_{2i}^{p} & 0 \\ b_s\hat{\boldsymbol{\mathcal{B}}}_j^{p}\boldsymbol{C}_{2i}^{p} & 0 \end{bmatrix}, \mathbb{B}_{esj}^{p} = \begin{bmatrix} a_s\hat{\boldsymbol{\mathcal{B}}}_j^{p} \\ b_s\hat{\boldsymbol{\mathcal{B}}}_j^{p} \end{bmatrix}
\end{aligned} \tag{58}
$$

Furthermore, the filter gains are given by

$$
\hat{\boldsymbol{A}}_j^{p} = (\boldsymbol{U}^{p})^{-1}\hat{\boldsymbol{\mathcal{A}}}_j^{p}, \quad \hat{\boldsymbol{B}}_j^{p} = (\boldsymbol{U}^{p})^{-1}\hat{\boldsymbol{\mathcal{B}}}_j^{p}, \quad \hat{\boldsymbol{C}}_j^{p} = \hat{\boldsymbol{\mathcal{C}}}_j^{p} \tag{59}
$$

Proof. According to Theorem 2, a feasible solution must satisfy the condition $\tilde{\boldsymbol{\Psi}}_{88}^{p} < 0$. Thus, it is easy to verify that $-\mathrm{sym}(\boldsymbol{F}_2^{p}) < 0$ and $\mathrm{sym}((\boldsymbol{U}^{p})) < 0$. It follows that $\boldsymbol{U}^{p}$ is nonsingular. Moreover, using the fact that

$$
\mathbb{A}_{sj}^{p} = \boldsymbol{F}_s^{p}\boldsymbol{A}_{sj}^{p},\ \mathbb{A}_{dsj}^{p} = \boldsymbol{F}_s^{p}\boldsymbol{A}_{dsj}^{p},\ \mathbb{B}_{esj}^{p} = \boldsymbol{F}_s^{p}\boldsymbol{B}_{esj}^{p},\ s = 1,2
$$

it can be verified that (57) is equivalent to (24). Hence, according to Theorem 1, System (19) is stochastically mean-squared stable and extended dissipative. □

4. Optimization-Based Filter Design Algorithm

An important aspect of the filter design lies in the choice of the parameters a_s and b_s ,$s = 1,2$. Additionally, the coupling terms $a_s\hat{\boldsymbol{\mathcal{A}}}_j^{p}$, $b_s\hat{\boldsymbol{\mathcal{A}}}_j^{p}$, $a_s\hat{\boldsymbol{\mathcal{B}}}_j^{p}$, and $b_s\hat{\boldsymbol{\mathcal{B}}}_j^{p}$ in $\tilde{\boldsymbol{\Gamma}}_{ij}^{p}$ can be reduced to linear ones if these parameters are determined a priori. Therefore, finding suitable a_s and b_s is a natural approach to obtaining the optimized gains $\hat{\boldsymbol{\mathcal{A}}}_j^{p}$ and $\hat{\boldsymbol{\mathcal{B}}}_j^{p}$. To achieve this, we constructed one of the following optimization problems according to the performance to be taken into account:

(i) Dissipative/passivity performances:

$$
\begin{aligned}
&\min \gamma \\
&\text{subject to (22), (23)} - \text{(57);}
\end{aligned} \tag{60}
$$

(ii) $H_\infty/l_2 - l_\infty$ performances:

$$
\begin{aligned}
&\min \gamma^2 \\
&\text{subject to (22), (23)} - \text{(57).}
\end{aligned} \tag{61}
$$

Through the particle swarm optimization algorithm (PSO) [60], we give the following solving method for the above optimization problems (60) or (61). In the PSO algorithm, the parameters are chosen arbitrarily by specifying the number of particles, and the decision values corresponding to the parameters that need to be evaluated using the position vector for the l_{th} particle, as described below:

$$\mathbf{X}^l = \begin{bmatrix} a_1^l & a_2^l & b_1^l & b_2^l \end{bmatrix}^T$$

where l denotes the swarm size (count of particles). Each iteration k of the algorithm updates the velocity for every particle in the swarm, and the positions are determined by using the following equations:

$$\begin{cases} \mathbf{X}_{k+1}^{(l)} & = \mathbf{v}_k^{(l)} + c_1 r_1 \left(\mathbf{X}_{pk}^{(l)} - \mathbf{X}_k^{(l)} \right) + c_2 r_2 \left(\mathbf{X}_{gk}^{(l)} - \mathbf{X}_k^{(l)} \right) \\ \mathbf{v}_{k+1}^{(l)} & = \mathbf{X}_k^{(l)} + \mathbf{v}_{k+1}^{(l)} \end{cases} \tag{62}$$

In these equations, $\mathbf{X}_k^{(l)}$ represents the current candidate solution encoded according to the position of the l_{th} particle during iteration k and $\mathbf{X}_{k+1}^{(l)}$ represents the updated particle position. $\mathbf{X}_k^{(l)} \in [\mathbf{X}_L, \mathbf{X}_U]$ with $\mathbf{X}_L$ and $\mathbf{X}_U$ denote the lower and upper bounds of $\mathbf{X}$. $\mathbf{v}_{k+1}^{(l)}$ and $\mathbf{v}_k^{(l)}$ are the respective velocities of the new and old particles. $\mathbf{X}_{pk}^{(l)}$ denotes the best position, which the l_{th} particle attained in the past. $\mathbf{X}_{gk}^{(l)}$ denotes the best position between the neighbors of every particle and is also referred to as the "gbest". Parameters r_1 and r_2 correspond to two random numbers in the range $(0, 1)$. Parameters c_1 and c_2 refer to the coefficients of acceleration, which determine how far a particle will reach during a single generation. Below is a detailed Algorithm 1 of the proposed approach:

Algorithm 1 Determining the minimum performances and filter gains.

1: **(Step 1)**: This step specifies the PSO population size n_p, the particle dimension d, and the maximum number of iterations it.
2: **(Step 2)**: Initialize the particle's position and velocity.
3: **(Step 3)**: Determine the parameters of $\mathbf{X}^{(l)}$ and calculate the fitness value corresponding to these parameters according to the equations (62).
4: **(Step 4)**: Compare each particle's fitness value with the best (minimized) value during its historical search, along with its position, which is stored as $\mathbf{X}_{pk}^{(l)}$. Furthermore, the "gbest" particle position corresponding to the minimized fitness value in the population is saved as $\mathbf{X}_{gk}^{(l)}$.
5: **(Step 5)**: As a result of (62), update the position of each particle, as well as its velocity.
6: **(Step 6)**: If the number of iterations reaches the maximum number of iterations it, then proceed to Step 7. Otherwise, go to Step 3.
7: **(Step 7)**: $\mathbf{X}_{pk}^{(l)}$ is defined as the optimal solution desired, and the optimal parameters are calculated.

5. Numerical Applications

This section of the paper details the computational framework used in the proposed method and illustrates its usefulness and advantages by employing the nonlinear single-link robot arm and lower limbs systems.

5.1. Computational Framework and Algorithm

Computational experiments were conducted using the `Matlab` programming language and a computer with the following characteristics: (i) (OS) Windows 10 Enterprise for 64 bits; (ii) (RAM) 8 gigabytes; (iii) (processor) Intel(R) Core(TM) i7-4790T CPU @ 2.70 gigahertz.

After a rigorous process of designing the filter, the detailed procedure is summarized in Algorithm 1 for determining the minimum performances and filter gains. This algorithm was applied using the `Yalmip` software in conjunction with the `mosek` optimization toolbox.

5.2. Single-Link Robot Arm System

Consider the single-link robot arm system described in [34] and stated as

$$\begin{cases} x_1(k+1) &= x_2(k) \\ x_2(k+1) &= -\dfrac{m(\bar{r}(k))glT_e}{J(\bar{r}(k))}\sin(x_1(k)) + (1 - \dfrac{D(\bar{r}(t))T_e}{J(\bar{r}(t))})x_2(k) + \dfrac{T_e}{J(\bar{r}(t))}u(k) \end{cases} \quad (63)$$

with $x_1(k)$, $x_2(k)$, $m(\bar{r}(k))$, $J(\bar{r}(k))$, $D(\bar{r}(k))$, and l being, respectively, the angle position, angle velocity, masses, moment of inertia, damping, and length of the robot arm, respectively. As in [34], the system can be converted into a Markov switching fuzzy system in the form of (4) with $T_s = 0.1s$, and the related transition-probability matrix is defined as:

$$\Pi = \begin{bmatrix} 0.5 & 0.25 & 0.25 \\ 2/14 & 9/14 & 3/14 \\ 0.35 & 0.15 & 0.5 \end{bmatrix}.$$

The considered model is described by the following system matrices:

$$\begin{aligned} &A_1^1 = \begin{bmatrix} 1 & 0.1 \\ -0.49 & 0.7 \end{bmatrix}, A_1^2 = \begin{bmatrix} 1 & 0.1 \\ -0.43 & 0.66 \end{bmatrix}, A_1^3 = \begin{bmatrix} 1 & 0.1 \\ -0.55 & 0.68 \end{bmatrix}, \\ &A_2^1 = \begin{bmatrix} 1 & 0.1 \\ -0.42 & 0.7 \end{bmatrix}, A_2^2 = \begin{bmatrix} 1 & 0.1 \\ -0.25v0.66 & \end{bmatrix}, A_2^3 = \begin{bmatrix} 1 & 0.1 \\ -0.29 & 0.68 \end{bmatrix} \\ &B_{wi}^p = \begin{bmatrix} 0 \\ 0.1 \end{bmatrix}, C_{2i}^p = \begin{bmatrix} 1 & 0 \end{bmatrix}, C_{1i}^p = \begin{bmatrix} 0 & 1 \end{bmatrix}, D_{1i}^p = 0, \quad p = 1,2,3. \end{aligned} \quad (64)$$

The membership functions are defined as follows:

$$h_1(x_1(k)) = \begin{cases} \dfrac{\sin(x_1(k)) - \beta_0 x_1(k)}{(1-\beta_0)x_1(k)}, & x_1(k) \neq 0 \\ 1 & x_1(k) = 0 \end{cases}$$

$$h_2(x_1(k)) = \begin{cases} \dfrac{x_1(k) - \sin(x_1(k))}{(1-\beta_0)x_1(k)}, & x_1(k) \neq 0 \\ 0 & x_1(k) = 0 \end{cases}$$

with $\beta_0 = 10^{-2}/\pi$.

Next, we give the MFs of the filter. Define a nonlinear function as $l(\hat{x}_1(k)) = \hat{x}_1^2(k)$. For $\hat{x}_1(k) \in [\pi - 0.01, \pi + 0.01]$, we have $l(\hat{x}_1(k)) \in [l_{min}(\hat{x}_1(k)), l_{max}(\hat{x}_1(k))]$ with $l_{min}(\hat{x}_1(k)) = 0$ and $l_{max}(\hat{x}_1(k)) = (\pi - 0.01)^2$. The MFs of the filter are given as $\nu_1(\hat{x}_1(k) = \dfrac{l(\hat{x}_1(k)) - \breve{l}_{min}(\hat{x}_1(k))}{\breve{l}_{max}(\hat{x}_1(k)) - \breve{l}_{min}(\hat{x}_1(k))}$ and $\nu_2(\hat{x}_1(k) = \dfrac{\breve{l}_{max}(\hat{x}_1(k)) - l(\hat{x}_1(k))}{\breve{l}_{max}(\hat{x}_1(k)) - \breve{l}_{min}(\hat{x}_1(k))}$ with $\breve{l}_{min}(\hat{x}_1(k)) = l_{min}(\hat{x}_1(k)) - 0.1$ and $\breve{l}_{max}(\hat{x}_1(k)) = \pi^2 + 0.1$.

For $\tau_m = 1$, $\tau_M = 5$, $\boldsymbol{\Theta} = 3$, $\zeta = 0.8$, $\varrho_1 = 0.15$, $\varrho_2 = 1/0.15$, $\boldsymbol{a}(k) = \tanh(0.6y(k))$, and $\boldsymbol{G} = 0.6$, Algorithm 1 is performed to design filters satisfying the extended dissipative performance from four aspects, namely H_∞, $l_2 - l_\infty$, passive, and dissipative performances. For the given cases, Table 2 lists the minimum allowable values for a_s, b_s, $(s = 1,2)$, and γ.

Table 2. Minimal values for different performances.

Performance	Ω_1	Ω_2	Ω_3	Ω_4	a_1^*	a_2^*	b_1^*	b_2^*	γ^*
H_∞	-1	0	γ^2	0	14.3165	-8.4295	14.5129	-9.4800	0.7327
passivity	0	1	γ	0	10.0000	-8.0000	14.5663	-10.0000	0.1201
dissipativity	-1	1	γ	0	10.6384	-8.0000	12.7780	-10.0000	0.1238

Then, we were concerned with different cases to conclude our discussion:

Case I: passivity filtering:

$$\begin{bmatrix} \hat{A}_1^1 & \hat{A}_2^1 & \hat{B}_1^1 & \hat{B}_2^1 \\ \hat{A}_1^2 & \hat{A}_2^2 & \hat{B}_1^2 & \hat{B}_2^2 \\ \hat{A}_1^3 & \hat{A}_2^3 & \hat{B}_1^3 & \hat{B}_2^3 \end{bmatrix} = \left[\begin{array}{cc|cc|c|c} 0.76238 & 0.1182 & 0.77373 & 0.12784 & 0.24699 & 0.25917 \\ -0.85348 & 0.31808 & -0.70568 & 0.31865 & 0.076161 & 0.20964 \\ ine0.62687 & 0.078345 & 0.68218 & 0.059552 & 0.33982 & 0.42384 \\ -1.4991 & 0.26763 & -1.3146 & 0.20334 & 0.40852 & 1.2533 \\ ine0.68872 & 0.085852 & 0.70391 & 0.10369 & 0.28929 & 0.39769 \\ -1.2972 & 0.2118 & -0.92588 & 0.17481 & 0.078596 & 0.83558 \end{array}\right]$$

$$\begin{bmatrix} \hat{C}_1^1 & \hat{C}_2^1 \\ \hat{C}_1^2 & \hat{C}_2^2 \\ \hat{C}_1^3 & \hat{C}_2^3 \end{bmatrix} = \left[\begin{array}{cc|cc} -0.38676 & 0.10875 & -0.38987 & 0.10831 \\ ine-0.57935 & 0.11218 & -0.58156 & 0.11422 \\ ine-0.47562 & 0.087605 & -0.49095 & 0.084071 \end{array}\right] \tag{65}$$

Case II: dissipativity filtering:

$$\begin{bmatrix} \hat{A}_1^1 & \hat{A}_2^1 & \hat{B}_1^1 & \hat{B}_2^1 \\ \hat{A}_1^2 & \hat{A}_2^2 & \hat{B}_1^2 & \hat{B}_2^2 \\ \hat{A}_1^3 & \hat{A}_2^3 & \hat{B}_1^3 & \hat{B}_2^3 \end{bmatrix} = \left[\begin{array}{cc|cc|c|c} 0.75035 & 0.10162 & 0.76625 & 0.115 & 0.26128 & 0.27117 \\ -0.67407 & 0.53154 & -0.51703 & 0.53359 & 0.076572 & 0.19088 \\ ine0.56972 & 0.052941 & 0.63612 & 0.027462 & 0.34737 & 0.47697 \\ -1.3854 & 0.46117 & -1.216 & 0.38875 & 0.45019 & 1.3471 \\ ine0.60712 & 0.041119 & 0.61428 & 0.053424 & 0.30315 & 0.52321 \\ -1.4628 & 0.3626 & -1.1198 & 0.28418 & 0.19734 & 1.4148 \end{array}\right]$$

$$\begin{bmatrix} \hat{C}_1^1 & \hat{C}_2^1 \\ \hat{C}_1^2 & \hat{C}_2^2 \\ \hat{C}_1^3 & \hat{C}_2^3 \end{bmatrix} = \left[\begin{array}{cc|cc} -0.32311 & 0.15403 & -0.32593 & 0.15345 \\ ine-0.61589 & 0.16114 & -0.61878 & 0.16324 \\ ine-0.4861 & 0.088156 & -0.50649 & 0.085026 \end{array}\right] \tag{66}$$

Case III: H_∞ filtering:

$$\begin{bmatrix} \hat{A}_1^1 & \hat{A}_2^1 & \hat{B}_1^1 & \hat{B}_2^1 \\ \hat{A}_1^2 & \hat{A}_2^2 & \hat{B}_1^2 & \hat{B}_2^2 \\ \hat{A}_1^3 & \hat{A}_2^3 & \hat{B}_1^3 & \hat{B}_2^3 \end{bmatrix} = \left[\begin{array}{cc|cc|cc} 0.8641 & 0.074582 & 0.88691 & 0.092459 & 0.15007 & 0.15425 \\ -0.44428 & 0.7861 & -0.30547 & 0.78572 & -0.038087 & -0.02165 \\ ine0.64921 & -0.017533 & 0.83116 & 0.06009 & 0.2362 & 0.31973 \\ -0.45812 & 0.74357 & -0.091925 & 0.70545 & 0.017246 & 0.1741 \\ ine0.33701 & -0.15945 & 0.24163 & -0.10596 & 0.1781 & 1.7443 \\ -1.0908 & 0.47433 & -1.3547 & 0.57936 & -0.42549 & 2.786 \end{array}\right]$$

$$\begin{bmatrix} \hat{C}_1^1 & \hat{C}_2^1 \\ \hat{C}_1^2 & \hat{C}_2^2 \\ \hat{C}_1^3 & \hat{C}_2^3 \end{bmatrix} = \left[\begin{array}{cc|cc} 0.00036846 & 0.63323 & 0.014015 & 0.63422 \\ ine-0.070462 & 0.57629 & 5.1593.10^{-6} & 0.47536 \\ ine-0.16548 & 0.51958 & 0.042926 & 0.52235 \end{array}\right] \tag{67}$$

Case IV: H_∞ filtering in [34]: In this case, we used the filter proposed in [34] with a constant threshold $\sigma = 0.5$.

For the first three cases, the gains were calculated by employing the minimal values recorded in Table 2.

5.2.1. Results and Graphical Plots

Given the initial conditions $x(0) = \hat{x}(0) = \begin{bmatrix} 0 & 0 \end{bmatrix}^T$, the numerical simulations were performed for $w(k) = 8sin^2(k)e^{-0.3k}$. For different cases, the simulation results are depicted in Figures 2–5, where the switching signal $\bar{r}(k)$ and Bernoulli distribution are, respectively, depicted in Figure 6a,b.

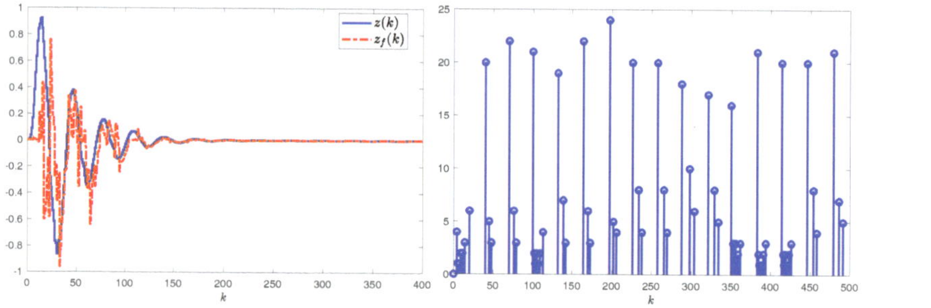

(**a**) (**b**)

Figure 2. $z(k)$ and its estimation $\hat{z}(k)$ (**a**) and the release instants and release interval (**b**), for Case I.

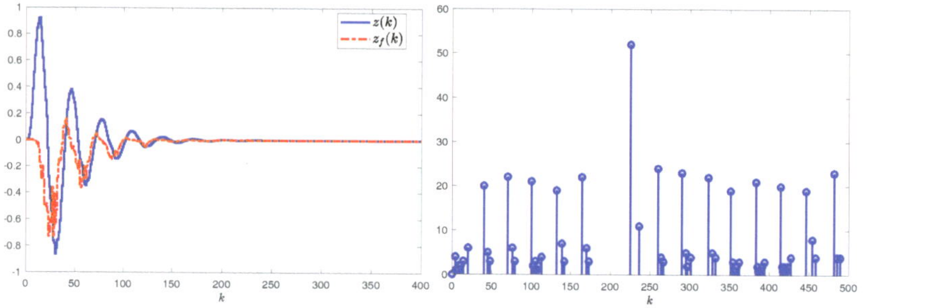

(**a**) (**b**)

Figure 3. $z(k)$ and its estimation $\hat{z}(k)$ (**a**) and the release instants and release interval (**b**), for Case II.

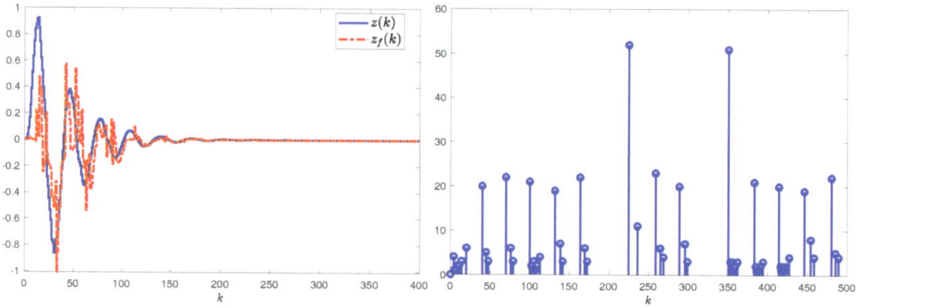

(**a**) (**b**)

Figure 4. $z(k)$ and its estimation $\hat{z}(k)$ (**a**) and the release instants and release interval (**b**), for Case III.

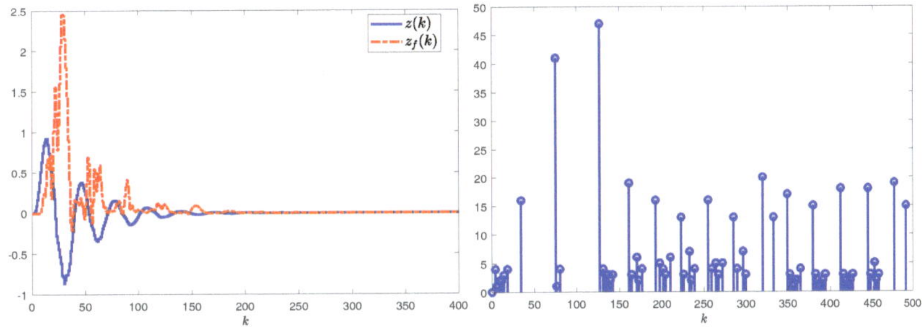

(**a**) (**b**)

Figure 5. $z(k)$ and its estimation $z_f(k)$ (**a**) and the release instants and release interval (**b**), for Case IV.

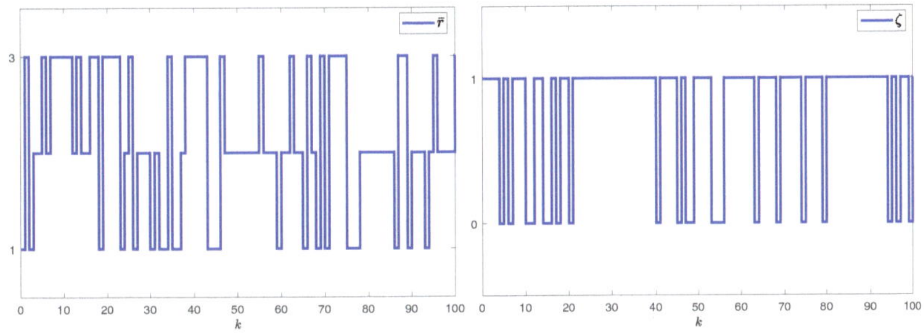

(**a**) (**b**)

Figure 6. A possible sequence of the system mode (**a**) and the states of measurements (**b**).

5.2.2. Comparative Explanations

Figures 2a–5a display the output $z(k)$ and its estimation $\hat{z}(k)$. Figures 2b–5b show the event-triggered release instants and the intervals. According to the time-triggered schemes, the percentages of transmitted data for different cases are shown in Table 3. Moreover, the table provides an overview of the comparison using a quantitative analysis, where the deviations of the state error $\hat{e}(k)$ are investigated by computing the integral squared error (ISE) and the integral absolute error (IAE) for different filters. In light of our results, we can substantiate that our method effectively mitigates the waste of computational resources and communication channels, with the smallest total deviation of $\hat{e}(k)$. To evaluate the merit of the proposed approach, the adaptive ET scheme proposed by [48] was applied to the system, as shown in Figure 7b. Using the AET mechanism described in [48], the adaptive variable showed clearly a negative value, which is inaccurate.

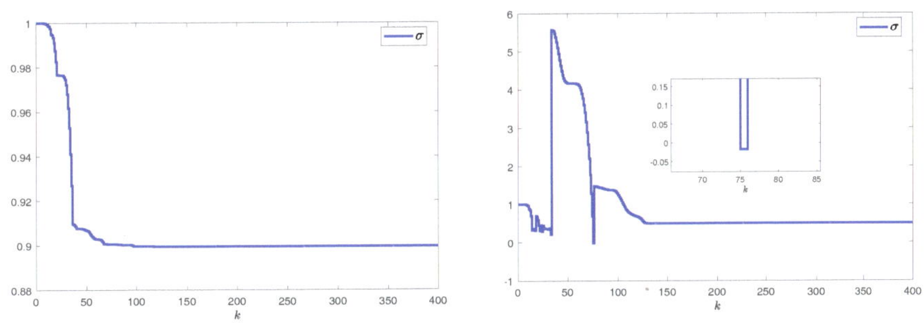

(**a**) (**b**)

Figure 7. Adaptive variable using the proposed method (**a**) and the AET mechanism in [48] (**b**).

Table 3. Data transmission rate, ISE, and IAE for different cases.

Filter Gains	(65)	(66)	(67)	[34]
Data transmission rate	14.172	13.573	12.575	18.563
ISE	1.5152	1.4895	0.94308	8.7833
IAE	2.9001	2.8546	2.4904	5.8065

5.3. A Lower Limbs Rehabilitation System

As shown in Figure 8, this example uses an isokinetic rehabilitation system for the lower limbs, which can assist the handicapped in Hail is described by the following differential nonlinear equation [61]:

$$\Gamma_m = J_m\ddot{\theta} + f_m\theta - k_m\frac{\dot{\theta}}{|\dot{\theta}|}\dot{\theta}^2 - Fl - a\cos(\theta) - b\sin(\theta) \tag{68}$$

Γ_m, J_m, and f_m, respectively, represent the controlled variables that determine the torque, inertia, and viscous friction. Moreover, the variables $\ddot{\theta}$, $\dot{\theta}$, and θ correspond to the angular position, velocity, and acceleration of the mobile part. The torque delivered by the patient is described by F times l; the potential energy of the system is described by $a\cos(\theta)$; the centrifugal force coefficient is described by $b\sin(\theta)$, while the centrifugal force coefficient needs to be added or subtracted according to the velocity sign. Define $x(t) = [x_1(t),\ x_2(t)] = [\dot{\theta},\ \theta]$, $u(t) = \Gamma_m$, and $w(t) = F$. A discrete-time model of System (68) can be obtained by applying Euler's discretization:

$$\begin{cases} x_1(k+1) &= (1+T_s g(k_m, x_1(k)))x_1(k) + T_s f(x_2(k))x_2(k) + \dfrac{T_s}{J_m}u(k) + \dfrac{T_s l}{J_m}w(k) \\ x_2(k+1) &= x_2(k) + T_s(k) \end{cases} \tag{69}$$

with $g(k_m, x_1(k)) = \dfrac{k_m|x_1(k)| - f_m}{J_m}$ and $f(x_2(k)) = \dfrac{a\cos(x_2(k)) - b\sin(x_2(k))}{J_m x_2(k)}$.

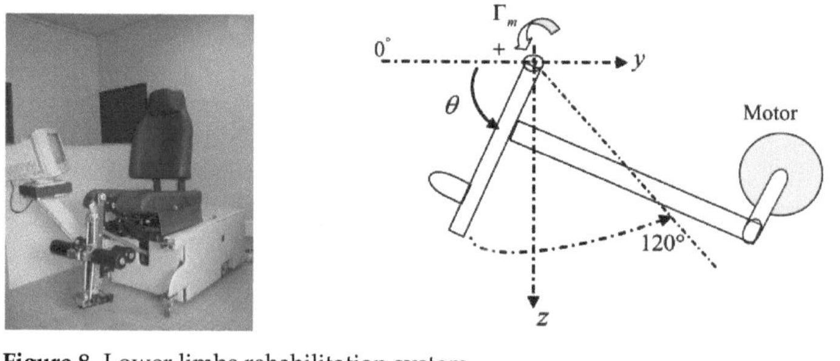

Figure 8. Lower limbs rehabilitation system.

A list of the parameters of the system can be found in Table 4. As a further consideration, it was assumed that parameter J_m has two different modes, listed in the table. Given the uncertainty associated with the parameter k_m, it is evident that the IT-2 fuzzy system should be adopted to model a nonlinear system (68). Assume that $x_1(k) \in [-2\pi,\ 2\pi]$ and $x_2(k) \in [\pi/180,\ 2\pi/3]$; we can obtain $g_{max} = -3.0651$ and $g_{min} = 11.806$, and $g(x_1(k)) \in [\bar{g}_{max}, \bar{g}_{min}] = [-4,\ 12]$.

By adopting the sector nonlinearity method, the membership functions of the corresponding IT-2 fuzzy model are defined as

$$\mu_1(x_1(k)) = m_1(x_1(k))h_1(x_2(k)), \qquad \mu_2(x_1(k)) = m_2(x_1(k))h_2(x_2(k)),$$
$$\mu_3(x_1(k)) = m_1(x_1(k))h_1(x_2(k)), \qquad \mu_4(x_1(k)) = m_2(x_1(k))h_2(x_2(k))$$

with $m_i(x_1(k)) = \underline{m}_i(x_1(k))\sin^2(x_1(k)) + \bar{m}_i(x_1(k))\cos^2(x_1(k))$, $\underline{m}_1(x_1(k)) = \frac{\bar{g}_{max} - g(80, x_1(k))}{\bar{g}_{max} - \bar{g}_{min}}$, $\bar{m}_1(x_1(k)) = \frac{\bar{g}_{max} - g(70, x_1(k))}{\bar{g}_{max} - \bar{g}_{min}}$, $\underline{m}_2(x_1(k)) = 1 - \bar{m}_1(x_1(k))$, $\bar{m}_2(x_1(k)) = 1 - \underline{m}_1(x_1(k)), h_1(x_2(k)) = \frac{187.35 - f(x_2(k))}{187.75}$, and $h_2(x_2(k)) = 1 - h_1(x_2(k))$.

Thus, the dynamical model of (68) can be described by the Markov switching fuzzy system in the form of (4) with $T_s = 0.025$ s, and the transition probability rate matrix and the system matrices are defined as:

$$\begin{aligned}
A_1^1 &= \begin{bmatrix} -0.2557 & -6.1348 \\ 0.0250 & 1.0000 \end{bmatrix}, A_2^1 = \begin{bmatrix} -0.2561 & -1.3733 \\ 0.0250 & 1.0000 \end{bmatrix}, A_3^1 = \begin{bmatrix} 0.0706 & -6.0071 \\ 0.0250 & 1.0000 \end{bmatrix}, \\
A_4^1 &= \begin{bmatrix} 0.0706 & -1.4156 \\ 0.0250 & 1.0000 \end{bmatrix}, A_1^2 = \begin{bmatrix} -0.2488 & -6.0116 \\ 0.0250 & 1.0000 \end{bmatrix}, A_2^2 = \begin{bmatrix} -0.2487 & -1.4191 \\ 0.0250 & 1.00001 \end{bmatrix}, \\
A_3^2 &= \begin{bmatrix} 0.0649 & -5.8815 \\ 0.0250 & 1.0000 \end{bmatrix}, A_4^2 = \begin{bmatrix} 0.0644 & -1.4777 \\ 0.0250 & 1.0000 \end{bmatrix}, B_i^1 = \begin{bmatrix} 0.00037 \\ 0 \end{bmatrix}, B_i^2 = \begin{bmatrix} 0.000355 \\ 0 \end{bmatrix}, \\
\Pi &= \begin{bmatrix} 0.7 & 0.3 \\ 0.2 & 0.8 \end{bmatrix}
\end{aligned} \tag{70}$$

For $\tau_m = 2$, $\tau_M = 6$, $\Theta = 5$, $\bar{\zeta} = 0.85$, $\varrho_1 = 0.15$, and $\varrho_2 = 1/0.15$, Algorithm 1 was performed, and the optimal values of the $l_2 - l_\infty$ performance were obtained as: $a_1^* = 2.0581$, $a_2^* = 2.5176$, $b_1^* - 0.1309$ $b_2^* = -0.1744$, and $\gamma^* = 7.223^{-4}$. The filter matrices are designed as

$$\begin{aligned}
&\begin{bmatrix} \hat{A}_1^1 & \hat{A}_2^1 & \hat{A}_3^1 & \hat{A}_4^1 \\ \hat{A}_1^2 & \hat{A}_2^2 & \hat{A}_3^2 & \hat{A}_4^2 \end{bmatrix} = \\
&\left[\begin{array}{cc|cc|cc|cc} 0.073469 & -3.4221 & 0.083413 & -0.6332 & 0.20643 & -3.2698 & 0.21936 & -0.73011 \\ 0.01021 & 0.926 & 0.0076208 & 0.97167 & 0.011397 & 0.92152 & 0.012629 & 0.97586 \\ ine - 0.060213 & -2.8422 & -0.092627 & -0.75619 & 0.17697 & -2.6329 & 0.19582 & -0.79572 \\ 0.013459 & 0.93971 & 0.010413 & 0.96712 & 0.014624 & 0.93769 & 0.018435 & 0.97071 \end{array}\right] \\
&\begin{bmatrix} \hat{B}_1^1 & \hat{B}_2^1 & \hat{B}_3^1 & \hat{B}_4^1 \\ \hat{B}_1^2 & \hat{B}_2^2 & \hat{B}_3^2 & \hat{B}_4^2 \end{bmatrix} = \\
&\left[\begin{array}{c|c|c|c} -9.5398 & 9.3138 & -8.6087 & 6.7673 \\ -0.0038308 & 0.32511 & -0.042507 & 0.38779 \\ ine - 7.5654 & 6.6368 & -6.6323 & 5.2521 \\ 0.041711 & 0.28217 & 0.019727 & 0.29285 \end{array}\right] \\
&\begin{bmatrix} \hat{C}_1^1 & \hat{C}_2^1 & \hat{C}_3^1 & \hat{C}_4^1 \\ \hat{C}_1^2 & \hat{C}_2^2 & \hat{C}_3^2 & \hat{C}_4^2 \end{bmatrix} = \\
&\left[\begin{array}{cc|cc|cc|cc} 0.029365 & 0.015436 & 0.029365 & 0.015436 & 0.029365 & 0.015436 & 0.029365 & 0.015436 \\ ine0.044923 & 0.026018 & 0.044923 & 0.026018 & 0.044923 & 0.026018 & 0.044923 & 0.026018 \end{array}\right]
\end{aligned} \tag{71}$$

According to the above parameters, the simulation results are presented in Figures 9–11, where Figure 11b illustrates, respectively, the random variables $\bar{r}(k)$ and $\zeta(k)$. As an intuitive method for evaluating the communication performance of the designed filter, Figures 9a–11a record the trajectories of the system's and filter's outputs, the triggering instants and intervals of each system state component, as well as the dynamic threshold parameters for the different trigger mechanisms suggested in [35,48,53,54]. As can be seen in these figures, the implemented filter for different ETMs can guarantee the convergence of the system's states despite uncertainties, external disturbances, and cyber-attacks. Based on Table 5, it appears that the application of the AETM in this article may further optimize communication resources and reduce bandwidth usage because the number of data packets transmitted with an acceptable IAE was significantly less than the number of data packets transmitted using the ETMs in [35,48,53,54,62].

Table 4. Markov mode parameters of the system.

Parameter	Physical Meaning	Value	Mode 1	Mode 2	Unit
a	Gravitational coefficient	110	-	-	(N)
b	Gravitational coefficient	31	-	-	(N)
l	Arm's length	0.5	-	-	(m)
f_m	Viscous friction	103.6	-	-	$(\mathrm{N(rad/s)}^{-1})$
k_m	Coriolis coefficient	[70, 80]	-	-	$(\mathrm{Nm(rad/s)}^{-1})$
J_m	inertia	-	33.8	35.2	$(\mathrm{kg\ m}^{-2})$

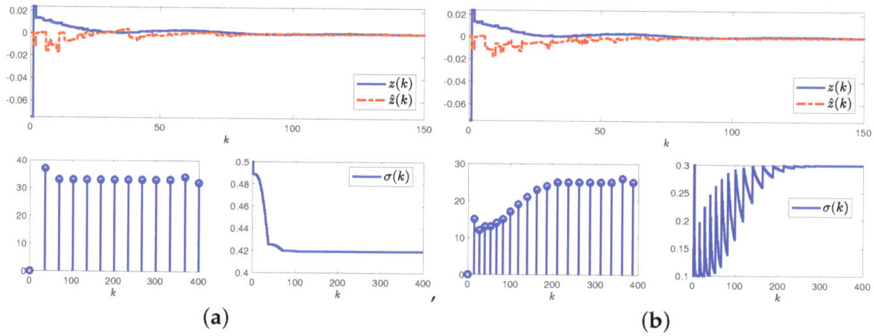

Figure 9. Simulation results based on the proposed AETM (**a**) and simulation results based on the AETM in [35] (**b**).

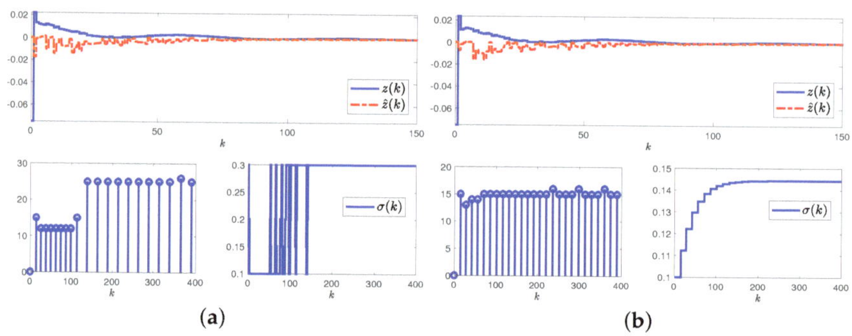

Figure 10. Simulation results based on the AETM in [48] (**a**) and simulation results based on the AETM in [53] (**b**).

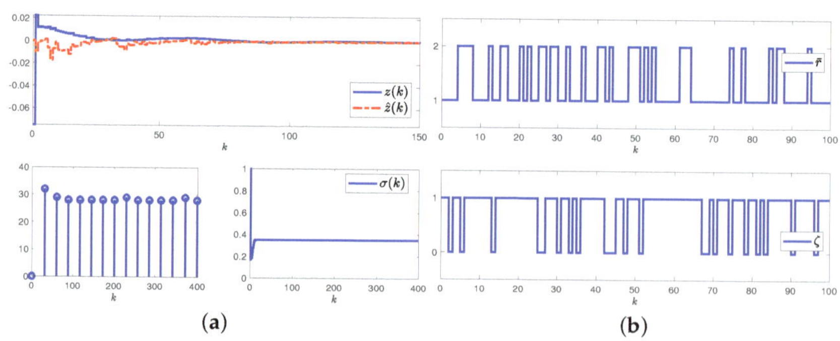

Figure 11. Simulation results based on the AETM in [54] (**a**) and a possible sequence of the system mode and the states of measurements (**b**).

Table 5. Number and rate of transmitted data packets and IAE for different event-triggered mechanisms.

Event-Triggered Mechanism	Packets Transmitted	Data Transmission Rate (%)	IAE (%)
Proposed	12	2.99	1.69
[35]	20	4.98	1.74
[48]	14	3.49	1.67
[53]	19	4.74	1.81
[54]	26	6.48	1.67

5.4. Comparative Explanations

The dynamic- and adaptive-event-triggered filtering problem for nonlinear networks has recently been the subject of numerous studies in the literature, so several methods are comparable to what we presented here. There are, however, some differences from our approaches as follows:

- To handle nonlinear network systems, a H_∞ linear filter was designed in [26] using Type-1 fuzzy models, which may lead to conservative results. As an alternative, we took a more general approach based on a Markovian jump Type-2 fuzzy model to design an IT2 fuzzy filter with extended dissipativity performance.
- Different from [49,62], the filter design method described in [26] was based on an improved matrix decoupling approach that uses appropriate selected scalars. The selection of these parameters can be achieved either by a numerical analysis or by using meta-heuristic techniques, such as the PSO method addressed in this study.

6. Conclusions, Limitations, and Future Work

Some conclusions, possibilities for future developments, and limitations of the filtering strategy are presented in this section.

6.1. Concluding Remarks

For a class of nonlinear discrete-time systems described by Markov jumping IT-2 fuzzy models, a novel adaptive-event-triggered extended dissipativity-based filtering problem was developed. This study successfully addressed the hypothesis of perturbations and the random occurrence of cyber-attacks. As a means of reducing the unnecessary use of limited communication resources, a new event-triggered scheme with an adaptive triggering scheme was used to determine whether or not the current measurements needed to be transmitted. Furthermore, delay-dependent conditions were developed for the filter design in order to ensure that the error system would be stochastically stable with extended dissipativity performance. Using the upper and lower membership functions, less conservative results were obtained. Furthermore, an optimization-based algorithm (PSO) was used for the determination of the filter gains and to achieve the best-possible performance. Finally, this paper concluded with two examples that illustrated how effective the method proposed here can be.

6.2. Limitations

Despite the power and performance of the synthesized technique, the latter imposes that the Markov chain transition probability is known as a delicate task in practice. Equally important, the synchronization between modes of the filter and model is a task that may be difficult to achieve in practice.

6.3. Future Work

Further research that may be focused on in the context of the present study is related to the state estimation of singular systems based on the dynamic-event-triggered communication protocol [56,63]. Moreover, the determination of partial and time-varying transition probabilities in our framework is an open problem to be addressed in the future.

Author Contributions: Writing—review & editing, M.A.R.; Project administration, M.K. All authors have read and agreed to the published version of the manuscript.

Funding: This research was funded by the Scientific Naser Al-Rasheed Research Chair at the University of Ha'il—Saudi Arabia through Project Number SCR-22 102.

Data Availability Statement: The data are contained within the article.

Acknowledgments: The authors thank the Reviewers for their constructive comments, which helped to obtain an improved version of the article.

Conflicts of Interest: The authors declare no conflict of interest.

References

1. Hou, L.; Chen, D.; He, C. Finite-time nonfragile dissipative control for discrete-time neural networks with Markovian jumps and mixed time-delays. *Complexity* **2019**, *2019*, 5748923. [CrossRef]
2. Sathishkumar, M.; Sakthivel, R.; Alzahrani, F.; Kaviarasan, B.; Ren, Y. Mixed H_∞ and passivity-based resilient controller for nonhomogeneous Markovian jump systems. *Nonlinear Anal. Hybrid Syst.* **2019**, *31*, 86–99. . [CrossRef]
3. Cheng, J.; Park, J.H.; Cao, J.; Qi, W. A hidden mode observation approach to finite-time SOFC of Markovian switching systems with quantization. *Nonlinear Dyn.* **2020**, *100*, 509–521. [CrossRef]
4. Cheng, J.; Park, J.H.; Zhao, X.; Karimi, H.R.; Cao, J. Quantized Nonstationary Filtering of Networked Markov Switching RSNSs: A Multiple Hierarchical Structure Strategy. *IEEE Trans. Autom. Control.* **2020**, *65*, 4816–4823. . [CrossRef]
5. Li, H.; Shi, P.; Yao, D.; Wu, L. Observer-based adaptive sliding mode control for nonlinear Markovian jump systems. *Automatica* **2016**, *64*, 133–142. [CrossRef]
6. Ma, Y.; Jia, X.; Zhang, Q. Robust finite-time non-fragile memory H_∞ control for discrete-time singular Markovian jumping systems subject to actuator saturation. *J. Frankl. Inst.* **2017**, *354*, 8256–8282. [CrossRef]
7. Wu, B.; Zhao, Y. Dissipative Control for Fuzzy Singular Markov Jump Systems With State-Dependent Noise and Asynchronous Modes. *IEEE Access* **2021**, *9*, 25691–25702. [CrossRef]
8. Qi, W.; Park, J.; Cheng, J.; Chen, X. Stochastic stability and L_1-gain analysis for positive nonlinear semi-Markovian jump systems with time-varying delay via TS fuzzy model approach. *Fuzzy Sets Syst.* **2019**, *371*, 110–122. [CrossRef]
9. Kavikumar, R.; Sakthivel, R.; Kwon, O.; Kaviarasan, B. Reliable non-fragile memory state feedback controller design for fuzzy Markovian jump systems. *Nonlinear Anal. Hybrid Syst.* **2020**, *35*, 100828. [CrossRef]
10. Tian, Y.; Wang, Z. Asynchronous Extended Dissipative Filtering for T-S Fuzzy Markov Jump Systems. *IEEE Trans. Syst. Man Cybern. Syst.* **2022**, *52*, 3915–3925. [CrossRef]
11. Li, J.; Zhang, Q.; Yan, X.G.; Spurgeon, S.K. Integral sliding mode control for Markovian jump T-S fuzzy descriptor systems based on the super-twisting algorithm. *Iet Control. Theory Appl.* **2017**, *11*, 1134–1143. [CrossRef]
12. Lv, C.; Chen, J.; Lv, X.; Zhang, Z. Finite-time H_∞ control for interval Type-2 fuzzy singular systems via switched fuzzy models and static output feedback. *Nonlinear Anal. Hybrid Syst.* **2022**, *45*, 101206. [CrossRef]
13. Zhao, T.; Wei, Z.; Dian, S.; Xiao, J. Observer-based H_∞ controller design for interval Type-2 T-S fuzzy systems. *Neurocomputing* **2016**, *177*, 9–25. [CrossRef]
14. Song, W.; Tong, S. Fuzzy decentralized output feedback event-triggered control for interval Type-2 fuzzy systems with saturated inputs. *Inf. Sci.* **2021**, *575*, 639–653. [CrossRef]
15. Alshammari, O.; Kchaou, M.; Jerbi, H.; Ben Aoun, S.; Leiva, V. A Fuzzy Design for a Sliding Mode Observer-Based Control Scheme of Takagi–Sugeno Markov Jump Systems under Imperfect Premise Matching with Bio-Economic and Industrial Applications. *Mathematics* **2022**, *10*, 3309. [CrossRef]
16. Wang, S.; Feng, J.; Zhang, H. Robust fault tolerant control for a class of networked control systems with state delay and stochastic actuator failures. *Int. J. Adapt. Control. Signal Process.* **2014**, *28*, 798–811. [CrossRef]
17. Latrech, C.; Kchaou, M.; Guéguen, H. Networked non-fragile H_∞ static output feedback control design for vehicle dynamics stability: A descriptor approach. *Eur. J. Control.* **2018**, *40*, 13–26. . [CrossRef]
18. Wang, B.; Chen, W.; Wang, J.; Zhang, B.; Zhang, Z.; Qiu, X. Cooperative Tracking Control of Multiagent Systems: A Heterogeneous Coupling Network and Intermittent Communication Framework. *IEEE Trans. Cybern.* **2019**, *49*, 4308–4320. [CrossRef]
19. Li, Q.; Wang, Z.; Sheng, W.; Alsaadi, F.E.; Alsaadi, F.E. Dynamic event-triggered mechanism for H_∞ non-fragile state estimation of complex networks under randomly occurring sensor saturations. *Inf. Sci.* **2020**, *509*, 304–316. [CrossRef]
20. Chen, M.; Karimi, H.R.; Sun, J. Observer-based finite time H_∞ control of nonlinear discrete time-varying systems with an adaptive-event-triggered Mechanism. *J. Frankl. Inst.* **2020**, *357*, 11668–11689. [CrossRef]
21. Chu, X.; Li, M. H_∞ non-fragile observer-based dynamic-event-triggered sliding mode control for nonlinear networked systems with sensor saturation and dead-zone input. *ISA Trans.* **2019**, *94*, 93 – 107. [CrossRef] [PubMed]
22. Qi, W.; Zong, G.; Zheng, W.X. Adaptive Event-Triggered SMC for Stochastic Switching Systems With Semi-Markov Process and Application to Boost Converter Circuit Model. *IEEE Trans. Circuits Syst. I: Regul. Pap.* **2021**, *68*, 786–796. [CrossRef]
23. Zhang, Y.; Shi, P.; Agarwal, R.K.; Shi, Y. Event-Based Dissipative Analysis for Discrete Time-Delay Singular Jump Neural Networks. *IEEE Trans. Neural Netw. Learn. Syst.* **2020**, *31*, 1232–1241. [CrossRef]

24. Wang, X.L.; Yang, G.H. Event-triggered fault detection for discrete-time T-S fuzzy systems. *ISA Trans.* **2018**, *76*, 18–30. [CrossRef] [PubMed]
25. Song, X.; Men, Y.; Zhou, J.; Zhao, J.; Shen, H. Event-triggered H_∞ control for networked discrete-time Markov jump systems with repeated scalar nonlinearities. *Appl. Math. Comput.* **2017**, *298*, 123–132. [CrossRef]
26. Chen, Q.X.; Chang, X.H. Resilient filter of nonlinear network systems with dynamic event-triggered mechanism and hybrid cyber-attack. *Appl. Math. Comput.* **2022**, *434*, 127419. [CrossRef]
27. Tian, E.; Wang, Z.; Zou, L.; Yue, D. Probabilistic-constrained filtering for a class of nonlinear systems with improved static event-triggered communication. *Int. J. Robust Nonlinear Control.* **2019**, *29*, 1484–1498. [CrossRef]
28. Liu, J.; Zhang, N.; Li, Y.; Xie, X. H_∞ filter design for discrete-time networked systems with adaptive-event-triggered mechanism and hybrid cyber-attacks. *J. Frankl. Inst.* **2021**, *358*, 9325–9345. [CrossRef]
29. Dong, S.; Fang, M.; Shi, P.; Wu, Z.G.; Zhang, D. Dissipativity-Based Control for Fuzzy Systems With Asynchronous Modes and Intermittent Measurements. *IEEE Trans. Cybern.* **2020**, *50*, 2389–2399. [CrossRef]
30. Liu, J.; Yin, T.; Cao, J.; Yue, D.; Karimi, H.R. Security Control for T-S Fuzzy Systems With Adaptive Event-Triggered Mechanism and Multiple Cyber-Attacks. *IEEE Trans. Syst. Man Cybern. Syst.* **2021**, *51*, 6544–6554. [CrossRef]
31. Hsieh, C.S. Robust two-stage Kalman filters for systems with unknown inputs. *IEEE Trans. Autom. Control.* **2000**, *45*, 2374–2378. [CrossRef]
32. Jwo, D.J.; Wang, S.H. Adaptive Fuzzy Strong Tracking Extended Kalman Filtering for GPS Navigation. *IEEE Sens. J.* **2007**, *7*, 778–789. [CrossRef]
33. Hamid, K.R.; Talukder, A.; Islam, A.K.M.E. Implementation of Fuzzy Aided Kalman Filter for Tracking a Moving Object in Two-Dimensional Space. *Int. J. Fuzzy Log. Intell. Syst.* **2018**, *18*, 85–96. [CrossRef]
34. Zhang, M.; Shi, P.; Liu, Z.; Ma, L.; Su, H. H_∞ filtering for discrete-time switched fuzzy systems with randomly occurring time-varying delay and packet dropouts. *Signal Process.* **2018**, *143*, 320–327. [CrossRef]
35. Wang, H.; Xue, A. Adaptive event-triggered H_∞ filtering for discrete-time delayed neural networks with randomly occurring missing measurements. *Signal Process.* **2018**, *153*, 221–230. [CrossRef]
36. Zhang, J.; Peng, C. Event-triggered H_∞ filtering for networked Takagi–Sugeno fuzzy systems with asynchronous constraints. *IET Signal Process.* **2015**, *9*, 403–411. [CrossRef]
37. Zhang, Y.; Shi, P.; Agarwal, R.K.; Shi, Y. Event-based mixed H_∞ and passive filtering for discrete singular stochastic systems. *Int. J. Control.* **2020**, *93*, 2407–2415. [CrossRef]
38. Tian, Y.; Wang, Z. Finite-Time Extended Dissipative Filtering for Singular T-S Fuzzy Systems With Nonhomogeneous Markov Jumps. *IEEE Trans. Cybern.* **2022**, *52*, 4574–4584. [CrossRef]
39. Shi, P.; Su, X.; Li, F. Dissipativity-Based Filtering for Fuzzy Switched Systems with Stochastic Perturbation. *IEEE Trans. Autom. Control.* **2016**, *61*, 1694–1699. [CrossRef]
40. Cheng, J.; Park, J.H.; Chadli, M. Peak-to-peak fuzzy filtering of nonlinear discrete-time systems with Markov communication protocol. *Inf. Sci.* **2022**, *607*, 361–376. [CrossRef]
41. Song, J.; Chang, X.H. Finite-Time Peak-To-Peak Filtering for Nonlinear Singular System. *IEEE Trans. Circuits Syst. I: Express Briefs* **2022**, *69*, 4369–4373. [CrossRef]
42. Kchaou, M.; Regaieg, M.A.; Al-Hajjaji, A. Quantized asynchronous extended dissipative observer-based sliding mode control for Markovian jump TS fuzzy systems. *J. Frankl. Inst.* **2022**, *359*, 9636–9665. [CrossRef]
43. Liu, J.; Gu, Y.; Cao, J.; Fei, S. Distributed event-triggered H_∞ filtering over sensor networks with sensor saturations and cyber-attacks. *ISA Trans.* **2018**, *81*, 63–75. [CrossRef] [PubMed]
44. Zhang, J.; Christofides, P.D.; He, X.; Wu, Z.; Zhang, Z.; Zhou, D. Event-triggered filtering and intermittent fault detection for time-varying systems with stochastic parameter uncertainty and sensor saturation. *Int. J. Robust Nonlinear Control.* **2018**, *28*, 4666–4680. [CrossRef]
45. Wu, Y.; Cheng, J.; Wu, Z.G. Fuzzy-Affine-Model-Based Filtering Design With Memory-Based Dynamic Event-Triggered Protocol. *IEEE Trans. Fuzzy Syst.* **2022**, 1–11. [CrossRef]
46. Peng, C.; Zhang, J.; Yan, H. Adaptive Event-Triggering H_∞ Load Frequency Control for Network-Based Power Systems. *IEEE Trans. Ind. Electron.* **2018**, *65*, 1685–1694. [CrossRef]
47. Gu, Z.; Yue, D.; Liu, J.; Ding, Z. H_∞ tracking control of nonlinear networked systems with a novel adaptive-event-triggered communication scheme. *J. Frankl. Inst.* **2017**, *354*, 3540–3553. [CrossRef]
48. Zhang, Z.; Liang, H.; Wu, C.; Ahn, C.K. Adaptive Event-Triggered Output Feedback Fuzzy Control for Nonlinear Networked Systems With Packet Dropouts and Actuator Failure. *IEEE Trans. Fuzzy Syst.* **2019**, *27*, 1793–1806. [CrossRef]
49. Chang, X.H.; Liu, Y. Robust $\mathcal{H}_\infty$ Filtering for Vehicle Sideslip Angle With Quantization and Data Dropouts. *IEEE Trans. Veh. Technol.* **2020**, *69*, 10435–10445. [CrossRef]
50. Xia, W.; Zheng, W.X.; Xu, S. Event-triggered filter design for Markovian jump delay systems with nonlinear perturbation using quantized measurement. *Int. J. Robust Nonlinear Control.* **2019**, *29*, 4644–4664. [CrossRef]
51. Li, F.; Cao, X.; Zhou, C.; Yang, C. Event-triggered asynchronous sliding mode control of CSTR based on Markov model. *J. Frankl. Inst.* **2021**, *358*, 4687–4704. [CrossRef]
52. Chu, X.; Li, M. H_∞ observer-based event-triggered sliding mode control for a class of discrete-time nonlinear networked systems with quantizations. *ISA Trans.* **2018**, *79*, 13–26. [CrossRef] [PubMed]

53. Ahmad, I.; Ge, X.; Han, Q.L. Decentralized Dynamic Event-Triggered Communication and Active Suspension Control of In-Wheel Motor Driven Electric Vehicles with Dynamic Damping. *IEEE/CAA J. Autom. Sin.* **2021**, *8*, 971–986. [CrossRef]
54. Pan, Y.; Yang, G.H. Event-based output tracking control for fuzzy networked control systems with network-induced delays. *Appl. Math. Comput.* **2019**, *346*, 513–530. [CrossRef]
55. Shen, H.; Jiao, S.; Huo, S.; Chen, M.; Li, J.; Chen, B. On energy-to-peak filtering for semi-Markov jump singular systems with unideal measurements. *Signal Process.* **2018**, *144*, 127–133. [CrossRef]
56. Tao, J.; Xiao, Z.; Li, Z.; Wu, J.; Lu, R.; Shi, P.; Wang, X. Dynamic Event-Triggered State Estimation for Markov Jump Neural Networks With Partially Unknown Probabilities. *IEEE Trans. Neural Netw. Learn. Syst.* **2022**, *33*, 7438–7447. [CrossRef]
57. Zhang, B.; Zheng, W.X.; Xu, S. Filtering of Markovian Jump Delay Systems Based on a New Performance Index. *IEEE Trans. Circuits Syst. Regul. Pap.* **2013**, *60*, 1250–1263. [CrossRef]
58. Shen, Y.; Wu, Z.G.; Shi, P.; Shu, Z.; Karimi, H.R. H_∞ control of Markov jump time-delay systems under asynchronous controller and quantizer. *Automatica* **2019**, *99*, 352–360. [CrossRef]
59. Chen, J.; Lu, J.; Xu, S. Summation inequality and its application to stability analysis for time-delay systems. *IET Control. Theory Appl.* **2016**, *10*, 391–395. [CrossRef]
60. Hamidi, F.; Aloui, M.; Jerbi, H.; Kchaou, M.; Abbassi, R.; Popescu, D.; Ben Aoun, S.; Dimon, C. Chaotic Particle Swarm Optimisation for Enlarging the Domain of Attraction of Polynomial Nonlinear Systems. *Electronics* **2020**, *9*, 1704. [CrossRef]
61. Moughamir, S.; Zaytoon, J.; Manamanni, N.; Afilal, L. A system approach for control development of lower-limbs training machines. *Control. Eng. Pract.* **2002**, *10*, 287–299. [CrossRef]
62. Kaviarasan, B.; Kwon, O.M.; Park, M.J.; Sakthivel, R. Reduced-order filtering for semi-Markovian jump systems against randomly occurring false data injection attacks. *Appl. Math. Comput.* **2023**, *444*, 127832. [CrossRef]
63. Wu, Z.; Chen, J.; Zhang, X.; Xiao, Z.; Tao, J.; Wang, X. Dynamic event-triggered synchronization of complex networks with switching topologies: Asynchronous observer-based case. *Appl. Math. Comput.* **2022**, *435*, 127413. [CrossRef]

mathematics

MDPI

Article

Redundancy Allocation of Components with Time-Dependent Failure Rates

Enrico Zio [1,2] and Hadi Gholinezhad [3,*]

1 Center of Research on Risk and Crises (CRC), MINES-Paris, PSL University, 75272 Paris, France; enrico.zio@polimi.it
2 Energy Department, Politecnico di Milano, 20133 Milan, Italy
3 Department of Industrial Engineering, University of Gonabad, Gonabad 96919-57678, Iran
* Correspondence: gholinezhad.hadi@gonabad.ac.ir

Abstract: The Redundancy Allocation Problem (RAP) is well-known in the field of reliability optimization. In this paper, RAP is investigated assuming that the distribution of the time to failure of the components has the form of an Erlang distribution with a time-dependent rate parameter and considering that the choice of redundancy for each subsystem can be none, active, standby or mixed. A genetic algorithm is used to solve the problem of optimal allocation. To analyze the effect of the time dependence, some numerical examples are worked out. Then, a case study of RAP from the literature is analyzed. The obtained results show that time dependence of the failure time distribution parameters can lead to significant differences in the optimal redundancy allocation.

Keywords: reliability optimization; redundancy allocation problem; redundancy strategies; Erlang distribution; time-dependent failure rates; genetic algorithm

MSC: 60E05; 90B25

Citation: Zio, E.; Gholinezhad, H. Redundancy Allocation of Components with Time-Dependent Failure Rates. *Mathematics* **2023**, *11*, 3534. https://doi.org/10.3390/math11163534

Academic Editors: Gurami Tsitsiashvili and Alexander Bochkov

Received: 3 July 2023
Revised: 11 August 2023
Accepted: 12 August 2023
Published: 16 August 2023

1. Introduction

The reliability of a system can be improved at design by allocating redundant components to the subsystems. In a redundant system, when a working component fails, the failed component function is taken over by a redundant one which can provide the same function. The problem to do this optimally is called Redundancy Allocation Problem (RAP), which amounts to optimally selecting the redundant components to be allocated to the subsystems. The choice of the redundancy allocation strategy is driven by objectives, and/or subject to constraints, such as reliability, cost, volume and weight, depending on the system application. Then, RAP can be defined as the problem of selecting the number and type of the redundant components to allocate to the system, so to optimize a specific objective, such as system reliability, under given constraints, such as total system cost and weight.

The two redundancy strategies commonly considered in RAPs are "active" and "standby", although one called "mixed" has been introduced relatively recently [1]. In the active redundancy strategy, all components start to work at time zero and a redundant component might be found already failed before it is called to operation. In the standby redundancy strategy, one component is active and the others (at least one component) are standby. The standby redundancy strategy comes in the three types: cold, warm and hot [2]. In cold-standby, redundant components do not fail before they are used. In warm-standby, redundant components are subject to failure even when they are dormant, but at a rate lower than that of the main component which is working. Finally, in hot-standby, redundant components are subject to failure even while dormant, at a rate equal to that of the main components. Without loss of generality of the analysis, among the standby strategies, the present paper considers the cold-standby redundancy strategy, only for simplicity of illustration. In the mixed redundancy strategy, more than one component can be active while some others (one or more) are standby. The mixed redundancy strategy tries to combine the advantages of both active and standby strategies. Indeed, it is a general form of the active and standby strategies. The mixed redundancy strategy can be used in most practical cases, but the active and standby redundancy strategies are also used in certain cases, due to their simple forms. In the standby and mixed redundancy strategies, a switching system has to be used to activate a standby redundant component.

Some researchers considered RAP with the same redundancy strategy for all subsystems in the system [3–34]. Following Coit [35], some researchers have turned to consider the redundancy strategy for each subsystem, either active or standby, as a decision variable that can be optimally determined by an optimization model [36–43]. The mixed redundancy strategy has also been considered as an alternative for the subsystems [44–52].

Other aspects come realistically into play in a RAP; these include the objective function, the states of the components and that of the entire system, the type of the components to be allocated to each subsystem, the reparability of the components, and so on. For example, RAPs may be considered with a single objective function, e.g., aiming to maximizing system reliability [1,3–11,27,28,30,31,35–37,39–41,44,46–48,53–55], minimizing total system cost [12–15] or maximizing/minimizing other parameters [16,29]. Alternatively, RAPs may consider multi-objective functions [18–25,32,34,38,42,45,50,56–59].

Regarding the components, they might be represented as binary or multi-state. In the binary state representation, the components can only be totally healthy or completely failed [1,3–13,17–23,25–35,40,41,43–48,53,55]; in the multi-state, the components might have other states, intermediate between these two [14–16,24,60–65]. The type of the components in the subsystems can be characterized from different viewpoints. For instance, allocation of either identical [1,3–7,12,18–20,35,36,40,45,46,53,54] or non-identical components to each subsystem might be allowed [8–11,13–17,21,22,25,27,28,33,37,41,44,48,49,62]. Moreover, the components characteristics, including their cost, weight and performance function, might be deterministic or uncertain functions [66]. From another aspect, the components can be repairable [14,15,20,24,26,30,56] or non-repairable [1,3–7,9,11,13,16–19,21,22,27–29,31,35–37,40,41,43–48,53–55]. In this regard, a comprehensive categorization of reliability optimization problems is provided in [67].

This paper considers the RAP for a system with identical binary components in each subsystem. The redundancy strategy for each subsystem is taken as the decision variable that can be active, standby or mixed, and the objective function requires system reliability to be maximized. Also, the distribution of the time to failure of the components is assumed to have the form of an Erlang distribution, whose rate parameter ($\lambda(t)$) is not constant in time. To the best of our knowledge, no RAP with these novel realistic properties has ever been investigated in the literature.

The failure of the components allocated to a system may be due to degradation or due to occurrence of some shocks. In the case where component failure is due to shock occurrence, the Erlang distribution can be considered suitable for modeling the stochastic time to failure of the components. If the distribution of the time to failure of a component is considered as an Erlang distribution with a shape parameter of k, it means that the failure of the component is due to the occurrence of some shocks, so that the component will fail as soon as the kth shock occurs. So, the problem is of practical interest, for example for electronic components which often fail due to some shocks. In the Erlang distribution, the rate parameter indicates the rate of occurrence of the shocks. The rate parameter is usually assumed to be constant, to describe the failure of components subject to shocks occurring at a constant rate. But the occurrence rate of the shocks may not be constant in time and, for this reason, in this paper the rate parameter is considered time-dependent, $\lambda(t)$.

The main contribution of this paper is to propose a model in which the rate parameter of the time to failure distribution of the components is time-dependent and the redundancy allocation strategy for the subsystems can be mixed or other redundancy strategies.

The rest of the paper is structured as follows. In Section 2, the mathematical formulation of the RAP is described in detail. A solution method based on a genetic algorithm is proposed in Section 3. A numerical example is elaborated in Section 4 with and without time dependence, to demonstrate the effect of the time dependence of the rate parameter. Then, a case study taken from literature is considered. Finally, the conclusions and the suggestions for future work are given in Section 5.

2. Materials and Methods

2.1. Problem Formulation

A single-objective RAP is considered, in which the distribution of the time to failure of any component has the form of an Erlang distribution. The main specificities are the time-dependent parameter λ(t) in the time to failure distribution function and the redundancy strategy of any subsystem that can select among active, standby, mixed or no redundancy strategies. Figure 1 depicts the structure of a system composed of a series of five subsystems.

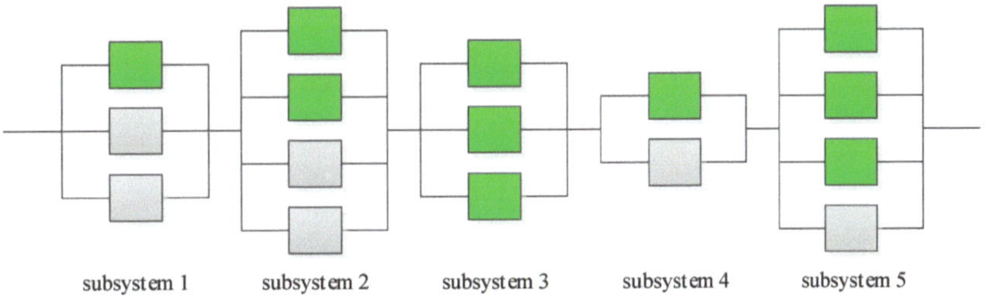

Figure 1. Structure of a system of redundant subsystems.

A number of redundant components can be allocated to each subsystem, each of which may be active or standby. In Figure 1, the active components are shown in green and the standby ones are shown in grey. The purpose of allocating the redundant components is to increase the reliability of the subsystems and consequently to increase the reliability of the system. But, the existence of some physical and/or economical restrictions makes it impossible to allocate any desired number of components to the subsystems. Therefore, the goal is to determine the best choices including determining the type and number of active and standby components allocated to each subsystem so that the reliability of the system is maximized, under given constraints.

The number of active and standby components allocated to each subsystem determines the redundancy strategy. If no redundant component is allocated to a subsystem, the subsystem has no redundancy strategy. The redundancy strategy of a subsystem is active if all of the components allocated to the subsystem are active. If one of the components allocated to a subsystem is active and the rest are standby, the redundancy strategy is standby, and if more than one allocated component is active and one or more components are standby, the redundancy strategy is mixed. Since the number of active and standby components allocated to each subsystem is a decision variable, the redundancy strategy is also a decision variable that is determined during the optimization process.

In Figure 1, the redundancy strategy for each subsystem can be determined according to the number of active and standby components allocated to the subsystem. Accordingly, the redundancy strategy for subsystems 1 and 4 is standby, that for subsystems 2 and 5 is mixed, and the one for subsystem 3 is active. Before addressing the mathematical model, the assumptions considered and the notations used are presented below.

2.1.1. Notation

i	index for subsystems $(i = 1, 2, \ldots, s)$
s	number of subsystems
$n_{A,i}$	number of active components allocated to subsystem i
$n_{S,i}$	number of standby components allocated to subsystem i
n_A, n_S	vectors of $[n_{A,i}]$ and $[n_{S,i}]$, $(n_A = [n_{A,1}, n_{A,2}, \ldots, n_{A,s}], n_S = [n_{S,1}, n_{S,2}, \ldots, n_{S,s}])$
n_i	number of components allocated to subsystem i, $(n_i = n_{A,i} + n_{S,i})$
$n_{max,i}$	maximum value allowed for n_i
ARS_i	redundancy strategy assigned for subsystem i, $(ARS_i \in \{NR, Active, Standby, Mixed\})$
$r_i(t)$	reliability at time t of the component (all identical) allocated to subsystem i
$R_{sys}(t; n_A, n_S)$	system reliability at time t due to the vectors n_A and n_S
$R_{i,ARS_i}(t)$	reliability of subsystem i at time t with redundancy strategy ARS_i
k_i	shape parameter of the Erlang distribution for the component allocated to subsystem i
$\lambda_i(t)$	rate parameter of the Erlang distribution for the component allocated to subsystem i
c_i, w_i	cost and weight of the component allocated to subsystem i
$c_{switch,i}$, $w_{switch,i}$	cost and weight of the switching system used in subsystem i (if present)
W, C	maximum allowed amount for the system weight and cost
$\rho_i(t)$	switching system reliability at time t for subsystem i (if present)
$f_i^{(j)}(t)$	pdf of the jth failure time of standby components allocated to subsystem i at time t
$f_i^{Max,n_{A,i}}(t)$	pdf of time to failure of the last active component allocated to subsystem i at time t
$f_i(t)$	pdf of the time to failure of the component allocated to subsystem i at time t
$F_i(t)$	cdf of the time to failure of the component allocated to subsystem i at time t
λ_{a-b}	average rate of occurrence of events between times a and b $(\lambda_{a-b} = \int_a^b \lambda(t)dt)$
$N(t)$	number of events occurred until time t.

2.1.2. Assumptions

As previously mentioned, the objective is to maximize the reliability of a system under a given number of constraints. The following assumptions are made:

- The components and the entire system are binary, i.e., they can be either totally healthy or completely failed.
- Components failures are independent events that individually cause no damage to the system.
- The redundancy strategy for each subsystem is a decision variable that can be selected to be active, standby, mixed or none.
- The distribution of the time to failure of the components has the form of an Erlang distribution with a time-dependent parameter $\lambda(t)$.
- The components are non-repairable and there is no preventive maintenance.
- The components of each subsystem are identical, i.e., mixing of component types is not allowed.

2.1.3. Mathematical Model

The mathematical model for the considered problem can be formulated as follows:

$$max\ R_{sys}(t; n_A, n_S) \tag{1}$$

$$s.t.$$

$$\sum_{i=1}^{s} c_i.n_i + \sum_{\substack{i \in Standby \\ or\ Mixed}} c_{switch,i} \leq C \tag{2}$$

$$\sum_{i=1}^{s} w_i.n_i + \sum_{\substack{i \in Standby \\ or\ Mixed}} w_{switch,i} \leq W \tag{3}$$

$$1 \leq n_i \leq n_{max,i} \tag{4}$$

Equation (1) expresses the objective of maximizing the system reliability by determining the type and the number of active and standby components allocated to each subsystem of the system. Constraints (2) and (3) express the maximum cost and weight for the entire system, respectively, whereas constraint set (4) considers the maximum number of components that can be allocated to each subsystem. The reliability of a system composed of a series of subsystems can be calculated as in Equation (5):

$$R_{sys}(t; n_A, n_S) = \prod_{i \in NR} R_{i,NR}(t) \prod_{i \in Active} R_{i,Active}(t) \prod_{i \in Standby} R_{i,Standby}(t) \prod_{i \in Mixed} R_{i,Mixed}(t) \tag{5}$$

where *NR* represents subsystems with no redundancy strategy, whereas *Active*, *Standby* and *Mixed* represent those with active, standby and mixed redundancy strategies, respectively. As previously mentioned, the redundancy strategy for each subsystem is determined according to the number of active and standby components allocated. Therefore, the number of active and standby components allocated to each subsystem determines in which part of Equation (5) the reliability calculation of the subsystem should be placed. After determining the redundancy strategy for each subsystem, the reliability of the subsystems at time t is calculated with one of the following equations:

$$R_{i,NR}(t) = r_i(t) \tag{6}$$

$$R_{i,Active}(t) = 1 - (1 - r_i(t))^{n_i} \tag{7}$$

$$R_{i,Standby}(t) = r_i(t) + \sum_{j=1}^{n_i-1} \int_0^t \left(\rho_i(u) f_i^{(j)}(u) r_i(t-u)\right) du \tag{8}$$

$$\begin{aligned} R_{i,Mixed}(t) &= 1 - (1 - r_i(t))^{n_{A,i}} + \int_0^t \left(\rho_i(u) f_i^{Max,n_{A,i}}(u) r_i(t-u)\right) du \\ &+ \sum_{j=1}^{n_{S,i}-1} \int_0^t \int_v^t \left(f_i^{Max,n_{A,i}}(v) \rho_i(u) f_i^{(j)}(u-v) r_i(t-u)\right) du dv \end{aligned} \tag{9}$$

In these equations, $r_i(t)$ is the reliability of the components used in subsystem i at time t, n_i is the number of components allocated to subsystem i, $\rho_i(t)$ is the reliability of the switching system at time t, $f_i^{(j)}(t)$ is the pdf of the time to failure of the jth component in subsystem i in standby and mixed redundancy strategies, whereas $n_{A,i}$ and $n_{S,i}$ represent the numbers of active and standby components assigned to subsystem i, respectively, in the case a mixed redundancy strategy is used (in this case $n_{is} = n_{A,i} + n_{S,i}$), and $f_i^{Max,n_{Ai}}(t)$ is the pdf of the time to failure of the last active component in subsystem i in the mixed redundancy strategy.

In Equation (6), the reliability of subsystem i with no redundancy strategy is equal to the reliability of the only component allocated. In Equation (7), the reliability of subsystem i at time t with an active redundancy strategy is equal to the probability that at least one of the allocated components can remain healthy until time t. In Equation (8), the reliability of subsystem i at time t with standby redundancy strategy consists of two parts. The first part is the reliability of the active component allocated and the second one is equal to the sum of the probabilities that the jth standby component fails at time u (a time before t) and the next standby component is activated by the switching system at this time and can be healthy until time t. In Equation (9), the reliability of subsystem i at time t with a mixed redundancy strategy consist of three parts. The first part is equal to the probability that at least one of the active components remains healthy until time t. The second part is equal to the probability that the last active component fails at time u (a time before t) and the first standby component is activated by the switching system at this time and can remain healthy until time t. The third part is equal to the sum of the probabilities that the last active component fails at time v (a time before t) and, then, the standby components will be activated one by one after the failure of the previous one, until the jth standby component fails at time u (a time between v and t), and the next standby component is activated by the switching system at this time and can remain healthy until time t.

In Equation (9), $f_i^{Max,n_{A,i}}(t)$ can be calculated using Equation (10):

$$f_i^{Max,n_{A,i}}(t) = n_{Ai}(F_i(t))^{n_{A,i}-1}f_i(t) \tag{10}$$

For example, if $n_{A,i} = 3$, $f_i^{Max,n_{A,i}}(t)$ can be calculated as in Equation (11):

$$f_i^{Max,3}(t) = f_i(t)F_i(t)F_i(t) + F_i(t)f_i(t)F_i(t) + F_i(t)F_i(t)f_i(t) = 3(F_i(t))^{3-1}f_i(t) \tag{11}$$

If the time to failure of a component in subsystem i follows an Erlang distribution, then its reliability at time t is given by Equation (12):

$$r_i(t) = \int_t^{\infty} f_i(t)dt = \sum_{j=0}^{k_i-1} \frac{e^{-\lambda_i t}(\lambda_i t)^j}{j!} \tag{12}$$

When the time to failure of a generic component follows an Erlang distribution with shape parameter k, this component is, then, exposed to shocks that follow a Poisson distribution, and fails at the occurrence of the kth shock. The reliability of the component at time t is, therefore, the probability that less than k shocks occur until time t.

Considering that the times to failure of the components follow Erlang distribution, the reliability of subsystem i with no redundancy strategy and with an active redundancy strategy is obtained as Equations (13) and (14), respectively:

$$R_{i,NR}(t) = \sum_{j=0}^{k_i-1} \frac{e^{-\lambda_i t}(\lambda_i t)^j}{j!} \tag{13}$$

$$R_{i,Active}(t) = 1 - \left(1 - \sum_{j=0}^{k_i-1} \frac{e^{-\lambda_i t}(\lambda_i t)^j}{j!}\right)^{n_i} \tag{14}$$

If the redundancy strategy of subsystem i is standby, considering that $\rho_i(t)$ is a non-increasing function, we have:

$$\begin{aligned} R_{i,Standby}(t) &\geq r_i(t) + \sum_{j=1}^{n_i-1} \rho_i(t) \int_0^t \left(f_i^{(j)}(u) r_i(t-u) \right) du \\ &= \sum_{j=0}^{k_i-1} \frac{e^{-\lambda_i t}(\lambda_i t)^j}{j!} + \rho_i(t) \sum_{j=k_i}^{n_i k_i - 1} \frac{e^{-\lambda_i t}(\lambda_i t)^j}{j!} \end{aligned} \tag{15}$$

The reliability formulation for a subsystem with a mixed redundancy strategy is slightly more complicated than the others. In Equation (9), we can rewrite $r_i(t)$ and $F_i(t)$ as in Equation (12). Also, we know that for a component with Erlang time to failure distribution of shape parameter k_i, $f_i(t)dt$ is the probability that $k_i - 1$ shocks occur before time t and the k_ith shock occurs at time t. So, we can write $f_i(t)dt$ so as to explicitly represent that $k_i - 1$ shocks occur before time t and one shock occurs between time t and $t + dt$:

$$f_i(t)dt = \frac{e^{-\lambda_i t}(\lambda_i t)^{k_i-1}}{(k_i-1)!} . \frac{e^{-\lambda_i dt}(\lambda_i dt)^1}{1!} \tag{16}$$

For $f_i^{(j)}(t)$, the pdf of the time to failure of the jth component in subsystem i, we have:

$$f_i^{(j)}(t)dt = \frac{e^{-\lambda_i t}(\lambda_i t)^{j.k_i-1}}{(j.k_i-1)!} . \frac{e^{-\lambda_i dt}(\lambda_i dt)^1}{1!} \tag{17}$$

This means that $f_i^{(j)}(t)dt$ is the probability that $j.k_i - 1$ shocks occur before time t and one shock occurs between time t and $t + dt$. So, if the redundancy strategy of subsystem i is mixed, the reliability of subsystem i, whose components times to failure follow an Erlang distribution, can be calculated as in Equation (18):

$$\begin{aligned} R_{i,Mixed}(t) &= 1 - \left(1 - \sum_{j=0}^{k_i-1} \frac{e^{-\lambda_i t}(\lambda_i t)^j}{j!} \right)^{n_{A,i}} \\ &+ \int_0^t \left(\rho_i(u) n_{A,i} \left(1 - \sum_{j=0}^{k_i-1} \frac{e^{-\lambda_i u}(\lambda_i u)^j}{j!} \right)^{n_{A,i}-1} \frac{e^{-\lambda_i u}(\lambda_i u)^{k_i-1}}{(k_i-1)!} . \frac{e^{-\lambda_i du}(\lambda_i)^1}{1!} \sum_{j=0}^{k_i-1} \frac{e^{-\lambda_i (t-u)}(\lambda_i (t-u))^j}{j!} \right) du \\ &+ \sum_{j=1}^{n_{S,i}-1} \int_0^t \int_v^t \left(\rho_i(u) n_{A,i} \left(1 - \sum_{j=0}^{k_i-1} \frac{e^{-\lambda_i v}(\lambda_i v)^j}{j!} \right)^{n_{A,i}-1} \frac{e^{-\lambda_i v}(\lambda_i v)^{k_i-1}}{(k_i-1)!} . \frac{e^{-\lambda_i dv}(\lambda_i)^1}{1!} . \frac{e^{-\lambda_i (u-v)}(\lambda_i (u-v))^{j.k_i-1}}{(j.k_i-1)!} . \frac{e^{-\lambda_i du}(\lambda_i)^1}{1!} \sum_{j=0}^{k_i-1} \frac{e^{-\lambda_i (t-u)}(\lambda_i (t-u))^j}{j!} \right) dvdu \end{aligned} \tag{18}$$

In this paper, it is assumed that the rate parameter is not constant, but a function of time, i.e., we have $\lambda(t)$ instead of λ. So, the components will be exposed to shocks with a variable rate over time. For a nonhomogeneous Poisson process with rate $\lambda(t)$, we have:

$$P\{N(b) - N(a) = j\} = \frac{e^{-\lambda_{a-b}}(\lambda_{a-b})^j}{j!} \tag{19}$$

where

$$\lambda_{a-b} = \int_a^b \lambda(t)dt \tag{20}$$

and $N(t)$ is the number of events occurring prior to time t. Therefore, the reliability at time t for a component exposed to the shocks that follow a nonhomogeneous Poisson distribution, i.e., the probability that less than k shocks occur until time t, can be calculated as in Equation (21):

$$r(t) = P\{N(t) - N(0) < k\} = \sum_{j=0}^{k-1} \frac{e^{-\lambda_{0-t}}(\lambda_{0-t})^j}{j!} \tag{21}$$

Now, if the rate parameter is time-dependent (λ(t)), by replacing $\lambda_i t$ by $\lambda_{i,0-t}$ in Equations (13)–(15) and (18), the reliability of the subsystems with no redundancy strategy, and of those with active, standby and mixed redundancy strategies can be written as Equations (22)–(25), respectively:

$$R_{i,NR}(t) = \sum_{j=0}^{k_i-1} \frac{e^{-\lambda_{i,0-t}} (\lambda_{i,0-t})^j}{j!} \tag{22}$$

$$R_{i,Active}(t) = 1 - \left(1 - \sum_{j=0}^{k_i-1} \frac{e^{-\lambda_{i,0-t}} (\lambda_{i,0-t})^j}{j!}\right)^{n_i} \tag{23}$$

$$R_{i,Standby}(t) = \sum_{j=0}^{k_i-1} \frac{e^{-\lambda_{i,0-t}} (\lambda_{i,0-t})^j}{j!} + \rho_i(t) \sum_{j=k_i}^{n_i k_i-1} \frac{e^{-\lambda_{i,0-t}} (\lambda_{i,0-t})^j}{j!} \tag{24}$$

$$\begin{aligned} R_{i,Mixed}(t) = {} & 1 - \left(1 - \sum_{j=0}^{k_i-1} \frac{e^{-\lambda_{i,0-t}} (\lambda_{i,0-t})^j}{j!}\right)^{n_{A,i}} \\ & + \int_0^t \left(\rho_i(u) n_{A,i} \left(1 - \sum_{j=0}^{k_i-1} \frac{e^{-\lambda_{i,0-u}} (\lambda_{i,0-u})^j}{j!}\right)^{n_{A,i}-1} \frac{e^{-\lambda_{i,0-u}} (\lambda_{i,0-u})^{k_i-1}}{(k_i-1)!} \cdot \frac{e^{-\lambda_{i,u-u+du}} (\lambda_i(u))^1}{1!} \sum_{j=0}^{k_i-1} \frac{e^{-\lambda_{i,u-t}} (\lambda_{i,u-t})^j}{j!}\right) du \\ & + \sum_{j=1}^{n_{S,i}-1} \int_0^t \int_v^t \left(\rho_i(u) n_{A,i} \left(1 - \sum_{j=0}^{k_i-1} \frac{e^{-\lambda_{i,0-v}} (\lambda_{i,0-v})^j}{j!}\right)^{n_{A,i}-1} \frac{e^{-\lambda_{i,0-v}} (\lambda_{i,0-v})^{k_i-1}}{(k_i-1)!} \cdot \frac{e^{-\lambda_{i,v-v+dv}} (\lambda_i(v))^1}{1!} \cdot \frac{e^{-\lambda_{i,v-u}} (\lambda_{i,v-u})^{j.k_i-1}}{(j.k_i-1)!} \cdot \frac{e^{-\lambda_{i,u-u+du}} (\lambda_i(u))^1}{1!} \sum_{j=0}^{k_i-1} \frac{e^{-\lambda_{i,u-t}} (\lambda_{i,u-t})^j}{j!}\right) dv du \end{aligned} \tag{25}$$

2.2. *RAP Solution Method*

RAP is a problem of NP-hard class [68], and a meta-heuristic algorithm is here developed to search for the optimal solution. The genetic algorithm (GA) is a well-known meta-heuristic method for solving combinatorial optimization problems [69] and, indeed, it has been successfully employed for solving optimization problems in various fields [1,4–6,8,24,30,38,41–47,56]. In the following, the deployment of the algorithm for the solution of the proposed RAP is briefly described.

2.2.1. Solution Encoding (Chromosome)

Solution encoding in the GA used is a 3 × s matrix, in which s represents the number of the subsystems in the system. For each subsystem, the first row is the type of selected component and the second and third rows give the numbers of allocated active and standby components, respectively. There is no need to add a row for the redundancy strategy, as it may be determined using the second and third rows: if the number of standby components is equal to zero, then the redundancy strategy is active (or no redundancy strategy, in case the number of active components is equal to one); if the number of active components is equal to one and there are some standby components, then the redundancy strategy is standby; finally, if the number of active components is larger than one and there are also some standby components, then the redundancy strategy is mixed. Figure 2 shows an example for the solution encoding with s = 6 subsystems. In the Figure, the active components are shown in green and the standby ones are shown in grey. As it can be seen in the Figure, the redundancy strategy for subsystems 1 and 4 is standby, that for subsystems 2 and 6 is mixed, and that for subsystems 3 and 5 is active. Figure 3 represents the structure of the system.

$$\begin{array}{r} \text{component type} \\ \text{number of active components} \\ \text{number of standby components} \end{array} \begin{pmatrix} 3 & 2 & 4 & 2 & 1 & 1 \\ 1 & 2 & 3 & 1 & 2 & 2 \\ 2 & 1 & 0 & 3 & 0 & 2 \end{pmatrix}$$

Figure 2. Chromosome representation.

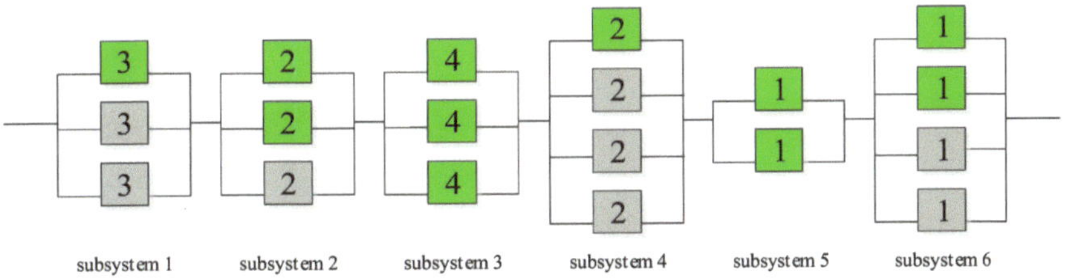

Figure 3. Structure of the system corresponding to the solution in Figure 2.

2.2.2. Initial Population

The initial population is composed of the *nPop* solution matrix, whose entries are generated randomly. The solution matrices should be generated in such a way that in each column (subsystem), the first element is an integer number between 1 and the number of components types available for the subsystem, and the second and third elements are integer numbers so that their summation is less than or equal to the maximum number of components allowed for the subsystem.

2.2.3. Fitness Function

The fitness function is defined as the system reliability minus the penalty for constraints violation. Indeed, if a solution goes beyond the constraints, a large enough amount of penalty is deducted from the fitness function to guarantee that the final optimal solution is feasible. The amount of the penalty is equal to the product of the total amount of violations of the constraints and a fixed number. This penalty also allows the algorithm to search in the infeasible space for maintaining appropriate diversity in the search.

2.2.4. Selection

At each iteration, new offspring are generated by implementation of the crossover and mutation operators. The roulette wheel selection method is used to select the parents [8,40]. In the roulette wheel selection method, selection of the parents occurs based on their fitness function value: in other words, a solution is selected with probability proportional to its fitness function. After the selection process, the crossover and mutation operators are performed.

2.2.5. Crossover Operator

The crossover operator activates at a prespecified rate, *rc*. To explore a large variety of solutions, use has been made of different types of single-point, double-point, uniform and max–min crossover operators [1,4,44], each with a predetermined probability.

2.2.6. Mutation Operator

The mutation operator is implemented at a prespecified rate, *rm*. The main purpose of using this operator is to not get caught in the local optimal solutions. In this paper, use has been made of the random and max–min mutations [1,4,44], each with a predetermined probability.

To create the next generation, the offspring generated by implementation of the crossover and mutation operators are combined with the current population and the *nPop* best ones are selected. These best solutions, then, form the next-generation population.

2.2.7. Stopping Criteria

The GA is terminated after a prespecified number of iterations, *MaxIt*.

3. Results

As a numerical example, the new optimization model proposed in this paper is applied to three simple systems composed of series subsystems, for each of which some component choices are available for selection. The first system consists of 6 subsystems and the second one consists of 15 subsystems. The objective is to maximize system reliability at a mission time t = 100 (in arbitrary units). The third example is a case study introduced in [6]. The system is a pharmaceutical plant containing 10 series subsystems as follows:

1. Weighting machine, 2. Sifter machine, 3. Mass machine, 4. Granulator, 5. Fluid bed dryer, 6. Octagonal blender, 7. Rotary compression machine, 8. Coating machine, 9. Air compressor, 10. Strip packing machine.

In this work, we extend this case study so that the rate parameters for the time to failure of the components are time-dependent. The mission time is equal to t = 1000.

The rate parameter of the Erlang distribution is assumed to be in the form of a bathtub curve [70–74], as shown in Figure 4. As can be seen, the rate parameter first decreases over time, representing infant mortality, i.e., the process for which defective items fail early and the occurrence rate of the shocks decreases over time as the defective items are weeded out of the population. Then, the rate parameter remains constant for a period of time, as fully random shocks are the cause of failure, independent of time. Finally, the rate parameter increases, due to the aging process, for which the components are more likely to fail as time goes on.

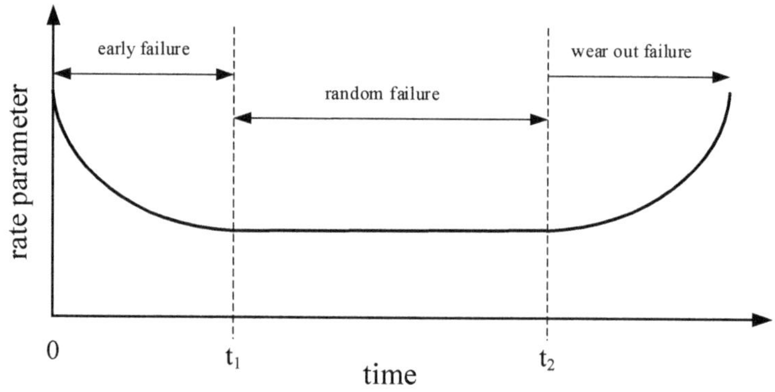

Figure 4. Rate parameter (bathtub curve).

We consider the mathematical form of the rate parameter as in Equation (26) to describe all three parts, decreasing, constant and increasing:

$$\lambda(t) = c\lambda_0(\lambda_0 t)^{\alpha-1} \tag{26}$$

In the above equation, c is a positive constant parameter. Different values of the parameter α allow representing the different parts of the rate parameter: a value of $\alpha < 1$ indicates a rate parameter decreasing over time; $\alpha = 1$ indicates a constant rate parameter over time; $\alpha > 1$ indicates a rate parameter increasing over time. Then, the rate parameter for the components in Figure 4 can be expressed by Equation (27), in which α_1 is a value less than one and α_2 is a value larger than one:

$$\lambda(t) = \begin{cases} c_1\lambda_0(\lambda_0 t)^{\alpha_1-1} & ; 0 \le t \le t_1 \\ \lambda_0 & ; t_1 \le t \le t_2 \\ c_2\lambda_0(\lambda_0 t)^{\alpha_2-1} & ; t \ge t_2 \end{cases} \tag{27}$$

In this equation, the values of c_1 and c_2 should be determined so that the rate function is continuous at points t_1 and t_2. For this purpose, their values should be considered as given in Equation (28):

$$\lambda(t) = \begin{cases} (\lambda_0 t_1)^{1-\alpha_1} \lambda_0 (\lambda_0 t)^{\alpha_1 - 1} = \lambda_0 (\frac{t}{t_1})^{\alpha_1 - 1} & ; 0 \leq t \leq t_1 \\ \lambda_0 & ; t_1 \leq t \leq t_2 \\ (\lambda_0 t_2)^{1-\alpha_2} \lambda_0 (\lambda_0 t)^{\alpha_2 - 1} = \lambda_0 (\frac{t}{t_2})^{\alpha_2 - 1} & ; t \geq t_2 \end{cases} \tag{28}$$

According to Equation (28), λ_{0-t} can be calculated as Equation (29) to be used in Equations (22)–(25).

$$\lambda_{0-t} = \int_0^t \lambda(t)dt = \begin{cases} \int_0^t \lambda_0 (\frac{u}{t_1})^{\alpha_1 - 1} du = \frac{\lambda_0 t_1}{\alpha_1} (\frac{t}{t_1})^{\alpha_1} & ; 0 \leq t \leq t_1 \\ \int_0^{t_1} \lambda_0 (\frac{u}{t_1})^{\alpha_1 - 1} du + \int_{t_1}^t \lambda_0 du = \frac{\lambda_0 t_1}{\alpha_1} + \lambda_0 (t - t_1) & ; t_1 \leq t \leq t_2 \\ \int_0^{t_1} \lambda_0 (\frac{u}{t_1})^{\alpha_1 - 1} du + \int_{t_1}^{t_2} \lambda_0 du + \int_{t_2}^t \lambda_0 (\frac{u}{t_2})^{\alpha_2 - 1} du = \frac{\lambda_0 t_1}{\alpha_1} + \lambda_0 (t_2 - t_1) + \frac{\lambda_0 t_2}{\alpha_2} \left((\frac{t}{t_2})^{\alpha_2} - 1 \right) & ; t \geq t_2 \end{cases} \tag{29}$$

The parameters α_1, α_2, t_1 and t_2 must be determined for each component type allocated in each subsystem. The components characteristics, including cost, weight and rate parameters, for the numerical examples are presented in Tables A1–A3 in Appendix A. The values of C, W, reliability of the switching system at mission time and the maximum number of components that can be allocated to each subsystem are provided in Table A4 (in Appendix A) for the examples considered.

In order to tune the parameters of the GA, the Taguchi method has been used. For this purpose, three levels have been considered for population size, crossover rate and mutation rate.

The results obtained by the implementation of the Taguchi method are shown in Figure 5. The considered levels, as well as the best values obtained by this method for population size, crossover rate and mutation rate are shown in Table 1. The number of iterations is set to be *MaxIt* = 100 for all the examples.

Table 1. The results obtained by the implementation of Taguchi method.

	Level	nPop	rc	rm
	1	100	0.2	0.2
	2	200	0.3	0.3
	3	300	0.4	0.4
Best level	Example 1	200	0.4	0.2
	Example 2	300	0.3	0.3
	Example 3	200	0.4	0.3

As the GA is a stochastic algorithm, 20 trials of the algorithm are performed for each example and the best solution found amongst them is selected as the optimal solution. The trend of the algorithm towards the optimal solution in the 20 trials is shown in Figures A1–A3 in Appendix A. In order to show the robustness of the algorithm, the values of the minimum, maximum, average, standard deviation and coefficient of variation values of the reliability, weight and cost of the system in the optimal solutions obtained by different trials are reported in Table 2. The values of the reliability, weight and cost of the system in the optimal solutions obtained by different trials are reported in Tables A5–A7 in Appendix A.

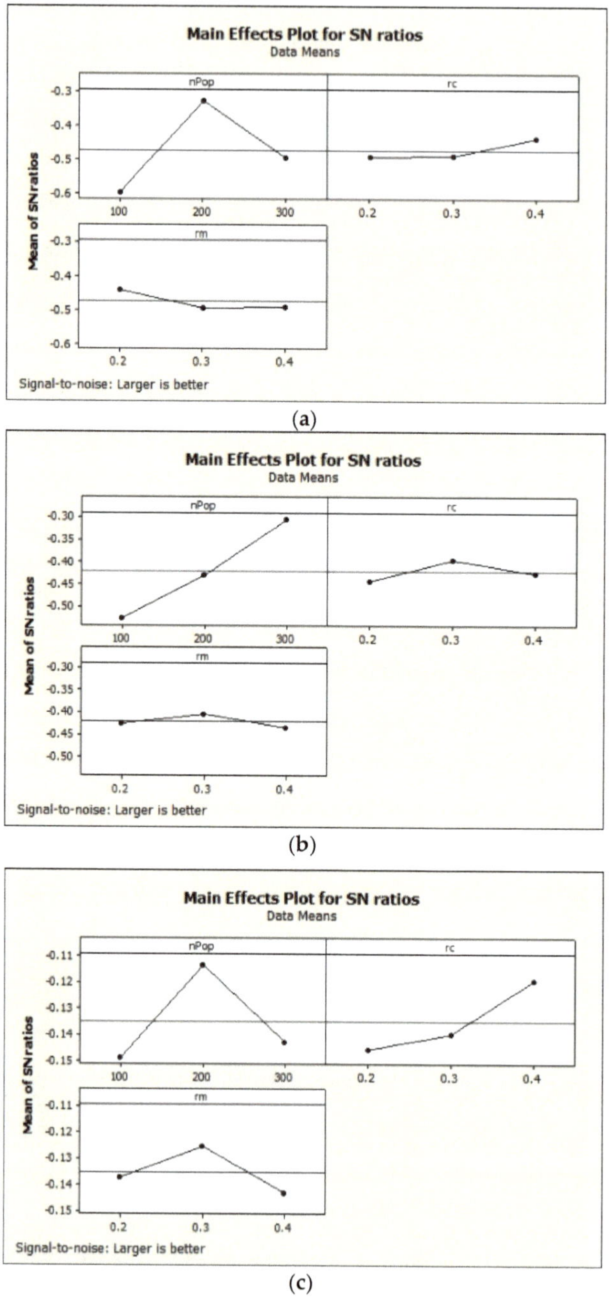

Figure 5. The results obtained by the implementation of Taguchi method: (**a**) first example; (**b**) second example; (**c**) third example.

Table 2. The results obtained by different trials.

	First Example	System Characteristics Reliability	Cost	Weight
First example	Minimum	0.9284	37	65
	Maximum	0.9702	50	70
	Average	0.9571	45.15	68.4
	Standard deviation	0.0131	3.7852	1.4283
	Coefficient of variation	0.0137	0.0838	0.0209
Second example	Minimum	0.9337	200	265
	Maximum	0.9707	306	393
	Average	0.9522	255.8	334.7
	Standard deviation	0.0102	26.5643	36.5392
	Coefficient of variation	0.0107	0.1038	0.1092
Third example	Minimum	0.9846	424	509
	Maximum	0.9896	478	519
	Average	0.9882	457.95	515.3
	Standard deviation	0.0014	13.1813	3.2879
	Coefficient of variation	0.0015	0.0288	0.0064

To the best of our knowledge, there is no benchmark problem to compare with the RAP presented in this paper. So, to show the effect of time dependence of the rate parameter, the above-mentioned examples have also been considered with a constant rate parameter (i.e., useful lifetime), of the same value for all components. The best solutions obtained for the examples by considering a constant rate parameter (λ) and by considering a time-dependent rate parameter ($\lambda(t)$) (i.e., also initial early failure and final wear-out parts of the lifetime) are given in Tables 3–5.

Table 3. Optimal solution with time dependence and without time dependence; first example.

Subsystem	With Time Dependence					Without Time Dependence				
	z_i	n_{Ai}	n_{Si}	Red. Strategy	Reliability	z_i	n_{Ai}	n_{Si}	Red. Strategy	Reliability
1	3	2	3	Mixed	0.9938	2	2	1	Mixed	0.9994
2	1	1	1	Standby	0.9977	1	1	1	Standby	0.9987
3	4	2	0	Active	0.9939	4	1	1	Standby	0.9986
4	2	1	1	Standby	0.9909	4	2	1	Mixed	0.9981
5	1	2	2	Mixed	0.9959	1	1	2	Standby	0.9953
6	1	1	1	Standby	0.9977	1	1	1	Standby	0.9980
System reliability				0.9702				0.9877		
System cost				45				37		
System weight				70				70		

Table 4. Optimal solution with time dependence and without time dependence; second example.

Subsystem	With Time Dependence					Without Time Dependence				
	z_i	n_{Ai}	n_{Si}	Red. Strategy	Reliability	z_i	n_{Ai}	n_{Si}	Red. Strategy	Reliability
1	1	3	1	Mixed	0.99997	7	5	1	Mixed	0.99788
2	2	2	1	Mixed	0.99835	9	4	2	Mixed	0.99938
3	3	2	2	Mixed	0.99999	3	1	1	Standby	0.99977
4	9	3	3	Mixed	0.99912	6	2	3	Mixed	0.99995
5	9	2	3	Mixed	0.99428	3	2	2	Mixed	0.99876
6	8	2	2	Mixed	0.99916	1	3	2	Mixed	0.99888
7	10	3	2	Mixed	0.99968	4	3	3	Mixed	0.99755
8	5	2	3	Mixed	0.99717	3	3	4	Mixed	0.99764

Table 4. *Cont.*

Subsystem	With Time Dependence					Without Time Dependence				
	z_i	n_{Ai}	n_{Si}	Red. Strategy	Reliability	z_i	n_{Ai}	n_{Si}	Red. Strategy	Reliability
9	8	2	3	Mixed	0.99796	7	4	1	Mixed	0.99554
10	6	3	3	Mixed	0.99567	5	3	3	Mixed	0.99686
11	1	2	3	Mixed	0.99761	1	2	2	Mixed	0.99853
12	1	1	3	Standby	0.99979	10	3	2	Mixed	0.99915
13	2	2	3	Mixed	0.99877	2	2	3	Mixed	0.99921
14	9	2	3	Mixed	0.99365	8	3	3	Mixed	0.99493
15	5	2	3	Mixed	0.99914	5	2	3	Mixed	0.99921
System reliability				0.9707				0.9736		
System cost				272				254		
System weight				282				335		

Table 5. Optimal solution with time dependence and without time dependence; third example.

Subsystem	With Time Dependence					Without Time Dependence				
	z_i	n_{Ai}	n_{Si}	Red. Strategy	Reliability	z_i	n_{Ai}	n_{Si}	Red. Strategy	Reliability
1	3	2	1	Mixed	0.99956	3	2	1	Mixed	0.99980
2	1	2	1	Mixed	0.99928	4	2	1	Mixed	0.99999
3	2	2	1	Mixed	0.99950	2	2	1	Mixed	0.99966
4	3	2	1	Mixed	0.99957	4	1	1	Standby	0.99976
5	1	1	1	Standby	0.99740	1	2	1	Mixed	0.99987
6	3	2	1	Mixed	0.99918	3	2	1	Mixed	0.99982
7	2	2	1	Mixed	0.99923	3	2	0	Active	0.99999
8	3	2	1	Mixed	0.99924	1	2	1	Mixed	0.99998
9	2	1	1	Mixed	0.99755	4	2	1	Mixed	0.99999
10	1	2	1	Mixed	0.99908	1	1	1	Standby	0.99981
System reliability			0.9896					0.9987		
System cost			454					480		
System weight			514					517		

4. Discussion

In this paper, a formulation for RAP is proposed in which the redundancy strategy for each subsystem is taken as the decision variable and the distribution of the time to failure of the components is assumed to have the form of an Erlang distribution, whose rate parameter ($\lambda(t)$) is not constant in time. Then, the genetic algorithm is proposed to search for the optimal solution.

As a numerical example, the proposed model was applied to three simple systems composed of series subsystems. As the GA is a stochastic algorithm, 20 trials of the algorithm were performed for each example. The trend of the algorithm towards the optimal solution was shown in Figures A1–A3. Due to the random nature of the GA, the trend of the algorithm towards the optimal solution varies in different trials. But, it can be seen that the differences are small, which shows the robustness of the algorithm in finding the optimal solution. It can also be seen that, in all three numerical examples, the different trials have converged to a similar solution, with only a slight difference, which shows the convergence of the algorithm to a same optimal solution.

Also, the results obtained by different trials including the values of the minimum, maximum, average, standard deviation and coefficient of variation of the reliability, the weight and cost of the system in the optimal solutions were reported in Table 2. As can be seen, the values of the coefficient of variation of the reliability, cost and weight of the system are small. This shows that the optimal solutions obtained by different trials are only slightly different from each other, which confirms the robustness of the algorithm in finding the optimal solution and its convergence to the optimal solution.

Since we did not find any benchmark problem to compare with the RAP presented in this paper, to show the effect of time dependence of the rate parameter, the examined examples have also been considered with a constant rate parameter, of the same value for all components. By comparing the best solution obtained for the problems with a constant rate parameter and with a time-dependent rate parameter (given in Tables 3–5), it can be seen that the structure of the obtained solutions, the type of components allocated to the subsystems and the redundancy strategy of the subsystems may be different.

For example, in the first numerical example, the type of the component allocated to the first subsystem is different in the cases of time-dependent rate parameter and constant rate parameter. In the case with time dependence, component type 3 is allocated to this subsystem, and in the case without time dependence, component type 2 is allocated to this subsystem. In this subsystem, the number of standby components is also different. In the case with time dependence, three standby components are allocated, and in the case without time dependence, one standby component is allocated to this subsystem. As another example, in the first numerical example, the redundancy strategy of the third subsystem is different in the two considered cases. In the case with time dependence, the redundancy strategy is active and, therefore, no switching system is required, but in the case without time dependence, the redundancy strategy is standby and, therefore, a switching system is required. These differences can also be seen in other cases.

In general, it can be seen that the structure of the optimal solution is different in the two cases. In other words, the optimal solution obtained in the case with time dependence is not optimal in the case without time dependence and vice versa. This issue is important in considering assumptions in designing a system. Because, if the rate parameter is variable over time, by considering it as constant, the real optimal solution cannot be found and, therefore, the system designed based on the considered assumptions is no longer optimal and its reliability in practice will be lower than what is theoretically expected. Therefore, to reach a real optimal solution, correct assumptions should be made.

In addition to the difference in the structure of the subsystems, there are also differences in the values of reliability, weight and cost in the obtained optimal solutions at the system level. For the cases studied, the obtained results show that, when the rate parameters are considered as time-dependent, the system reliability is lower. This is expected, because when the time-dependent rate parameters are considered for the components, the failure rates in the early and wear-out parts of the lifetime are higher than the failure rates in the useful part of the lifetime. Therefore, the probability of failure of the components is larger and, as a result, the reliability of the system is smaller than with constant failure rates. This decrease in reliability of the system in the case with time dependence is even accompanied by an increase in the cost of the system in some cases. For example, in the first numerical example, in the case with time dependence, the system reliability is 0.9702 and the system cost is 45, but in the case without time dependence, the system reliability is 0.9877 and the system cost is 37. This means that if the rate parameter is time-dependent, the system cost is higher than the case of a constant rate parameter but the system reliability is lower.

All of these confirm that it is important to consider the actual time-dependence of the rate parameter according to reality.

5. Conclusions

A redundancy allocation problem with multiple redundancy types has been investigated, in which the distribution of time to failure of the components has the form of an Erlang distribution with time-dependent parameter $\lambda(t)$. The mathematical model has been formulated and a genetic algorithm has been developed to find the optimal solution.

To demonstrate the effect of time dependence of the time to failure distribution, numerical examples have been worked out with and without time dependence of the rate parameter. In the cases studied, it has been shown that taking the time dependence of the parameter $\lambda(t)$ into account changes the reliability of the components and as a result, both the optimal redundancy structure of the subsystems of the system and the values of its reliability, cost and weight are different from those of the case with constant rate parameter.

In this paper, only identical components have been assumed to be allocated to each subsystem. However, mixing non-identical components might lead to improved system reliability; so, it is suggested that future studies examine the effect of mixing non-identical components. Another aspect to consider is the uncertainty in the characteristics and the time to failure distribution parameters of the components. Considering multi-state, repairable components, other pdfs for the times to failure of the components and/or preventive maintenance is another interesting area for future investigation.

Since the model presented in each research needs to be compared with the previous ones, it can be useful to define benchmark problems that can show the impact of the assumptions of each model well and be used in future research. Therefore, the definition of suitable benchmark problems to be used in the research in this field can also be a suggestion for future research.

Author Contributions: Conceptualization, E.Z. and H.G.; Investigation, E.Z. and H.G.; Methodology, H.G.; Software, H.G.; Supervision, E.Z.; Validation, E.Z. and H.G.; Writing—original draft, H.G.; Writing—review and editing, E.Z. All authors have read and agreed to the published version of the manuscript.

Funding: This research received no external funding.

Institutional Review Board Statement: Not applicable.

Informed Consent Statement: Not applicable.

Data Availability Statement: In this of article, no new data were created.

Conflicts of Interest: The authors declare no conflict of interest.

Appendix A

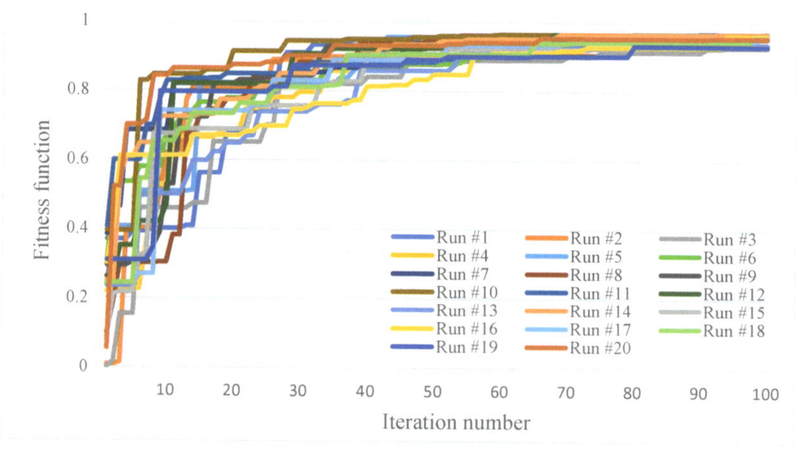

Figure A1. Trend of GA towards the optimal solution in 20 different trials; first example.

Table A1. Component data for the first numerical example.

	Component Type 1 (j = 1)								Component Type 2 (j = 2)								Component Type 3 (j = 3)								Component Type 4 (j = 4)							
i	λ_{0ij}	α_{1ij}	α_{2ij}	t_{1ij}	t_{2ij}	k_{ij}	c_{ij}	w_{ij}	λ_{0ij}	α_{1ij}	α_{2ij}	t_{1ij}	t_{2ij}	k_{ij}	c_{ij}	w_{ij}	λ_{0ij}	α_{1ij}	α_{2ij}	t_{1ij}	t_{2ij}	k_{ij}	c_{ij}	w_{ij}	λ_{0ij}	α_{1ij}	α_{2ij}	t_{1ij}	t_{2ij}	k_{ij}	c_{ij}	w_{ij}
1	0.052	0.3	3	10	90	6	2	4	0.007	0.1	2	12	90	2	2	4	0.049	0.4	2	10	100	4	3	2	0.081	0.3	3	10	80	4	3	4
2	0.017	0.5	2	15	120	4	2	5	0.110	0.4	4	18	100	4	3	5	0.124	0.3	3	5	75	6	2	6	0.046	0.1	5	5	75	4	4	3
3	0.043	0.4	4	8	110	5	3	4	0.056	0.6	3	20	110	6	4	5	0.028	0.7	4	10	95	3	3	5	0.004	0.5	4	15	100	2	2	4
4	0.026	0.4	3	12	80	4	4	6	0.001	0.3	5	8	80	1	3	8	0.004	0.6	3	12	85	1	3	4	0.009	0.4	2	20	120	2	3	6
5	0.023	0.6	5	6	70	3	2	4	0.077	0.8	2	15	85	4	2	6	0.133	0.3	3	7	90	5	2	5	0.122	0.7	3	17	85	7	4	5
6	0.011	0.7	7	9	95	3	4	5	0.083	0.2	3	14	95	5	5	4	0.035	0.2	2	15	110	3	2	6	0.043	0.3	3	14	90	4	5	8

Table A2. Component data for the second numerical example.

Subsystem	Component Type	λ_{0ij}	α_{1ij}	α_{2ij}	t_{1ij}	t_{2ij}	k_{ij}	c_{ij}	w_{ij}
i = 1	1	0.014	0.5	5	20	70	3	4	2
	2	0.034	0.4	6	7	80	5	4	3
	3	0.014	0.2	4	10	110	2	3	6
	4	0.052	0.2	2	7	85	7	5	4
	5	0.048	0.3	3	16	85	8	2	6
	6	0.013	0.4	3	6	75	2	3	4
	7	0.021	0.1	4	5	100	3	4	2
	8	0.045	0.3	3	19	80	7	5	5
	9	0.023	0.3	5	18	80	4	5	8
	10	0.071	0.7	5	5	90	7	2	4
i = 2	1	0.027	0.5	3	8	100	3	5	3
	2	0.017	0.4	4	16	105	2	3	3
	3	0.105	0.2	6	11	80	8	2	4
	4	0.019	0.8	3	19	80	3	2	6
	5	0.029	0.7	3	10	80	3	3	4
	6	0.049	0.5	4	19	80	6	4	5
	7	0.053	0.8	3	11	105	5	2	6
	8	0.046	0.1	5	14	90	7	5	8
	9	0.017	0.8	3	6	80	2	5	5
	10	0.023	0.7	6	10	80	3	4	3

Table A2. *Cont.*

Subsystem	Component Type	λ_{0ij}	α_{1ij}	α_{2ij}	t_{1ij}	t_{2ij}	k_{ij}	c_{ij}	w_{ij}
$i = 3$	1	0.071	0.6	5	10	85	8	2	8
	2	0.018	0.3	3	7	95	2	2	8
	3	0.022	0.6	3	15	85	2	5	2
	4	0.034	0.6	2	18	85	4	2	4
	5	0.021	0.5	2	12	70	2	5	8
	6	0.019	0.2	6	16	70	4	5	4
	7	0.071	0.6	4	15	95	8	3	7
	8	0.051	0.8	5	20	85	6	4	2
	9	0.063	0.6	5	11	105	7	5	6
	10	0.018	0.4	2	10	70	2	4	2
$i = 4$	1	0.021	0.8	6	6	105	2	2	7
	2	0.014	0.1	4	14	90	3	3	6
	3	0.025	0.1	2	13	95	5	2	3
	4	0.032	0.2	4	20	105	5	2	5
	5	0.047	0.3	3	11	75	4	3	2
	6	0.017	0.1	4	10	80	3	2	7
	7	0.049	0.1	4	5	75	4	5	6
	8	0.051	0.3	5	17	85	8	3	8
	9	0.021	0.7	6	20	75	3	2	7
	10	0.043	0.2	5	20	95	7	4	6
$i = 5$	1	0.073	0.5	3	10	85	8	5	3
	2	0.048	0.5	5	10	85	6	3	2
	3	0.022	0.2	4	6	105	3	2	3
	4	0.055	0.8	3	11	85	6	3	8
	5	0.054	0.2	6	11	95	8	5	7
	6	0.017	0.3	3	10	100	2	2	8
	7	0.032	0.6	5	6	95	4	3	2
	8	0.063	0.3	6	12	90	9	4	4
	9	0.051	0.7	6	10	90	6	4	8
	10	0.073	0.5	4	7	105	9	2	2

Table A2. *Cont.*

Subsystem	Component Type	λ_{0ij}	α_{1ij}	α_{2ij}	t_{1ij}	t_{2ij}	k_{ij}	c_{ij}	w_{ij}
$i = 6$	1	0.016	0.8	2	19	95	2	2	6
	2	0.027	0.1	3	9	90	5	3	8
	3	0.012	0.2	5	20	90	2	2	3
	4	0.009	0.2	6	8	70	2	4	7
	5	0.051	0.7	5	20	70	7	5	7
	6	0.029	0.5	4	19	110	4	4	3
	7	0.027	0.8	4	8	105	3	3	7
	8	0.017	0.6	4	15	75	2	4	7
	9	0.021	0.7	4	15	85	2	3	4
	10	0.031	0.1	6	5	85	4	3	5
$i = 7$	1	0.032	0.5	6	9	70	6	4	7
	2	0.022	0.2	2	9	100	3	3	6
	3	0.053	0.4	6	13	90	7	4	5
	4	0.017	0.7	2	14	95	2	5	5
	5	0.029	0.3	6	13	105	4	3	3
	6	0.034	0.1	4	13	90	7	4	8
	7	0.037	0.6	4	16	70	5	4	6
	8	0.051	0.3	6	20	75	8	3	8
	9	0.024	0.7	4	16	95	2	2	4
	10	0.015	0.4	2	20	90	2	3	5
$i = 8$	1	0.018	0.4	5	10	85	2	3	3
	2	0.023	0.7	4	20	75	3	3	6
	3	0.012	0.2	7	8	110	2	2	4
	4	0.024	0.2	4	20	80	5	3	2
	5	0.02	0.2	4	11	90	3	5	3
	6	0.058	0.6	6	19	95	7	2	5
	7	0.061	0.6	4	6	70	8	2	4
	8	0.024	0.3	4	20	85	3	2	6
	9	0.068	0.7	3	12	105	7	3	2
	10	0.073	0.5	5	14	105	8	2	7

Table A2. *Cont.*

Subsystem	Component Type	λ_{0ij}	α_{1ij}	α_{2ij}	t_{1ij}	t_{2ij}	k_{ij}	c_{ij}	w_{ij}
$i = 9$	1	0.049	0.7	2	8	85	5	5	7
	2	0.068	0.6	6	11	85	8	2	7
	3	0.024	0.6	6	15	70	4	3	5
	4	0.019	0.2	3	10	90	2	3	7
	5	0.024	0.7	6	10	80	3	5	5
	6	0.049	0.7	3	13	75	6	3	5
	7	0.035	0.2	3	20	110	6	5	5
	8	0.025	0.8	3	7	80	3	3	6
	9	0.049	0.8	4	5	75	6	2	8
	10	0.039	0.2	2	14	90	5	4	5
$i = 10$	1	0.028	0.1	5	20	90	7	4	2
	2	0.048	0.4	2	5	90	5	4	2
	3	0.019	0.2	2	20	105	3	3	4
	4	0.015	0.2	4	12	70	2	3	4
	5	0.018	0.3	4	6	85	2	5	5
	6	0.019	0.8	6	14	95	2	2	3
	7	0.058	0.3	6	15	80	8	3	8
	8	0.016	0.1	4	6	110	2	4	4
	9	0.051	0.5	6	11	70	9	2	6
	10	0.065	0.3	5	10	100	8	2	3
$i = 11$	1	0.033	0.6	5	14	80	4	3	3
	2	0.027	0.3	2	5	85	3	3	7
	3	0.035	0.4	4	15	90	5	2	8
	4	0.009	0.1	2	10	110	2	5	8
	5	0.038	0.2	6	17	85	6	2	3
	6	0.035	0.3	3	20	110	5	3	3
	7	0.029	0.1	6	11	75	6	2	3
	8	0.022	0.6	3	10	70	3	2	3
	9	0.058	0.6	4	18	90	7	2	8
	10	0.062	0.2	3	11	90	8	2	2

Table A2. *Cont.*

Subsystem	Component Type	λ_{0ij}	α_{1ij}	α_{2ij}	t_{1ij}	t_{2ij}	k_{ij}	c_{ij}	w_{ij}
$i = 12$	1	0.022	0.6	4	14	105	2	5	6
	2	0.041	0.2	5	18	85	7	2	6
	3	0.037	0.5	6	6	95	4	3	4
	4	0.023	0.5	6	6	105	2	3	4
	5	0.032	0.6	6	18	90	4	2	4
	6	0.059	0.7	3	6	80	7	5	3
	7	0.038	0.3	3	20	80	6	5	4
	8	0.054	0.4	4	20	90	8	2	7
	9	0.057	0.7	4	5	100	5	2	3
	10	0.015	0.5	2	8	85	2	3	8
$i = 13$	1	0.047	0.8	3	11	95	5	2	7
	2	0.019	0.5	3	12	75	2	5	5
	3	0.053	0.3	3	10	80	7	4	8
	4	0.039	0.6	2	11	70	4	5	3
	5	0.019	0.8	5	17	85	2	3	2
	6	0.048	0.4	2	11	90	6	2	6
	7	0.023	0.1	3	7	90	3	4	8
	8	0.034	0.1	4	6	80	5	3	7
	9	0.072	0.3	3	11	85	9	2	8
	10	0.051	0.8	2	20	85	5	4	3
$i = 14$	1	0.019	0.5	3	16	75	2	5	5
	2	0.038	0.6	2	7	85	4	5	7
	3	0.045	0.3	5	19	85	7	5	6
	4	0.033	0.2	6	9	95	4	2	7
	5	0.045	0.4	3	12	105	6	4	2
	6	0.024	0.2	3	9	80	3	4	5
	7	0.023	0.8	3	8	110	3	2	4
	8	0.048	0.3	3	14	90	7	3	3
	9	0.024	0.8	3	14	85	3	5	8
	10	0.041	0.2	2	6	70	6	4	7

Table A2. *Cont.*

Subsystem	Component Type	λ_{0ij}	α_{1ij}	α_{2ij}	t_{1ij}	t_{2ij}	k_{ij}	c_{ij}	w_{ij}
$i = 15$	1	0.021	0.8	4	14	75	2	4	2
	2	0.041	0.4	7	11	90	5	5	3
	3	0.043	0.5	3	6	70	5	4	8
	4	0.025	0.6	4	16	110	3	3	3
	5	0.012	0.2	2	18	95	2	5	6
	6	0.039	0.8	4	7	75	4	5	4
	7	0.047	0.3	5	9	90	5	2	6
	8	0.013	0.2	6	14	95	2	2	6
	9	0.015	0.6	4	18	90	2	4	4
	10	0.044	0.2	6	16	75	8	2	3

Table A3. Component data for the third example (pharmaceutical plant).

	Component Type 1								Component Type 2								Component Type 3								Component Type 4							
i	$\lambda_{0ij} \times 10^5$	α_{1ij}	α_{2ij}	t_{1ij}	t_{2ij}	k_{ij}	c_{ij}	w_{ij}	$\lambda_{0ij} \times 10^5$	α_{1ij}	α_{2ij}	t_{1ij}	t_{2ij}	k_{ij}	c_{ij}	w_{ij}	$\lambda_{0ij} \times 10^5$	α_{1ij}	α_{2ij}	t_{1ij}	t_{2ij}	k_{ij}	c_{ij}	w_{ij}	$\lambda_{0ij} \times 10^5$	α_{1ij}	α_{2ij}	t_{1ij}	t_{2ij}	k_{ij}	c_{ij}	w_{ij}
1	128	0.5	5	200	880	3	16	26	176	0.8	2	50	840	3	20	19	276	0.6	3	140	860	5	18	17	607	0.9	4	190	1060	8	21	24
2	195	0.7	6	120	790	4	14	20	78	0.3	4	90	870	2	21	23	92	0.7	2	100	1000	2	13	26	430	0.1	4	80	1010	9	19	18
3	419	0.6	3	220	1050	7	20	24	473	0.8	5	150	1090	7	15	17	368	0.2	6	80	720	8	16	26	455	0.8	2	240	1010	7	17	27
4	480	0.5	4	90	850	7	19	22	388	0.7	6	120	990	6	17	24	394	0.9	7	80	750	8	20	17	159	0.1	8	90	780	5	19	21
5	495	0.7	6	80	820	8	14	19	173	0.5	7	70	990	3	18	22	97	0.5	7	150	900	2	19	19	88	0.9	5	190	840	2	18	24
6	101	0.2	6	150	850	3	15	23	106	0.1	2	210	940	4	13	26	194	0.2	4	70	980	4	20	19	68	0.2	6	210	930	2	14	22
7	68	0.3	7	180	830	2	14	26	245	0.5	7	160	800	5	15	18	241	0.2	8	240	760	8	21	23	455	0.9	3	100	820	7	14	22
8	194	0.2	4	220	1070	5	18	23	189	0.2	5	220	920	5	14	26	205	0.3	4	70	870	4	16	20	190	0.3	4	90	1100	4	20	25
9	265	0.8	6	50	780	5	21	27	415	0.8	5	240	830	7	15	19	293	0.7	3	210	1000	5	16	26	429	0.3	5	250	820	9	20	18
10	99	0.1	2	170	1050	4	14	18	597	0.6	4	90	840	9	15	25	213	0.7	6	150	910	4	21	23	452	0.2	4	120	990	8	18	25

Table A4. Values of C, W, $n_{max,i}$ and $\rho_i(t)$ for the examples.

	C	W	$n_{max,i}$	ρ_i (t)
First example	50	70	6	0.99
Second example	310	400	6	0.99
Third example	480	519	6	0.99

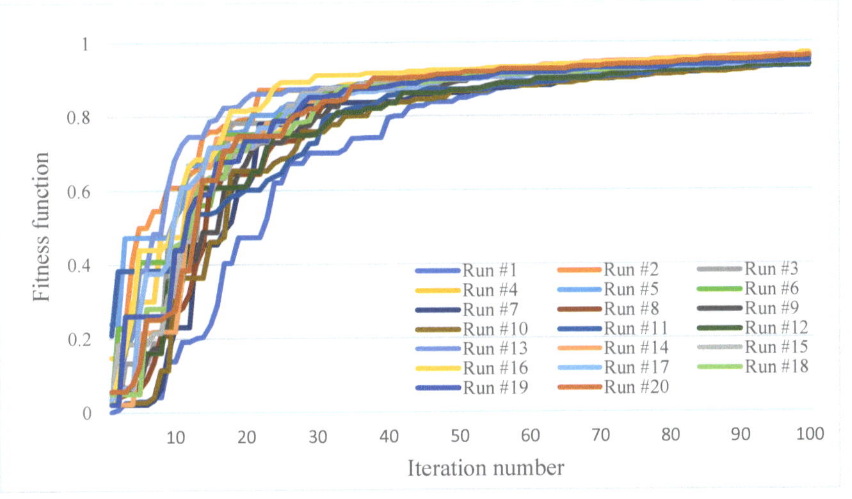

Figure A2. Trend of GA towards the optimal solution in 20 different trials; second example.

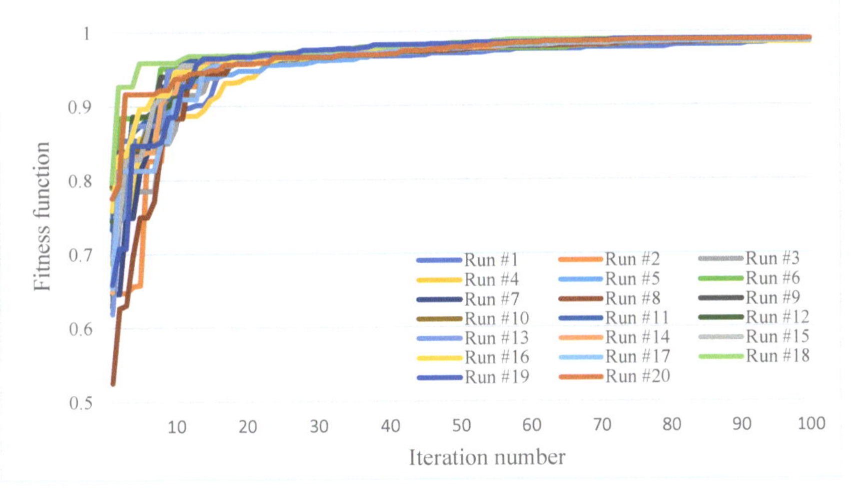

Figure A3. Trend of GA towards the optimal solution in 20 different trials; third example.

Table A5. Details of the optimal solutions obtained by different trials; first example.

Iteration Number	System Characteristics		
	Reliability	Cost	Weight
1	0.9612	45	68
2	0.9541	37	70
3	0.9284	45	70
4	0.9578	39	68
5	0.9692	48	69
6	0.9702	45	70
7	0.9619	49	66
8	0.9662	46	68
9	0.9695	45	70
10	0.9652	50	70
11	0.9673	49	68
12	0.9669	49	68
13	0.9376	39	68
14	0.9689	48	69
15	0.934	47	66
16	0.9624	38	68
17	0.9549	47	65
18	0.9577	45	70
19	0.9315	46	69
20	0.9562	46	68

Table A6. Details of the optimal solutions obtained by different trials; second example.

Iteration Number	System Characteristics		
	Reliability	Cost	Weight
1	0.9626	200	265
2	0.9506	306	367
3	0.9614	259	302
4	0.9557	251	380
5	0.9425	259	352
6	0.9584	250	337
7	0.9397	271	372
8	0.9574	303	319
9	0.9584	249	339
10	0.9351	242	325
11	0.9337	235	311
12	0.9384	237	298
13	0.9565	219	393
14	0.9643	248	333
15	0.9488	236	285
16	0.9707	272	282
17	0.946	240	323
18	0.9524	277	384
19	0.9493	302	384
20	0.9628	260	343

Table A7. Details of the optimal solutions obtained by different trials; third example.

Iteration Number	System Characteristics		
	Reliability	Cost	Weight
1	0.9866	458	519
2	0.9865	448	518
3	0.9881	466	517
4	0.9846	478	518
5	0.9896	454	514
6	0.9894	453	516
7	0.9889	469	513
8	0.9881	469	519
9	0.9886	464	509
10	0.9894	453	516
11	0.9889	469	513
12	0.9869	471	512
13	0.9864	429	518
14	0.9896	454	514
15	0.9868	451	519
16	0.9891	468	510
17	0.9869	424	509
18	0.9896	467	517
19	0.9896	461	519
20	0.9894	453	516

References

1. Ardakan, M.A.; Hamadani, A.Z. Reliability optimization of series–parallel systems with mixed redundancy strategy in subsystems. *Reliab. Eng. Syst. Saf.* **2014**, *130*, 132–139. [CrossRef]
2. Ebeling, C.E. *An Introduction to Reliability and Maintainability Engineering*; Tata McGraw-Hill Education: New York, NY, USA, 2004.
3. Hsieh, T.J. Hierarchical redundancy allocation for multi-level reliability systems employing a bacterial-inspired evolutionary algorithm. *Inf. Sci.* **2014**, *288*, 174–193. [CrossRef]
4. Abouei Ardakan, M.; Mirzaei, Z.; Zeinal Hamadani, A.; Elsayed, E.A. Reliability Optimization by Considering Time-Dependent Reliability for Components. *Qual. Reliab. Eng. Int.* **2017**, *33*, 1641–1654. [CrossRef]
5. Gupta, R.K.; Bhunia, A.K.; Roy, D. A GA based penalty function technique for solving constrained redundancy allocation problem of series system with interval valued reliability of components. *J. Comput. Appl. Math.* **2009**, *232*, 275–284. [CrossRef]
6. Mellal, M.A.; Zio, E. System reliability-redundancy allocation by evolutionary computation. In Proceedings of the 2017 2nd International Conference on System Reliability and Safety (ICSRS) 2017, Milan, Italy, 20–22 December 2017; pp. 15–19.
7. Coit, D.W. Cold-standby redundancy optimization for nonrepairable systems. *Iie Trans.* **2001**, *33*, 471–478. [CrossRef]
8. Zhao, J.; Zeng, S.; Guo, J.; Yang, C. Redundancy allocation with non-identical component and uncertainty. In Proceedings of the 2015 Annual Reliability and Maintainability Symposium (RAMS) 2015, Palm Harbor, FL, USA, 26–29 January 2015; pp. 1–6.
9. Liang, Y.C.; Smith, A.E. An ant colony optimization algorithm for the redundancy allocation problem (RAP). *IEEE Trans. Reliab.* **2004**, *53*, 417–423. [CrossRef]
10. Onishi, J.; Kimura, S.; James, R.J.; Nakagawa, Y. Solving the redundancy allocation problem with a mix of components using the improved surrogate constraint method. *IEEE Trans. Reliab.* **2007**, *56*, 94–101. [CrossRef]
11. Caserta, M.; Voß, S. An exact algorithm for the reliability redundancy allocation problem. *Eur. J. Oper. Res.* **2015**, *244*, 110–116. [CrossRef]
12. Ramirez-Marquez, J.E.; Coit, D.W. A heuristic for solving the redundancy allocation problem for multi-state series-parallel systems. *Reliab. Eng. Syst. Saf.* **2004**, *83*, 341–349. [CrossRef]
13. Sharma, V.K.; Agarwal, M.; Sen, K. Reliability evaluation and optimal design in heterogeneous multi-state series-parallel systems. *Inf. Sci.* **2011**, *181*, 362–378. [CrossRef]
14. Lai, C.M.; Yeh, W.C. Two-stage simplified swarm optimization for the redundancy allocation problem in a multi-state bridge system. *Reliab. Eng. Syst. Saf.* **2016**, *156*, 148–158. [CrossRef]
15. Li, Y.; Zio, E. A quantum-inspired evolutionary approach for non-homogeneous redundancy allocation in series-parallel multi-state systems. In Proceedings of the 2014 10th International Conference on Reliability, Maintainability and Safety (ICRMS) 2014, Guangzhou, China, 6–8 August 2014; pp. 526–532.
16. Wang, Y.; Li, L. A PSO algorithm for constrained redundancy allocation in multi-state systems with bridge topology. *Comput. Ind. Eng.* **2014**, *68*, 13–22. [CrossRef]
17. Hsieh, T.J.; Yeh, W.C. Penalty guided bees search for redundancy allocation problems with a mix of components in series–parallel systems. *Comput. Oper. Res.* **2012**, *39*, 2688–2704. [CrossRef]

18. Garg, H.; Rani, M.; Sharma, S.P.; Vishwakarma, Y. Bi-objective optimization of the reliability-redundancy allocation problem for series-parallel system. *J. Manuf. Syst.* **2014**, *33*, 335–347. [CrossRef]
19. Zhang, E.; Chen, Q. Multi-objective reliability redundancy allocation in an interval environment using particle swarm optimization. *Reliab. Eng. Syst. Saf.* **2016**, *145*, 83–92. [CrossRef]
20. Guo, J.; Wang, Z.; Zheng, M.; Wang, Y. Uncertain multiobjective redundancy allocation problem of repairable systems based on artificial bee colony algorithm. *Chin. J. Aeronaut.* **2014**, *27*, 1477–1487. [CrossRef]
21. Kulturel-Konak, S.; Coit, D.W. Determination of Pruned Pareto sets for the multi-objective system redundancy allocation problem. In Proceedings of the 2007 IEEE Symposium on Computational Intelligence in Multi-Criteria Decision-Making 2007, Honolulu, HI, USA, 1–5 April 2007; pp. 390–394.
22. Mariano, C.H.; Kuri-Morales, A.F. Complex componential approach for redundancy allocation problem solved by simulation-optimization framework. *J. Intell. Manuf.* **2014**, *25*, 661–680. [CrossRef]
23. Cao, D.; Murat, A.; Chinnam, R.B. Efficient exact optimization of multi-objective redundancy allocation problems in series-parallel systems. *Reliab. Eng. Syst. Saf.* **2013**, *111*, 154–163. [CrossRef]
24. Sun, M.X.; Li, Y.F.; Zio, E. On the optimal redundancy allocation for multi-state series–parallel systems under epistemic uncertainty. *Reliab. Eng. Syst. Saf.* **2019**, *192*, 106019. [CrossRef]
25. Valaei, M.R.; Behnamian, J. Allocation and sequencing in 1-out-of-N heterogeneous cold-standby systems: Multi-objective harmony search with dynamic parameters tuning. *Reliab. Eng. Syst. Saf.* **2017**, *157*, 78–86. [CrossRef]
26. Bei, X.; Chatwattanasiri, N.; Coit, D.W.; Zhu, X. Combined redundancy allocation and maintenance planning using a two-stage stochastic programming model for multiple component systems. *IEEE Trans. Reliab.* **2017**, *66*, 950–962. [CrossRef]
27. Dobani, E.R.; Ardakan, M.A.; Davari-Ardakani, H.; Juybari, M.N. RRAP-CM: A new reliability-redundancy allocation problem with heterogeneous components. *Reliab. Eng. Syst. Saf.* **2019**, *191*, 106563. [CrossRef]
28. Hsieh, T.J. Component mixing with a cold standby strategy for the redundancy allocation problem. *Reliab. Eng. Syst. Saf.* **2021**, *206*, 107290. [CrossRef]
29. Bhattacharyee, N.; Kumar, N.; Mahato, S.K.; Bhunia, A.K. Development of a blended particle swarm optimization to optimize mission design life of a series–parallel reliable system with time dependent component reliabilities in imprecise environments. *Soft Comput.* **2021**, *25*, 11745–11761. [CrossRef]
30. Wang, W.; Wu, Z.; Xiong, J.; Xu, Y. Redundancy optimization of cold-standby systems under periodic inspection and maintenance. *Reliab. Eng. Syst. Saf.* **2018**, *180*, 394–402. [CrossRef]
31. Mellal, M.A.; Zio, E. System reliability-redundancy optimization with cold-standby strategy by an enhanced nest cuckoo optimization algorithm. *Reliab. Eng. Syst. Saf.* **2020**, *201*, 106973. [CrossRef]
32. Yeh, W.C.; Su, Y.Z.; Gao, X.Z.; Hu, C.F.; Wang, J.; Huang, C.L. Simplified swarm optimization for bi-objection active reliability redundancy allocation problems. *Appl. Soft Comput.* **2021**, *106*, 107321. [CrossRef]
33. Yeh, W.C. BAT-based algorithm for finding all Pareto solutions of the series-parallel redundancy allocation problem with mixed components. *Reliab. Eng. Syst. Saf.* **2022**, *228*, 108795. [CrossRef]
34. Kundu, P. A multi-objective reliability-redundancy allocation problem with active redundancy and interval type-2 fuzzy parameters. *Oper. Res.* **2021**, *21*, 2433–2458. [CrossRef]
35. Coit, D.W. Maximization of system reliability with a choice of redundancy strategies. *IIE Trans.* **2003**, *35*, 535–543. [CrossRef]
36. Kong, X.; Gao, L.; Ouyang, H.; Li, S. Solving the redundancy allocation problem with multiple strategy choices using a new simplified particle swarm optimization. *Reliab. Eng. Syst. Saf.* **2015**, *144*, 147–158. [CrossRef]
37. Tavakkoli-Moghaddam, R.; Safari, J. A new mathematical model for a redundancy allocation problem with mixing components redundant and choice of redundancy strategies. *Appl. Math. Sci.* **2007**, *45*, 2221–2230.
38. Qiu, X.; Ali, S.; Yue, T.; Zhang, L. Reliability-redundancy-location allocation with maximum reliability and minimum cost using search techniques. *Inf. Softw. Technol.* **2017**, *82*, 36–54. [CrossRef]
39. Kim, H. Optimal reliability design of a system with k-out-of-n subsystems considering redundancy strategies. *Reliab. Eng. Syst. Saf.* **2017**, *167*, 572–582. [CrossRef]
40. Kim, H.; Kim, P. Reliability–redundancy allocation problem considering optimal redundancy strategy using parallel genetic algorithm. *Reliab. Eng. Syst. Saf.* **2017**, *159*, 153–160. [CrossRef]
41. Kim, H. Parallel genetic algorithm with a knowledge base for a redundancy allocation problem considering the sequence of heterogeneous components. *Expert Syst. Appl.* **2018**, *113*, 328–338. [CrossRef]
42. Wang, W.; Lin, M.; Fu, Y.; Luo, X.; Chen, H. Multi-objective optimization of reliability-redundancy allocation problem for multi-type production systems considering redundancy strategies. *Reliab. Eng. Syst. Saf.* **2020**, *193*, 106681. [CrossRef]
43. Li, X.Y.; Li, Y.F.; Huang, H.Z. Redundancy allocation problem of phased-mission system with non-exponential components and mixed redundancy strategy. *Reliab. Eng. Syst. Saf.* **2020**, *199*, 106903. [CrossRef]
44. Gholinezhad, H.; Hamadani, A.Z. A new model for the redundancy allocation problem with component mixing and mixed redundancy strategy. *Reliab. Eng. Syst. Saf.* **2017**, *164*, 66–73. [CrossRef]
45. Ardakan, M.A.; Hamadani, A.Z.; Alinaghian, M. Optimizing bi-objective redundancy allocation problem with a mixed redundancy strategy. *ISA transactions* **2015**, *55*, 116–128. [CrossRef] [PubMed]
46. Peiravi, A.; Karbasian, M.; Ardakan, M.A.; Coit, D.W. Reliability optimization of series-parallel systems with K-mixed redundancy strategy. *Reliab. Eng. Syst. Saf.* **2019**, *183*, 17–28. [CrossRef]

47. Peiravi, A.; Ardakan, M.A.; Zio, E. A new Markov-based model for reliability optimization problems with mixed redundancy strategy. *Reliab. Eng. Syst. Saf.* **2020**, *201*, 106987. [CrossRef]
48. Ouyang, Z.; Liu, Y.; Ruan, S.J.; Jiang, T. An improved particle swarm optimization algorithm for reliability-redundancy allocation problem with mixed redundancy strategy and heterogeneous components. *Reliab. Eng. Syst. Saf.* **2019**, *181*, 62–74. [CrossRef]
49. Reihaneh, M.; Ardakan, M.A.; Eskandarpour, M. An exact algorithm for the redundancy allocation problem with heterogeneous components under the mixed redundancy strategy. *Eur. J. Oper. Res.* **2022**, *297*, 1112–1125. [CrossRef]
50. Azizi, S.; Mohammadi, M. Strategy selection for multi-objective redundancy allocation problem in a k-out-of-n system considering the mean time to failure. *Opsearch* **2023**, *60*, 1021–1044. [CrossRef]
51. Kim, H. Parallel Genetic Algorithm with Knowledge Archives for the Redundancy Allocation Problem in a Mixed Redundant System. Available online: https://papers.ssrn.com/sol3/papers.cfm?abstract_id=4283392 (accessed on 22 November 2022).
52. Zhang, J.; Lv, H.; Hou, J. A novel general model for RAP and RRAP optimization of k-out-of-n: G systems with mixed redundancy strategy. *Reliab. Eng. Syst. Saf.* **2023**, *229*, 108843. [CrossRef]
53. Yeh, W.C. Solving cold-standby reliability redundancy allocation problems using a new swarm intelligence algorithm. *Appl. Soft Comput.* **2019**, *83*, 105582. [CrossRef]
54. Yeh, W.C. A novel boundary swarm optimization method for reliability redundancy allocation problems. *Reliab. Eng. Syst. Saf.* **2019**, *192*, 106060. [CrossRef]
55. Huang, X.; Coolen, F.P.; Coolen-Maturi, T. A heuristic survival signature based approach for reliability-redundancy allocation. *Reliab. Eng. Syst. Saf.* **2019**, *185*, 511–517. [CrossRef]
56. Lins, I.D.; Droguett, E.L. Redundancy allocation problems considering systems with imperfect repairs using multi-objective genetic algorithms and discrete event simulation. *Simul. Model. Pract. Theory* **2011**, *19*, 362–381. [CrossRef]
57. Marseguerra, M.; Zio, E.; Podofillini, L.; Coit, D.W. Optimal design of reliable network systems in presence of uncertainty. *IEEE Trans. Reliab.* **2005**, *54*, 243–253. [CrossRef]
58. Monalisa, P.; Satchidananda, D.; Kumar, J.A. Multi-objective artificial bee colony algorithm in redundancy allocation problem. *Int. J. Adv. Intell. Paradig.* **2023**, *25*, 24–50. [CrossRef]
59. Maji, A.; Duary, A.; Bhunia, A.K.; Mondal, S.K. *A Redundancy Allocation Problem for a Series-Parallel System with Multiple Choice Technologies Considering Fuzzy Sense with Ambiguity and Vagueness*; Springer: Berlin/Heidelberg, Germany, 2022. [CrossRef]
60. Yeh, C.T.; Fiondella, L. Optimal redundancy allocation to maximize multi-state computer network reliability subject to correlated failures. *Reliab. Eng. Syst. Saf.* **2017**, *166*, 138–150. [CrossRef]
61. Li, Y.F.; Zhang, H. The methods for exactly solving redundancy allocation optimization for multi-state series–parallel systems. *Reliab. Eng. Syst. Saf.* **2022**, *221*, 108340. [CrossRef]
62. Xu, Y.; Pi, D.; Yang, S.; Chen, Y. A novel discrete bat algorithm for heterogeneous redundancy allocation of multi-state systems subject to probabilistic common-cause failure. *Reliab. Eng. Syst. Saf.* **2021**, *208*, 107338. [CrossRef]
63. Li, J.; Wang, G.; Zhou, H.; Chen, H. Redundancy allocation optimization for multi-state system with hierarchical performance requirements. *Proc. Inst. Mech. Eng. Part O J. Risk Reliab.* **2022**. [CrossRef]
64. Blokus, A. *Multistate System Reliability with Dependencies*; Academic Press: Cambridge, MA, USA, 2020.
65. Yingkui, G.; Jing, L. Multi-state system reliability: A new and systematic review. *Procedia Eng.* **2012**, *29*, 531–536. [CrossRef]
66. Paramanik, R.; Mahato, S.K.; Bhattacharyee, N. Optimization of Redundancy Allocation Problem Using Quantum Particle Swarm Optimization Algorithm Under Uncertain Environment. In *Advances in Reliability, Failure and Risk Analysis*; Springer Nature: Singapore, 2023; pp. 177–197.
67. Coit, D.W.; Zio, E. The evolution of system reliability optimization. *Reliab. Eng. Syst. Saf.* **2019**, *192*, 106259. [CrossRef]
68. Chern, M.S. On the computational complexity of reliability redundancy allocation in a series system. *Oper. Res. Lett.* **1992**, *11*, 309–315. [CrossRef]
69. Holland, J.H. *Adaptation in Natural and Artificial Systems: An Introductory Analysis with Applications to Biology, Control, and Artificial Intelligence*; MIT Press: Cambridge, MA, USA, 1992.
70. Scheidegger, A.; Leitao, J.P.; Scholten, L. Statistical failure models for water distribution pipes–A review from a unified perspective. *Water Res.* **2015**, *83*, 237–247. [CrossRef]
71. Romaniuk, M.; Hryniewicz, O. Estimation of maintenance costs of a pipeline for a U-shaped hazard rate function in the imprecise setting. *Eksploat. I Niezawodn.* **2020**, *22*, 352–362. [CrossRef]
72. Al-nefaie, A.H.; Ragab, I.E. A Novel Lifetime Model with A Bathtub-Shaped Hazard Rate: Properties & Applications. *J. Appl. Sci. Eng.* **2022**, *26*, 413–421.
73. Bai, B.; Li, Z.; Wu, Q.; Zhou, C.; Zhang, J. Fault data screening and failure rate prediction framework-based bathtub curve on industrial robots. *Ind. Robot. Int. J. Robot. Res. Appl.* **2020**, *47*, 867–880. [CrossRef]
74. Ahsan, S.; Lemma, T.A.; Gebremariam, M.A. Reliability analysis of gas turbine engine by means of bathtub-shaped failure rate distribution. *Process Saf. Prog.* **2020**, *39*, e12115. [CrossRef]

mathematics MDPI

Article

Gibbs Distribution and the Repairman Problem

Hassan Chetouani and Nikolaos Limnios *

Applied Mathematics Laboratory, Université de Technologie de Compiègne, Sorbonne University Alliance, 60203 Compiègne, France; hchetoua@gmail.com
* Correspondence: nlimnios@utc.fr

Abstract: In this paper, we obtain weak convergence results for a family of Gibbs measures depending on the parameter $\theta > 0$ in the following form $dP_\theta(x) = Z_\theta \exp(-H_\theta(x)/\theta)dQ(x)$, where we show that the limit distribution is concentrated in the set of the global minima of the limit Gibbs potential. We also give an explicit calculus for the limit distribution. Here, we use the above as an alternative to Lyapunov's function or to direct methods for stationary probability convergence and apply it to the *repairman problem*. Finally, we illustrate this method with a numerical example.

Keywords: Gibbs measure; Gibbs potential; Laplace's method; weak convergence; repairman problem

MSC: 60E99; 60F99; 60G99; 60J99

1. Introduction

In this paper, we study the weak convergence of a family of Gibbs probability measures via Laplace's method ([1–4]). This method is interpreted as a weak convergence of probability measures, which was used by Hwang [5]. Yu. Kaniovski and G. Pflug [6] investigated Laplace's method to study the limit stationary distribution of the birth and death process.

In this paper, we suppose that the Gibbs potential H_θ depends on a parameter θ, and Q is a probability measure on the Euclidean space $\mathbb{R}^r$, such that it dominates P_θ for any θ. To show the tightness of the family

$$\frac{dP_\theta}{dQ}(x) = Z_\theta \exp(-H_\theta(x)/\theta), \quad \theta > 0, \tag{1}$$

we give some additional conditions for the probability measure Q and for the limit Gibbs potential H of H_θ when $\theta \to 0$. Under these conditions, the limit distribution P of P_θ, as $\theta \to 0$, is concentrated on the set of the global minima of the limit Gibbs potential, so we give an explicit calculus for the limit probability P. We use these results to prove that the stationary probability of the repairman problem converges to some probability measure as its state space goes to infinity.

The repairman problem was introduced very early in queuing theory, and in the 1960s, important results concerning stochastic approximations, in particular diffusion approximation, were given by Iglehart [7], and later on, more detailed results on averaging and diffusion approximation were given by Korolyuk [8], see also [4,9,10]. The repairman problem is as follows. Consider n identical devices working independently and simultaneously with lifetimes exponentially distributed with parameter $\lambda > 0$. It is also supposed that we have the possibility of repairing $r \le n$ failed devices at one time, where the service times are independent and exponentially distributed with mean value $1/\mu$. Suppose that we have a stock of m separate devices of the same type assumed as not broken while waiting for replacements. As soon as a device breaks down, we replace it with another identical

Citation: Chetouani, H.; Limnios, N. Gibbs Distribution and the Repairman Problem. *Mathematics* **2023**, *11*, 4120. https://doi.org/10.3390/math11194120

Academic Editors: Alexander Bochkov and Gurami Tsitsiashvili

Received: 4 September 2023
Revised: 26 September 2023
Accepted: 28 September 2023
Published: 29 September 2023

one [7,8,11]. This is a special case of the birth and death process, but with interesting features, see, e.g., [4,8,12,13].

The aim of this paper is to generalize the Hwang theorem [5] and apply it to obtain the weak convergence of the stationary distribution of the repairman problem. The method used here is that proposed by Kaniovski and Pflug in [6], where the stationary distribution is written in Gibbs form, and then we apply the weak convergence of the last stationary distribution.

Hence, we propose an alternative to Lyapunov's function ([14]) or to direct methods ([11]) for stationary probability convergence in a series scheme (i.e., a functional setting). For weak convergence, see, e.g., [1–3,15].

Section 2 presents the problem setting and notation. Section 3 presents the tightness of the family of probability measures $\{P_\theta, \theta > 0\}$, the limit probability concentration on the set of global minima of the limit Gibbs potential, and the main results and an explicit calculus of the limit distribution. Section 4 presents an application, where we study the stationary distribution of the repairman problem with limited service and spare devices [8], which has a Gibbs representation, using the results of Sections 3. It is proven that the limit stationary distribution is concentrated and uniformly distributed on the set of global minima of the limit Gibbs potential. Finally, in Section 5, we present some conclusions and perspectives.

2. Problem Statement and Existence of the Limit Probability

Let Q be a fixed probability measure on $(\mathbb{R}^r, \mathcal{B}_r)$, where $r \in \mathbb{N}^* = \mathbb{N}\backslash\{0\}$, and $\mathcal{B}_r$ is the Borel σ–algebra of the Euclidean space $\mathbb{R}^r$, with the scalar product denoted by $\langle \cdot , \cdot \rangle$ and the induced Euclidean norm denoted by $||\cdot||$. Let $(H_\theta(\cdot))_{\theta>0}$ be a family of real-valued functions defined on $\mathbb{R}^r$, and $H(\cdot)$ be a real-valued continuous function defined on $\mathbb{R}^r$ with a finite global minimum denoted by

$$H^* := \min_{y \in \mathbb{R}^r} H(y).$$

Define the set of points where the global minimum of $H(\cdot)$ is reached as

$$N = \{x \mid H(x) = H^*\},$$

and denote the neighborhood of the set N by

$$N^\delta := \{x \mid H(x) - H^* < \delta\}, \quad \text{for all} \quad \delta > 0.$$

In what follows, we suppose that for any $\delta > 0$, we have

$$Q\{x \,|\, H(x) - H^* < \delta\} > 0, \tag{2}$$

and H is continuous on some neighborhood of its global minimum.

The space of integrable and bounded functions is equipped with the uniform norm

$$||H_\theta - H||_\infty := \sup_x |H_\theta(x) - H(x)|.$$

Laplace's method is used here to define a probability measure P on the set of the global minima of the limit Gibbs potential as weak convergence of the family $(P_\theta, \theta > 0)$, defined as

$$\frac{dP_\theta}{dQ}(x) = Z_\theta \, \exp\left(-\frac{H_\theta(x)}{\theta}\right), \quad \theta > 0,$$

where

$$Z_\theta = \left[\int \exp\left(-\frac{H_\theta(x)}{\theta}\right) dQ(x)\right]^{-1},$$

as $\theta \to 0$. If P_θ converges weakly to P, then P is the searched probability measure. We investigate this method with some additional conditions to show that the limit probability P exists and is explicitly defined on the set of global minima of the limit Gibbs potential.

Subsequently, when the integration domain is not indicated, it is supposed to be the whole space $\mathbb{R}^r$.

Let us suppose here that H_θ converges uniformly to H, i.e.,

$$\lim_{\theta \to 0} \|H_\theta - H\|_\infty = 0. \tag{3}$$

We have the following result, similar to Proposition 1 in [5], but here in a more general setting where the Gibbs potential depends on a parameter.

Proposition 1. *If P_θ is tight, then H has a global finite minimum.*

Proof. We provide the proof by contradiction. We suppose that the minimum of H exists and it is equal to 0 ($H^* = 0$), and that P_θ is not tight, i.e., there exists $\varepsilon_0 > 0$, and $\{P_{\theta_k},\ k = 0, 1, 2, \ldots\}$, $\theta_k \to 0$ as $k \to \infty$, such that for any compact set K in $\mathbb{R}^r$,

$$P_{\theta_k}(K^c) > \varepsilon_0, \quad \text{for all } k \geq 0. \tag{4}$$

Since H is continuous on a neighborhood of its global minimum, then there exists $\delta_0 > 0$ such that for all $\delta \in (0, \delta_0)$, $K = [H \leq \delta] := \{x |\ H(x) \leq \delta\}$ is a compact set. Hence,

$$\begin{aligned} P_{\theta_k}(K^c) &= P_{\theta_k}[H > \delta] \\ &= Z_{\theta_k} \int_{[H>\delta]} \exp\left(-\frac{H_{\theta_k}(x)}{\theta_k}\right) dQ(x) \\ &< Z_{\theta_k} \exp\left(-\frac{\delta}{\theta_k}\right) \int_{[H>\delta]} \exp\left(-\frac{H_{\theta_k}(x) - H(x)}{\theta_k}\right) dQ(x). \end{aligned}$$

The uniform convergence (3) implies that for $\delta > 0$, there exists k_0 such that for all $k \geq k_0$

$$-\frac{\delta}{2} < H_{\theta_k}(\cdot) - H(\cdot) < \frac{\delta}{2}.$$

Then, we obtain the following inequality

$$\begin{aligned} P_{\theta_k}[H > \delta] &< Z_{\theta_k} \exp\left(-\frac{\delta}{\theta_k}\right) \exp\left(\frac{\delta}{2\theta_k}\right) \\ &< \exp\left(-\frac{\delta}{2\theta_k}\right) \left[\int_{N^{\frac{\delta}{3}}} \exp(-H_{\theta_k}(x)/\theta_k) dQ(x)\right]^{-1} \\ &< \exp\left(-\frac{\delta}{2\theta_k}\right) \left[\int_{N^{\frac{\delta}{3}}} \exp\left(-\frac{H_{\theta_k}(x) - H(x)}{\theta_k}\right) \exp\left(-\frac{H(x)}{\theta_k}\right) dQ(x)\right]^{-1} \\ &< \exp\left(-\frac{\delta}{2\theta_k}\right) \left[\exp\left(-\frac{\delta}{3\theta_k}\right) \int_{N^{\frac{\delta}{3}}} \exp\left(-\frac{H_{\theta_k}(x) - H(x)}{\theta_k}\right) dQ(x)\right]^{-1}. \end{aligned}$$

Again, assumption (3) implies that, for $\delta > 0$, there exists k_1 such that, for all $k \geq k_1$,

$$-\frac{\delta}{8} < H_{\theta_k}(\cdot) - H(\cdot) < \frac{\delta}{8},$$

Finally, we obtain

$$P_{\theta_k}[H > \delta] < \frac{\exp\left(-\frac{\delta}{24\theta_k}\right)}{Q\left(N^{\frac{\delta}{3}}\right)},$$

By assumption (2), we have

$$Q\left(N^{\frac{\delta}{3}}\right) > 0.$$

Therefore, $P_{\theta_k}(K^c) = P_{\theta_k}[H > \delta]$ converges to 0 as $k \to +\infty$. Hence, a contradiction arises with hypothesis (4). □

Remark 1. *It is worth noting here that we cannot conclude from the above first inequality that* $Z_{\theta_k} \exp\left(-\frac{\delta}{2\theta_k}\right)$ *goes to zero, since* Z_{θ_k} *can go to infinity, as* θ_k *goes to zero.*

In the following, we suppose that the minimum H^* of the limit Gibbs potential H exists and is finite.

Proposition 2. *For all* $\varepsilon > 0$*, the following convergences hold*

$$P_\theta[|H_\theta - H^*| > \varepsilon] \to 0, \text{ and } P_\theta[H - H^* > \varepsilon] \to 0, \text{ as } \theta \to 0 \tag{5}$$

exponentially fast in $1/\theta$*.*

To prove Proposition 2, we need the following lemma:

Lemma 1. *For all* $\varepsilon > 0$*, there exists* θ_0 *such that for all* $\theta < \theta_0$

$$\left\{x \,|\, H(x) - H^* < \frac{\varepsilon}{2}\right\} \subset \{x \,|\, |H_\theta(x) - H^*| < \varepsilon\}.$$

Proof of Lemma 1. Let $x \in \left\{x \,|\, H(x) - H^* < \frac{\varepsilon}{2}\right\}$, then

$$|H_\theta(x) - H^*| \leq |H_\theta(x) - H(x)| + H(x) - H^*,$$

By the uniform convergence (3), there exists θ_0 such that for all $\theta \leq \theta_0$, we have

$$\|H_\theta - H\|_\infty < \frac{\varepsilon}{2},$$

Hence,

$$|H_\theta(x) - H^*| < \varepsilon,$$

i.e.,

$$\left\{x \,|\, H(x) - H^* < \frac{\varepsilon}{2}\right\} \subset \{x \,|\, |H_\theta(x) - H^*| < \varepsilon\}.$$

□

Proof of Proposition 2. Owing to Lemma 1, there exists θ_0 such that for all $\theta < \theta_0$,

$$\begin{aligned} P_\theta[|H_\theta - H^*| > \varepsilon] &\leq P_\theta[H - H^* > \varepsilon/2] \\ &\leq Z_\theta \int_{[H-H^*>\varepsilon/2]} \exp\left(-\frac{H_\theta(x)}{\theta}\right) dQ(x) \\ &\leq Z_\theta \exp\left(-\frac{H^*}{\theta}\right) \int_{[H-H^*>\varepsilon/2]} \exp\left(-\frac{H_\theta(x) - H(x)}{\theta}\right) \\ &\qquad \times \exp\left(-\frac{H(x) - H^*}{\theta}\right) dQ(x), \end{aligned}$$

By (3), there exists θ_0 such that for all $\theta < \theta_0$, we have

$$|H_\theta(\cdot) - H(\cdot)| < \frac{\varepsilon}{4},$$

Hence,

$$\begin{aligned}
P_\theta[H - H^* > \varepsilon/2] &\leq Z_\theta \exp\left(-\frac{H^*}{\theta}\right)\exp\left(\frac{-\varepsilon}{2\theta}\right)\exp\left(\frac{\varepsilon}{4\theta}\right) \\
&\leq \exp\left(\frac{-\varepsilon}{4\theta}\right)\left[\int_{[|H_\theta - H^*|<\frac{\varepsilon}{8}]} \exp\left(-\frac{H_\theta(x)-H^*}{\theta}\right) dQ(x)\right]^{-1} \\
&\leq \exp\left(\frac{-\varepsilon}{4\theta}\right)\exp\left(\frac{\varepsilon}{8\theta}\right) \\
&\leq \exp\left(\frac{-\varepsilon}{8\theta}\right).
\end{aligned}$$

The above inequality concludes the proof. □

Remark 2. *Proposition 2 means that the limit probability P is concentrated on the set of global minima of the limit Gibbs potential $H(\cdot)$.*

3. Characterization of the Limit Probability

The uniform convergence in (3) implies that for any $x,\ y\ \in \mathbb{R}^r$

$$H_\theta(x) - H_\theta(y) = H(x) - H(y) + \Delta_\theta(x,y),$$

with $\lim_{\theta\to 0} \Delta_\theta(x,y) = 0$.

In this section, we suppose that

$(A1)$ For all $x,\ y\ \in\ N$, $\lim_{\delta\to 0}\ \lim_{\|x-y\|<\delta}\ \Delta_\theta(x,y) = 0$.

$(A2)$ For all $x,\ y\ \in\ N$, $\dfrac{\Delta_\theta(x,y)}{\theta}$ is bounded as $\theta \to 0$.

3.1. Case When $Q(N) > 0$

Theorem 1. *If assumptions $(A1)$ and $(A2)$ are verified, then the limit probability P of P_θ as $\theta \to 0$ is concentrated on the set N and its density with respect to Q is*

$$f(x) = \begin{cases} \left[\int_N \exp(-\Delta(x,y))Q(dy)\right]^{-1}, & \text{if } x \in N, \\ 0, & \text{otherwise,} \end{cases}$$

where $\Delta(x,y) := \lim_{\theta\to 0} \frac{\Delta_\theta(x,y)}{\theta}$.

Proof. To prove Theorem 1, we must prove that the density

$$f_\theta(x) = Z_\theta\ \exp\left(-\frac{H_\theta(x)}{\theta}\right),$$

converged to $f(x)$.

For $x\ \in\ N$, we have

$$\begin{aligned}
f_\theta(x) &= \frac{\exp\left(-\frac{H_\theta(x)}{\theta}\right)}{\int \exp\left(-\frac{H_\theta(y)}{\theta}\right)Q(dy)} \\
&= \frac{1}{\int \exp\left(-\frac{H_\theta(y)-H_\theta(x)}{\theta}\right)Q(dy)}. \qquad (6)
\end{aligned}$$

The denominator in (6) can be written as follows:

$$\begin{aligned}\int \exp\Big(-\frac{H_\theta(y)-H_\theta(x)}{\theta}\Big)Q(dy) &= \int \exp\Big(-\frac{H(y)-H(x)+\Delta_\theta(y,x)}{\theta}\Big)Q(dy)\\ &= \int_N \exp\Big(-\frac{\Delta_\theta(y,x)}{\theta}\Big)Q(dy)\\ &+ \int_{N^c} \exp\Big(-\frac{H(y)-H(x)}{\theta}\Big)\times\\ &\qquad \exp\Big(-\frac{\Delta_\theta(y,x)}{\theta}\Big)Q(dy).\end{aligned}$$

Since $\frac{\Delta_\theta(x,y)}{\theta} \to \Delta(x,y)$ as $\theta \to 0$, then, by the dominated convergence theorem, we obtain

$$\int_N \exp\Big(-\frac{\Delta_\theta(y,x)}{\theta}\Big)Q(dy) \to \int_N \exp(-\Delta(y,x))Q(dy),$$

and

$$\int_{N^c} \exp\Big(-\frac{H(y)-H(x)}{\theta}\Big)\ \exp\Big(-\frac{\Delta_\theta(y,x)}{\theta}\Big)Q(dy) \to 0,$$

as $\theta \to 0$.

We can prove in the same way that for $x \notin N$, $f(x) = 0$. □

Corollary 1. *If $\Delta(\cdot,\cdot) = 0$, then the probability P is the uniform distribution on the set N, with density $1/Q(N)$.*

Example 1. *Let $Q(dx) = \frac{dx}{4}\mathbb{1}_{\{|x|\le 2\}}$ be a probability measure on $\mathbb{R}$ and*

$$H_\theta(x) = \begin{cases} x^2-\theta x, & 1<|x|<2,\\ 1-\theta x, & |x|\le 1,\\ 0, & \text{otherwise.}\end{cases}$$

We observe that H_θ converges uniformly to the function

$$H(x) = \begin{cases} x^2, & 1<|x|<2,\\ 1, & |x|\le 1,\\ 0, & \text{otherwise.}\end{cases}$$

The global minimum for $H(\cdot)$ is 1, and the set of its global minimum is

$$N = \{x \in \mathbb{R} \,|\, |x| \le 1\} = [-1;\ 1].$$

The Gibbs potential H is continuous on some neighborhood of the set of its global minima N and for all $x,\ y \in N$

$$\Delta(x,\ y) = -(x-y),$$

Then,

$$f(x) = \Big(\int_{-\infty}^{\infty} e^{x-y}\mathbb{1}_{\{|y|\le 1\}}\frac{dy}{4}\Big)^{-1}\mathbb{1}_{\{|x|\le 1\}} = \frac{4\,e^{-x}}{e-e^{-1}}\mathbb{1}_{\{|x|\le 1\}}.$$

3.2. Case When $Q(N) = 0$

In the sequel, we denote by $O(\theta)$, $\Delta(\theta,\delta)$ all quantities tending to 0 as $\theta \to 0$ or $\delta \to 0$, even if they are not equal, and by $o(\theta)$, all quantities such that $o(\theta)/\theta \to 0$ as $\theta \to 0$.

In what follows, we suppose that $N = \{x^1, x^2, \ldots, x^n, \ldots\}$ and there exists a continuous positive real function $\psi_i(\cdot)$ in the neighborhood of 0, with $\psi_i(0) = 0$, such that in the neighborhood of x^i,

$$H(x) = H^* + \psi_i(x - x^i).$$

The measure Q is also supposed to be absolutely continuous with respect to the Lebesgue measure, with continuous density f. We also define

$$P_\delta^{i,\theta} = P_\theta\left(V_i^\delta\right),$$

where $V_i^\delta = \{x \in \mathbb{R}^r \mid \|x - x^i\| < \delta\}$, and $V^\delta = \{x \in \mathbb{R}^r \mid \|x\| < \delta\}$.

Theorem 2. *Let assumptions $(A1)$ and $(A2)$ hold, and for all $i \in \mathbb{N}^*$, there exists $\alpha_i > 0$ such that*

$$\lim_{\delta \to 0} \sup_{\|x\| < \delta} \left| \frac{\psi_i(x)}{\|x\|^2} - \alpha_i \right| = 0,$$

Then, in the case of $f(x^i) \neq 0$, we obtain

$$\lim_{\theta \to 0} \lim_{\delta \to 0} \theta \log\left(\frac{P_\delta^{i,\theta}\ \alpha_i^{1/2}}{Z_\theta\ \exp(-H^*/\theta) f(x^i) \pi^{r/2} \theta^{1/2}} \right) = 0,$$

and, in the case of $f(x^i) = 0$, we obtain

$$P\left\{x^i\right\} = 0.$$

Proof. Let $V_i^\delta, i \in \mathbb{N}^*$ be open neighborhoods of $x^i, i \in \mathbb{N}^*$, with a radius less than $\delta > 0$. We have

$$\begin{aligned} P_\delta^{i,\theta} &= P_\theta\left(V_i^\delta\right) \\ &= Z_\theta \int_{V_i^\delta} \exp\left(-\frac{H_\theta(x)}{\theta}\right) f(x) dx \\ &= Z_\theta A_i(\theta, \delta), \end{aligned}$$

where

$$\begin{aligned} A_i(\theta, \delta) &= \int_{V_i^\delta} \exp\left(-\frac{H_\theta(x)}{\theta}\right) f(x) dx \\ &= \exp\left(-\frac{H^*}{\theta}\right) \int_{V_i^\delta} \exp\left(-\frac{H_\theta(x) - H^*}{\theta}\right) f(x) dx \\ &= \exp\left(-\frac{H^*}{\theta}\right) \int_{V_i^\delta} \exp\left(-\frac{H_\theta(x) - H(x)}{\theta}\right) \times \\ &\qquad \exp\left(-\frac{H(x) - H^*}{\theta}\right) f(x) dx \\ &= \exp\left(-\frac{H^*}{\theta}\right) \exp\left(-\frac{H_\theta(x^i) - H(x^i)}{\theta}\right) \times \\ &\qquad \int_{V_i^\delta} \exp\left(-\frac{\Delta_\theta(x, x^i)}{\theta}\right) \exp\left(-\frac{H(x) - H^*}{\theta}\right) f(x) dx. \end{aligned}$$

According to assumption $(A1)$, the following estimation holds

$$1 - O(\delta) \leq \exp\left(-\frac{\Delta_\theta(x, x^i)}{\theta}\right) \leq 1 + O(\delta).$$

It follows that

$$(1 - O(\delta))B_i(\theta,\delta) \leq A_i(\theta,\delta) \leq (1 + O(\delta))B_i(\theta,\delta),$$

where

$$B_i(\theta,\delta) = \exp\left(\frac{-H^*}{\theta}\right)\exp\left(-\frac{H_\theta(x^i) - H(x^i)}{\theta}\right)\int_{V_i^\delta}\exp\left(-\frac{H(x) - H^*}{\theta}\right)f(x)dx.$$

Define also

$$\begin{aligned} D_i(\theta,\delta) &= \int_{V_i^\delta}\exp\left(-\frac{H(x) - H^*}{\theta}\right)f(x)dx \\ &= \int_{V_i^\delta}\exp\left(-\frac{\psi_i(x - x^i)}{\theta}\right)f(x)dx. \end{aligned}$$

Now, setting $y = x - x^i$, $i \in \mathbb{N}^*$, we obtain

$$\begin{aligned} D_i(\theta,\delta) &= \int_{V^\delta}\exp\left(-\frac{\psi_i(y)}{\theta}\right)f(y + x^i)dy \\ &= \int_{V^\delta}\exp\left(-\frac{\alpha_i\,\|y\|^2 + o(\|y\|^2)}{\theta}\right)f(y + x^i)dy \end{aligned}$$

and

$$\left(1 - o(\delta^2)/\theta\right)F_i(\theta,\delta) \leq D_i(\theta,\delta) \leq \left(1 + o(\delta^2)/\theta\right)F_i(\theta,\delta),$$

where

$$F_i(\theta,\delta) = \int_{V^\delta}\exp\left(-\frac{\alpha_i\,\|y\|^2}{\theta}\right)f(y + x^i)dy.$$

We have

$$\begin{aligned}
F_i(\theta,\delta) &= \int_{V^\delta} \exp\Big(-\|y\|^2\Big) f(\sqrt{\frac{\theta}{\alpha_i}}y + x^i)\sqrt{\frac{\theta}{\alpha_i}}dy \\
&= \int_{V^{\frac{\delta}{\sqrt{\theta}}}} \exp\Big(-\|y\|^2\Big)\Big(f(x^i) + \Delta(\theta,y)\Big)\sqrt{\frac{\theta}{\alpha_i}}dy \\
&= \sqrt{\frac{\theta}{\alpha_i}} f(x^i) \int_{V^{\frac{\delta}{\sqrt{\theta}}}} \exp\Big(-\|y\|^2 dy\Big) + \sqrt{\frac{\theta}{\alpha_i}} \int_{V^{\frac{\delta}{\sqrt{\theta}}}} \exp\Big(-\|y\|^2\Big)\Delta(\theta,y)dy \\
&= \sqrt{\frac{\theta}{\alpha_i}} f(x^i) \int_{V^{\frac{\delta}{\sqrt{\theta}}}} \exp\Big(-\|y\|^2\Big)dy + \Delta(\theta,\delta) \\
&= \sqrt{\frac{\theta}{\alpha_i}} f(x^i) \int \exp\Big(-\|y\|^2\Big)dy(1 - O(\theta)) + \Delta(\theta,\delta) \\
&= \sqrt{\frac{\theta}{\alpha_i}} f(x^i) \Big(\int_{-\infty}^{+\infty} \exp\Big(-u^2\Big)du\Big)^r (1 - O(\theta)) + \Delta(\theta,\delta) \\
&= \sqrt{\frac{\theta}{\alpha_i}} f(x^i)\pi^{r/2}(1 - O(\theta)) + \Delta(\theta,\delta).
\end{aligned}$$

If $f(x^i) = 0$, then

$$\lim_{\delta\to 0}\lim_{\theta\to 0} P_\delta^{i,\theta} = 0,$$

and if $f(x^i) \neq 0$, then

$$\lim_{\delta\to 0}\lim_{\theta\to 0} \theta \log\left(\frac{P_\delta^{i,\theta}\ \alpha_i^{1/2}}{Z_\theta\ \exp(-H^*/\theta) f(x^i)\pi^{r/2}\theta^{1/2}}\right) = 0.$$

□

Corollary 2. *Under the assumption of Theorem 2, if for all $x^i \in N$, we have*

$$H_\theta\Big(x^i\Big) - H^* = a_i\theta + o(\theta), \tag{7}$$

where $a_i \in \mathbb{R}$, and if there exists $x^k \in N$ such that $f(x^k) \neq 0$, then

$$P\Big\{x^i\Big\} = \frac{f(x^i)\alpha_i^{-1/2}\exp(-a_i)}{\sum_{k=1}^{+\infty} f(x^k)\alpha_k^{-1/2}\exp(-a_k)}.$$

Proof. According to the proof of Theorem 2, we obtain

$$P_\theta\Big(V_i^\delta\Big) \asymp \pi^{r/2} Z_\theta \sqrt{\frac{\theta}{\alpha_i}} f(x^i)\exp\Big(\frac{-H^*}{\theta}\Big)\exp\Big(-\frac{H_\theta(x^i) - H(x^i)}{\theta}\Big), \tag{8}$$

and by (7),

$$\exp\Big(-\frac{H_\theta(x^i) - H(x^i)}{\theta}\Big) \asymp \exp(-a_i).$$

Then, we obtain that

$$\pi^{r/2} Z_\theta \sqrt{\theta}\exp\Big(\frac{-H^*}{\theta}\Big)\sum_{k\geq 1} f(x^k)\alpha_k^{-1/2}\exp(-a_k) \asymp 1,$$

Hence,

$$Z_\theta^{-1} \asymp \pi^{r/2}\sqrt{\theta}\exp\left(\frac{-H^*}{\theta}\right)\sum_{k\geq 1} f(x^k)\alpha_k^{-1/2}\exp(-a_k). \tag{9}$$

Finally, (8) and (9) give

$$P\left\{x^i\right\} = \frac{f(x^i)\alpha_i^{-1/2}\exp(-a_i)}{\sum_{k=1}^{+\infty} f(x^k)\alpha_k^{-1/2}\exp(-a_k)}.$$

Now, suppose that the function $H \in \mathcal{C}^2(\mathbb{R}^r)$, with Hessian matrix

$$A(x) = \left(\frac{\partial^2 H}{\partial x_i \partial x_j}(x)\right)_{1\leq i,j\leq r},$$

is invertible for all $x \in N$, then we have the following result. □

Corollary 3. *If $H \in \mathcal{C}^2(\mathbb{R}^r)$ and*

$$\sup_{x\in N}|H_\theta(x) - H^*| = o(\theta),$$

and there exists $k \in \mathbb{N}$ such that $f(x^k) \neq 0$, then

$$P\left\{x^i\right\} = \frac{f(x^i)\left|\det\left(A(x^i)\right)\right|^{-1/2}}{\sum_{k=1}^{+\infty} f(x^k)\left|\det\left(A(x^k)\right)\right|^{-1/2}}.$$

Proof. To prove Corollary 3, it is sufficient to put in Corollary 2,

$$a_i = 0, \text{ and } \alpha_i = \left|\det\left(A\left(x^i\right)\right)\right|,$$

for all x^i belonging to N. □

Remark 3. *Equation (9) implies that*

$$\sum_{k\geq 1} f(x^k)\alpha_k^{-1/2} < +\infty.$$

4. The Repairman Problem

Let us apply the previous results to obtain the limit distribution of the stationary one in the repairman problem [7,8] in the averaging scheme. This is an alternative method to Lyapunov's function method in the diffusion approximation scheme to prove convergence of stationary probabilities [11].

This system can be described by a Markov birth and death process $x_n(t),\ t > 0$, representing the number of failed components at time t, with state space [7,8,11]

$$E^n = \{0, 1, \dots, m+n\},$$

and jump intensities given by

$$\lambda_i = \begin{cases} n\lambda, & 0 \leq i \leq m \\ (m+n-i)\lambda, & m \leq i \leq m+n, \end{cases} \qquad \mu_i = \begin{cases} i\mu, & 0 \leq i \leq r \\ r\mu, & r \leq i \leq m+n. \end{cases}$$

In what follows, suppose that $m =: m_0 n$ and $r = r_0 n$, where m_0 and r_0 are constants.

Now, consider a Markov process $\nu_n(t)$ with state space $E_n = \{v_i = i/n \mid 0 \leq i \leq n+m\}$ and intensity functions given by

$$\lambda(v_i) = \begin{cases} \lambda, & 0 \leq v_i \leq m_0 \\ (m_0 + 1 - v_i)\lambda, & m_0 \leq v_i \leq m_0 + 1 \end{cases} \qquad \mu(v_i) = \begin{cases} v_i\mu, & 0 \leq v_i \leq r_0 \\ r_0\mu, & r_0 \leq v \leq m_0 + 1 \end{cases}$$

where $v_i = i/n$.

Then, we have

$$x_n(t) = n\nu_n(nt).$$

Consider now the normalized process defined by

$$\xi_n(t) := \nu_n(nt) = x_n(t)/n,$$

with the velocity of jumps defined by

$$C(v) := \lambda(v) - \mu(v).$$

This function enables us to classify the repairman problem depending on the position of the equilibrium point p defined by

$$C(p) = 0. \tag{10}$$

Under the condition $r_0 = \lambda/\mu < m_0$, the interval $V^* = [r_0, m_0]$ is an equilibrium set of a repairable system. Our main objective is to describe the stationary distribution of this repairable system on the equilibrium set V^*.

4.1. Gibbs Potential for the Stationary Distribution

The stationary distribution ρ^n of the process $x_n(t), t \geq 0$, may be written as follows:

$$\rho^n(k) = \rho_0^n \prod_{i=1}^{k} [\lambda(v_{i-1})/\mu(v_i)], \quad 1 \leq k \leq m+n, \tag{11}$$

where

$$\rho_0^n = \left[1 + \sum_{k=1}^{N} \prod_{i=1}^{k} [\lambda(v_{i-1})/\mu(v_i)]\right]^{-1}. \tag{12}$$

By using the Gibbs potential, the stationary distribution (11) is represented as follows (see [6]):

$$\rho^n(v_k) = \rho_0^n \exp[-NH_n(v_k)], \quad 1 \leq k \leq r,$$

with $N = m + n$ and where H_n is the Gibbs potential determined by the following relation

$$H_n(v_k) = -\frac{a}{n} \sum_{i=1}^{k} \ln[F(v_i)],$$

where $a = n/N = (m_0 + 1)^{-1}$ and $F(v) := \lambda(v)/\mu(v)$ is the kernel of the Gibbs potential H_n.

Then, the limit Gibbs potential is represented as

$$H(v) = -a \int_0^v \ln[F(u)]du, \quad 0 \leq v \leq 1 + m_0. \tag{13}$$

Therefore, in what follows and in the case when $r_0 = \lambda/\mu$, we will use the Gibbs potential for a stationary distribution with the kernel represented on the interval $(0, 1 + m_0]$ by

$$F(v) = \begin{cases} r_0/v, & 0 < vs. \geq r_0 \\ 1, & r_0 \leq vs. \leq m_0 \\ 1 + m_0 - v, & m_0 \leq vs. \leq 1 + m_0. \end{cases} \tag{14}$$

Alternatively, in explicit form on the left interval $[0, r_0]$

$$H(v) = av[\ln(v/r_0) - 1],$$

and on the right interval $[m_0, 1 + m_0]$

$$H(v) = a(m_0 + 1 - v)\ln(m_0 + 1 - v) - a(m_0 + r_0 - v), \quad m_0 \leq v \leq m_0 + 1.$$

Owing to (10), on the interval, $V^* = [r_0, m_0]$

$$H(v) = H(r_0) = -ar_0 = -r_0/(1 + m_0),$$

It is easy to verify that the set V^* is an equilibrium set for the limit Gibbs potential (13)

$$H^* = \min_{0 \leq v \leq 1 + m_0} H(v) = H(r_0) = -ar_0.$$

4.2. Limit Stationary Distribution

The stationary distribution of a repairable system induces a stationary distributed random variable x_n^*, with

$$\mathbb{P}\{x_n^* = k\} = \rho_k^n, \quad 0 \leq k \leq N.$$

In this example, Q is the probability measure and $Q(dx) = p^{-1}\mathbf{1}_{[r_0, m_0]}(x)dx$ on $[r_0, m_0]$, where $p = m_0 - r_0$.

We can show that

$$|\Delta_n(x, y)| \leq \frac{C}{n^2},$$

which means that

$$\lim_{n \to +\infty} n\Delta_n(x, y) = 0.$$

Then, the random variable x_n^* converges weakly to some random variable uniformly distributed on the equilibrium set V^* with density $f(x) = (m_0 - r_0)^{-1}\mathbf{1}_{[r_0, m_0]}(x)$.

Finally, we give the following numerical example (see Figure 1), which clearly shows that the limit stationary distribution is uniformly distributed on the equilibrium set V^*.

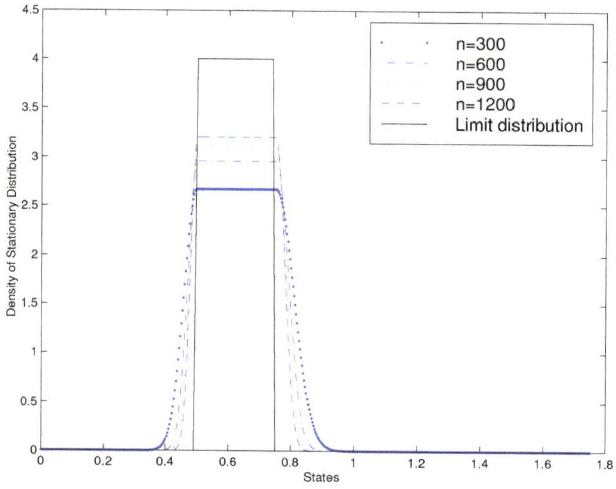

Figure 1. Convergence of the stationary distribution density for the repairman problem (with $\lambda = 0.5$, $\mu = 1$, $r_0 = \lambda/\mu = 0.5$ and $m_0 = 0.75$).

5. Concluding Remarks

We have presented the use of Gibbs distributions as a way to prove the convergence of the stationary distribution of the repairman problem in the averaging series scheme. The key contributions of this work are the weak convergence Gibbs distribution when the potential depends on a parameter θ and the transformation of the repairman stationary probabilities to Gibbs distributions. It will be of interest, on the one hand, to try to establish the same kind of results in a diffusion approximation scheme. Beyond the repairman problem, it would be of interest to apply this method to other similar problems. This problem must be rescaled in time when the finite state space becomes infinite. On the other hand, other work of interest can be performed in the direction of large deviations to provide a rate of convergence to the stationary distribution, see, e.g., [16].

Author Contributions: Methodology, H.C. and N.L. All authors have read and agreed to the published version of the manuscript.

Funding: This research received no external funding.

Data Availability Statement: Not applicable.

Acknowledgments: We are indebted to the three anonymous referees for their useful comments that greatly improved the presentation of this paper.

Conflicts of Interest: The authors declare no conflict of interest.

References

1. Billingsley, P. *Convergence of Probability Measures*; Wiley: Hoboken, NJ, USA, 1999.
2. Dudley, R.M. *Real Analysis and Probability*; Wadsworth: Belmont, CA, USA, 1989.
3. Ethier, S.N.; Kurtz, T.G. *Markov Processes: Characterization and Convergence*; Wiley: Hoboken, NJ, USA, 1986.
4. Koroliuk, V.S.; Limnios, N. *Stochastic Systems in Merging Phase Space*; World Scientific: Singapore, 2005.
5. Hwang, C.R. Laplace's Method Revisited: Weak Convergence of Probability Measures. *Ann. Probab.* **1980**, *6*, 1177–1182. [CrossRef]
6. Kaniovski, Y.; Pflug, G. *Limit Thoerems for Stationary Distributions of Birth-and-Death Processes*; Technical Report, IR–97–041; IIASA: Laxenburg, Austria, 1997.
7. Iglehart, D.L. Limit approximations for the many servers queue and repairman problem. *J. Appl. Probab.* **1965**, *2*, 429–441. [CrossRef]
8. Korolyuk, V.S. Maintenance Problems for Repairable Markov Systems. In *Statistical and Probabilistics Models in Reliability*; Ionescu, C., Limnios, N., Eds.; Birkhäuser: Boston, MA, USA, 1999.
9. Crampes, C.; Renault, J. The repairman problem revisited. *Ann. Econ. Stat.* **2016**, 7–24. [CrossRef]

10. Jain, M.; Shekhar, C.; Shukla, S. Queueing analysis of machine repair problem with controlled rates and working vacation under F-policy. *Proc. Natl. Acad. Sci. India Sect. Phys. Sci.* **2016**, *86*, 21–31. [CrossRef]
11. Chetouani, H.; Korolyuk, V.S. Stationary Distribution for Repairable Systems. *Appl. Stoch. Models Bus. Ind.* **2000**, *16*, 179–196. [CrossRef]
12. Gihman, I.I.; Skorohod, A.V. *Theory Stochastic Processes*; Springer: Berlin/Heidelberg, Germany, 1974; Volume 1–3.
13. Rogers, L.C.G.; Williams, D. *Diffusions, Markov Processes, and Martingales*; Volume 1: Foundations; Wiley: Chichester, UK, 1994.
14. Khasminskii, R. *Stability of the Solutions of Systems of Differential Equations under Random Disturbance of Their Parameters*; Nauka: Moscow, Russia, 1969.
15. Rudin, W. *Functional Analysis*; McGraw-Hill: New York, NY, USA, 1991.
16. Dupuis, P.; Laschos, V.; Ramanan, K. Large deviations for configurations generated by Gibbs distributions with energy functionals consisting of singular interaction and weakly confining potentials. *Electron. J. Probab.* **2020**, *25*, 1–41. [CrossRef]

mathematics MDPI

Article

On Positive Recurrence of the $M_n/GI/1/\infty$ Model

Alexander Veretennikov

Kharkevich Institute for Information Transmission Problems, Moscow 127051, Russia; ayv@iitp.ru

Abstract: Positive recurrence for a single-server queueing system is established under generalized service intensity conditions, without the assumption of the existence of a service density distribution function, but with a certain integral type lower bound as a sufficient condition. Positive recurrence implies the existence of the invariant distribution and a guaranteed slow convergence to it in the total variation metric.

Keywords: $M_n/GI/1/\infty$; positive recurrence; general service distribution function

MSC: 60K25; 90B22

Citation: Veretennikov, A. On Positive Recurrence of the $M_n/GI/1/\infty$ Model. *Mathematics* **2023**, *11*, 4514. https://doi.org/10.3390/math11214514

Academic Editors: Gurami Tsitsiashvili and Alexander Bochkov

Received: 23 September 2023
Revised: 23 October 2023
Accepted: 30 October 2023
Published: 1 November 2023

1. Introduction

The goal of this paper is to establish the **positive recurrence** of the model $M_n/G/1/\infty$ under certain assumptions. Intensity of service is assumed only partially at zero (as a lower left derivative value at zero of the "integrated intensity"); in addition, an integral type condition on the "integrated intensity" over intervals of some length is assumed.

By positive recurrence we mean the property of finite expectation of the time τ_0 to hit the regeneration state starting from any initial state with an estimate of such expected value (see (2) in what follows for the definition of τ_0), and not just a finite expectation of a regeneration period. For this aim, a new approach is suggested. For other more standard ways to establish positive recurrence for regenerative models, see [1,2], and also the references therein. While it may require conditions that may look too restrictive, the hidden main goal is to develop this new method. It looks likely that it may help establish better rates of convergence toward stationarity in the model under investigation. We also hope that it may help to make some progress in more involved queueing models such as Erlang–Sevastyanov, with an infinite number of servers. To the best of our knowledge, moment conditions on the service time under which positive recurrence is established in ([1], Chapter 2) and ([2], Chapter 5) are not applicable to infinite server Erlang–Sevastyanov-like systems; at least, the author is not aware of any progress in this direction so far. There is some well-known advice from Leonard Euler: if you see several possible paths toward the goal, then, in mathematics, you have to try all of them, not just one; some of them, or even all of them may turn out to be useful in some other adjacent areas or problems. Even though our conditions for positive recurrence are stronger than necessary for applications of other approaches, they may be more useful in some other particular situations. This was the main motivation for this work.

For the recent history of the topic see [3–11]. One of the reasons—although not the only one—why various versions of this system are so popular is because of their intrinsic links to important topics of mathematical insurance theory, see [5].

In this paper, we return to the less involved single-server system, $M_n/GI/1/\infty$, where the intensity of arrivals may only depend on the number of customers in the system, with the goal of reviewing conditions of its positive recurrence. An importance of this property may be highlighted, for example, by the publications [3,12] where the investigation of the model *assumes* that it is in the "steady state", which is a synonym for stationarity. As is well-known, positive recurrence along with some mild mixing or coupling properties guarantees the existence of a stationary regime of the system. One particular aspect of this issue is how to achieve bounds without assuming the existence of intensity of service in the $M/GI/1$ model and, more generally, in Erlang–Sevastyanov-type systems. Certain results in this direction were recently established in [13] for a slightly different model. Still, in [13] it is essential that the absolute continuous part of the distribution function F (in our notation) is non-degenerate; in the present paper, this is not required and the approach is different.

Please note that in such a model certain close results may be obtained by the methods of regenerative processes if it is assumed that the same distribution function F (see below) has enough moments. However, our conditions and methods are different. The main (moderate) hope of the author is that this approach may also be useful in studying ergodic properties of Erlang–Sevastyanov-type models, as happened with the earlier results and approaches based on the intensity of service as in [11], successfully applied in [14]. The present paper is an initial attempt in the program of developing tools that could help approach the problem outlined recently in [15].

The paper consists of the introduction in Section 1, the setting and the main result in Section 2, two simple auxiliary lemmata in Section 3, the proof of the main result in Section 4, and two simple examples in Section 5 for the comparison of sufficient conditions of Theorem 1 with conditions in terms of the intensity of service in the case when the latter does exist.

2. The Setting and Main Results

2.1. Definition of the Process

The model is as follows. There is one server with an incoming flow of customers or jobs; this flow is Poissonian with intensity λ_n where n is the number of customers in the system. If the server is idle, it immediately starts the service of the customer who arrives, unless the queue of waiting customers is not empty; in the latter case, it starts the service of one of them. If the server is busy, then the newly arrived customer goes to the queue where it waits until the server completes the earlier job(s). The buffer for the queue is unlimited (denumerable). The discipline of how the server chooses the next customer from the queue for serving is FIFO ("first in–first out"). All services are independent with the same distribution function F, and they are independent of the arrivals. A serve is a synonym of a completed job. It is assumed that the mean value $\int_0^\infty x dF(x) = \int_0^\infty (1 - F(x)dx) =: \mu^{-1}$ is finite. It is assumed that

$$\Lambda := \sup_n \lambda_n < \infty. \tag{1}$$

Also, it is assumed that

$$\lambda_n > 0, \quad \forall\, n \geq 0.$$

This will not be used except in one not-so-important remark. Therefore, we do not number this equation.

The following state space is convenient for the description of the stochastic process which describes the model. It will be convenient to identify the zero state $\{0\}$ with a zero couple $\{0,0\}$. Then the state space of the process is a union,

$$\mathcal{X} = \{(0,0)\} \cup \bigcup_{n=1}^{\infty} \{(n,x):\ x \geq 0\},$$

and the process itself is described at all times by a two-dimensional vector $X_t = (n_t, x_t)$, with $t \geq 0$, where $n_t = 0,1,\ldots$ stands for the number of the customers in the system including both in the server and in the queue; after the identification of $\{0\}$ with $\{0,0\}$ mentioned above, $x_t = 0$ in the case of $n_t = 0$ by definition; the second component x_t stands for the elapsed time of the current service. It is assumed that the initial value X_0 is any pair of non-negative values, $X_0 = (n, x)$, and the process evolves in time according to the provided description. By the construction, it is a Markov process in the state space $\mathcal{X}$.

We are interested in estimating the expectation of the stopping time

$$\tau_0 := \inf(t \geq 0 : X_t = (0, 0)), \tag{2}$$

where the process X_t starts from any initial state $X_0 = (n_0, x_0)$.

On some occasions, it will be convenient to write $n_t = n(X_t)$ for the first component of X_t and $x_t = x(X_t)$ for the second one. For any $X = (n, x)$ where $F(x) < 1$, the "integrated intensity" of service

$$dH(x) = (1 - F(x))^{-1} dF(x)$$

is defined by a Stiltjes integral

$$H(x) = \int_0^x (1 - F(s))^{-1} dF(s).$$

The integral $\int_0^x (1 - F(s))^{-1} dF(s)$ is assumed to be finite for any $x \geq 0$, and for the additional small part of the assumption see the next subsection. The intuitive meaning of the differential $dH(x_t)$ is the infinitesimal conditional probability of a job completion on the interval $(t, t + dt]$ under the condition that this job was not fulfilled by time t. If $dH(x_t) = \mu(x_t)dt$, then $\mu(x_t)$ is called intensity of service at x_t; however, we do not assume that $dH(\cdot)$ has to be absolutely continuous with respect to the Lebesgue measure. For simplicity of setting and proofs, in order to avoid possible singularities it is assumed that

$$\begin{gathered} F(0+) = 0, \quad \& \quad F(x) < 1, \quad \forall\, x \geq 0, \\ \text{and} \\ \tilde{F}(1) := \inf_{x \leq 1} (F(x + 1) - F(x)) > 0. \end{gathered} \tag{3}$$

2.2. Some Notation

1. The notation $\mathsf{P}_X = \mathsf{P}_{n,x}$ for the probability and $\mathsf{E}_X = \mathsf{E}_{n,x}$ for the expectation will be used. Both correspond to the initial value $X = (n, x)$ of the Markov process under consideration. We highlight that this is a standard notation from the theory of homogeneous Markov processes.
2. For a possibly discontinuous distribution function F, or for its integrated intensity H, integrals written like $\int_t^{t'} \ldots dF(s)$ will be understood as integrals

$$\int_t^{t'} \ldots dF(s) := \int_{(t,t']} \ldots dF(s),$$

   and likewise with $dH(s)$.
3. The following convention will be used, $dH(x) = 0$ if $(n, x) = (0, 0)$.

2.3. Main Result for $M_n/GI/1/\infty$

Recall that it is assumed that

$$H(t) < \infty, \quad \forall\, t \geq 0. \tag{4}$$

Let us also assume that there exists a constant r such that

$$r > 2(1 + \Lambda) \tag{5}$$

(so that $1/2 > (1 + \Lambda)/r$) where Λ was defined in (1), and such that

$$\forall\, 0 \le t \le 1, \quad \int_0^t (1+s)dH(s) \ge rt, \tag{6}$$

and, moreover,

$$\inf_{x\ge 1} \int_x^{x+\Delta} (1+s)dH(s) \ge r\Delta, \quad \forall\, \frac{1}{2} \le \Delta \le 1. \tag{7}$$

Let us highlight that the latter inequality is not supposed to hold for small values of Δ approaching zero, but only for $\Delta \in [1/2, 1]$. The increase of the integral $\int_x^{x+\Delta} (1+s)dH(s)$ for a fixed x as $\Delta \uparrow 1$ may be achieved due to a positive intensity $H' > 0$ if this derivative exists, and due to jumps of H, and also due to the increase of this function on sets of the Lebesgue measure zero like for the Cantor function.

Note that the process X_t may have no explosion on any bounded interval of time with probability one since the arrival intensities are bounded. In what follows $\|\mu_1 - \mu_2\|_{TV} = 2\sup_A(\mu_1(A) - \mu_2(A))$ is the distance in total variation between two probability measures; here the supremum is taken over all Borel measurable sets $A \in \mathcal{X}$.

Theorem 1. *Let assumptions (1)–(7) be satisfied. Then there exists $C > 0$ such that*

$$\mathsf{E}_{n,x}\tau_0 \le C(n+x+1). \tag{8}$$

Also, there exists a unique stationary measure μ, and, moreover, there exists $C > 0$ such that for any $t \ge 0$,

$$\|\mu_t^{n,x} - \mu\|_{TV} \le C\,\frac{(1+n+x)}{(1+t)}, \tag{9}$$

where $\mu_t^{n,x}$ is a marginal distribution of the process $(X_t,\ t \ge 0)$ with the initial data $X = (n, x)$, and constant C is the same as in (8).

Remark 1. *We highlight that the initial value of the second component here is arbitrary. Also, please note that this theorem is not a pure existence-type result because the constant C may be evaluated, as can be seen from the proof.*

3. Lemmata

Lemma 1. *Under assumption (3), for any $T > 0$ and for any $m \le n$,*

$$\begin{aligned} &\mathsf{P}_{n,0}(\textit{no less than } m \textit{ jobs completed on } [0,T]) \le F(T)^m, \\ &\& \\ &\mathsf{P}_{n,x}(\textit{no less than } m \textit{ jobs completed on } [0,T]) \le F(T)^{m-1}. \end{aligned} \tag{10}$$

Proof. The probability of no less than m completed jobs over time T for $m \le n$ is given by the repeated integral

$$\mathsf{P}_{n,x}(\text{no less than } m \text{ jobs completed over time } T)$$

$$= \int_0^T dF(x+t_1) \int_0^{T-t_1} dF(t_2) \ldots \int_0^{T-t_1-\ldots-t_{m-2}} dF(t_{m-1}) \int_0^{T-t_1-\ldots-t_{m-1}} dF(t_m)$$

$$\le (F(x+T) - F(x))F(T)^{m-1}.$$

Hence, we obtain both inequalities in (10), as required. Please note that assumption (3) implies that $\sup_{x\ge 0}(F(x+T) - F(x)) < 1$ for any $T > 0$, otherwise the distribution dF would be concentrated on a finite interval, which would contradict the assumption. □

Recall the notation $\Lambda = \sup_n \lambda_n$ and assumption (1).

Lemma 2. *Under assumption (3), for any n*

$$\inf_{x\le 1} \mathsf{P}_{n,x}(\tau_0 \le n) \ge p^n \exp(-n\Lambda) > 0, \quad \textit{with } p = \tilde F(1). \tag{11}$$

Proof. First, for any $n = 0, 1, \ldots$, and any x,

$$\mathsf{P}_{n,x}(\text{there are no arrivals on } [0,n]) \ge \exp(-n\Lambda).$$

Second, with probability no less than $p = \inf_{x\le 1}(F(x+1) - F(x)$ there is at least one completed job on $[0,1]$. The value p is positive according to assumption (3). After each jump down on $[0,1]$ — which is a stopping time — this argument may be repeated by induction n times. Note that since (n,x) is the initial value of the process, the server may not become idle until n jobs are completed. So, this gives the lower bound

$$\inf_{x\le 1} \mathsf{P}_{n,x}(\text{at least } n \text{ jobs completed on } [0,n]) \ge p^n.$$

By the multiplication of the two values $\exp(-n\Lambda)$ and p^n, due to the independence of the services and arrivals, the proof is completed. □

4. Proof of Theorem 1

0. The proof will be split into several steps. We shall consider the embedded Markov chain, namely, the process X_t at times $t = 0, 1, \ldots$, and it will be shown that *this* process hits some suitable compact around "zero state" $(0,0)$ in time which admits a finite expectation. From this property, the main result will follow. The reader is warned that after this first hit the definition of the embedded Markov chain will change, as further times may become random and possibly non-integer, see step 4 of this proof.

The main goal is to establish the bound (8), from which the estimate (9) follows as a corollary.

1. Let us choose $\varepsilon > 0$ so that

$$\varepsilon < \frac{1}{2} - \frac{(1+\Lambda)}{r} \tag{12}$$

(see condition (5)). NB: We highlight that this value ε will be fixed for what follows and will not tend to zero.

Once ε is chosen, let

$$M = M_\varepsilon := \left\lceil \frac{|\ln \varepsilon|}{|\ln \tilde F(1)|} \right\rceil, \tag{13}$$

(see (3) for the definition of $\tilde F(1)$) and let us choose $x(\varepsilon) \ge 2$ such that

$$\sup_{x \ge x(\varepsilon)} (F(x+1) - F(x)) \le \varepsilon. \tag{14}$$

Since $F(x) \uparrow 1$ as $x \uparrow \infty$, this is possible for any $\varepsilon > 0$. Let us introduce an auxiliary stopping time

$$\tau = \tau_\varepsilon := \inf(t = 0, 1, \ldots : \, n_t \le M_\varepsilon \;\&\; x_t \le x(\varepsilon)) \quad (\inf \varnothing = \infty). \tag{15}$$

Denote

$$K_\varepsilon = ((n,x) : n \le M_\varepsilon \;\&\; x \le x(\varepsilon)) \quad \& \quad L(n,x) = n + x + 1.$$

Function L will serve as a Lyapunov function outside the compact set K_ε.

First of all, we are going to estimate the first moment of τ, namely, to prove that there exists $C > 0$ such that

$$\mathsf{E}_{n,x}\tau \le CL(X), \quad X = (n,x). \tag{16}$$

Recall that

$$\tau_0 := \inf(t \ge 0 : \ X_t = (0,0)),$$

and highlight that the definition of $\tau = \tau_\varepsilon$ is quite different from that of τ_0.

Let $X_0 = X$. Applying the rules of stochastic calculus, we have for $X_t = (n_t, x_t) \neq (0,0)$,

$$\begin{aligned} dL(X_t) &= \lambda_{n_t}\left((n_t+2+x_t) - (n_t+1+x_t)\right)dt \\ &+ 1(n_t>0)((n_t+1+x_t+dt) - (n_t+1+x_t)) \\ &+ 1(n_t>0)(n_t - (n_t+1+x_t))dH(x_t) + dM_t, \end{aligned} \tag{17}$$

where M_t is a martingale (see, e.g., [16]). Indeed, let us comment on (17). Denote

$$\begin{aligned} dJ^1(t) &:= \lambda_{n_t}dt\left((n_t+2+x_t) - (n_t+1+x_t)\right) = \lambda_{n_t}dt, \\ dJ^2(t) &:= 1(n_t>0)((n_t+1+x_t+dt) - (n_t+1+x_t)) = 1(n_t>0)dt, \\ dJ^3(t) &:= 1(n_t>0)((n_t+1+x_t) - n_t)\,dH(x_t) = 1(n_t>0)(1+x_t)dH(x_t), \end{aligned}$$

and let $J^i(0) = 0$, $1 \le i \le 3$. At any non-random time $t \ge 0$, on the infinitesimal interval $(t, t+dt]$ there might occur the following events with corresponding probabilities, assuming that $n_t > 0$:

$$\begin{aligned} p_1(t,dt) &:= P(\text{one arrival and no completed jobs on } (t,t+dt]|\mathcal{F}_t) \\ &= \underbrace{\frac{1-F(x_t+dt)}{1-F(x_t)}}_{=1-dH(x_t)}\lambda_{n_t}dt; \end{aligned}$$

$$\begin{aligned} p_2(t,dt) &:= P(\text{one completed job and no arrivals on } (t,t+dt]|\mathcal{F}_t) \\ &= \underbrace{\frac{F(x_t+dt)-F(x_t)}{1-F(x_t)}}_{=dH(x_t)}(1-\lambda_{n_t}dt) = dH(x_t); \end{aligned}$$

$$p_3(t,dt) := P(\text{no arrivals and no completed jobs on } (t,t+dt]|\mathcal{F}_t) = 1-\lambda_{n_t}dt - dH(x_t);$$

$$p_4(t,dt) := P(\text{more than one arrival on } (t,t+dt]|\mathcal{F}_t) = o(dt) = 0;$$

$$p_5(t,dt) := P(\text{more than one completed job on } (t,t+dt]|\mathcal{F}_t) = (dH(x_t))(dH(0)) = 0;$$

$$\begin{aligned} p_6(t,dt) &:= P(\text{at least one arrival and at least one completed job on } (t,t+dt]|\mathcal{F}_t) \\ &= dH(x_t)\lambda_{n_t}dt = 0. \end{aligned}$$

Notice that we write $dH(x_t)\lambda_{n_t}dt = 0$ because when we integrate over a finite interval of time, the integral $\int \ldots dH(x_t)$ is finite anyway, while the multiplier dt is, in any case, $o(1)$. By a similar reason we claim $(dH(x_t))(dH(0)) = 0$ since the assumption $F(0+) = 0$ means that, in any case, $dH(0) = o(1)$.

So, by the full expectation formula we have for $n_t > 0$,

$$\begin{aligned}
&\mathsf{E}(L(X_{t+dt})|\mathcal{F}_t) \\
&= L(n_t+1, x_t+dt)p_1(t,dt) + L(n_t-1,0)p_2(t,dt) + L(n_t,x_t+dt)p_3(t,dt) \\
&= L(X_t) + (L(n_t+1,x_t) - L(n_t,x_t))\lambda_{n_t}dt \\
&+ 1(n_t>0)(L(n_t-1,0) - L(n_t,x_t))dH(x_t) + 1(n_t>0)dt \\
&= L(X_t) + \lambda_{n_t}dt - 1(n_t>0)(1+x_t)dH(x_t) + 1(n_t>0)dt.
\end{aligned}$$

Integrating from t_1 to t with $t > t_1$, we obtain,

$$\mathsf{E}(L(X_t) - \int_{t_1}^{t}(\lambda_{n_s} + 1(n_s>0))ds + \int_{t_1}^{t}(1+x_s)dH(x_s)|\mathcal{F}_{t_1}) = L(X_{t_1}). \tag{18}$$

Denote

$$M_t := L(X_t) - \int_{t_1}^{t}(\lambda_{n_s} + 1(n_s>0))ds + \int_{t_1}^{t}(1+x_s)dH(x_s),$$

and

$$I_{t_1,t} := \int_{t_1}^{t}(\lambda_{n_s} + 1(n_s>0))ds - \int_{t_1}^{t}(1+x_s)dH(x_s), \quad t_1 < t.$$

Notice that the assumption (1) straightforwardly implies that $\mathsf{E}M_t < \infty$ for any $t \geq 0$. Hence, we have for any $t_1 < t$,

$$\mathsf{E}(M_t|\mathcal{F}_{t_1}) = \mathsf{E}\big(L(X_t) - I_{t_1,t} - I_{0,t_1}|\mathcal{F}_{t_1}\big) = L(X_{t_1}) - I_{0,t_1} = M_{t_1}, \quad \text{a.s.}$$

So, as promised, M_t is a martingale. This justifies (17), as required.

The following bound will be established:

$$\mathsf{E}_{n,x}(J^1(1) + J^2(1) - J^3(1)) \leq -C, \tag{19}$$

with some $C > 0$, if $(n,x) \notin K_\varepsilon$.

First, we have,

$$J^1(1) + J^2(1) \leq \int_0^1 (1+\lambda_{n_t})dt \leq 1 + \Lambda. \tag{20}$$

To evaluate J^3 let us introduce a sequence of stopping times. Let

$$\gamma = \gamma_{n,x} := \inf(0 \leq t \leq 1 : x_t = 0) \quad (\inf(\emptyset) = 1).$$

To evaluate J^3, using the identity $1(t<\gamma)1(n_t>0) = 1(t<\gamma)$ which holds provided that $n > 0$, let us estimate,

$$\mathsf{E}_{n,x}J^3_{n,x}(1) = \mathsf{E}_{n,x}\int_0^1 1(n_t>0)(1+x_t)dH(x_t)$$

$$= \mathsf{E}_{n,x} \int_0^{\gamma} \overbrace{1(n_t > 0)}^{=1}(1 + x_t)dH(x_t) + \mathsf{E}_{n,x} \int_{\gamma}^{1} 1(n_t > 0)(1 + x_t)dH(x_t)$$

$$= \mathsf{E}_{n,x} \int_0^{\gamma} (1 + x + t)dH(x + t) + \mathsf{E}_{n,x} 1(\inf_{0 \le t \le 1} n_t > 0) \int_{\gamma}^{1} (1 + x_t)dH(x_t)$$

$$\underbrace{+ \mathsf{E}_{n,x} 1(\inf_{0 \le t \le 1} n_t = 0) \int_{\gamma}^{1} 1(n_t > 0)(1 + x_t)dH(x_t)}_{\ge 0}$$

$$\ge \mathsf{E}_{n,x} \int_0^{\gamma} (1 + x + t)dH(x + t)$$

$$+ \mathsf{E}_{n,x} 1(\inf_{0 \le t \le 1} n_t > 0) \int_{\gamma}^{1} (1 + x_t)dH(x_t) =: F^1 + F^2. \tag{21}$$

Let us estimate the term F^1. We have (recall that $\gamma \le 1$ by definition),

$$\begin{aligned} F^1 &= \mathsf{E}_{n,x} \int_0^{\gamma} (1 + x + t)dH(x + t) \\ &\ge \mathsf{E}_{n,x} 1(\gamma \ge 1/2) \underbrace{\int_0^{\gamma} (1 + x + t)dH(x + t)}_{\ge r\gamma} \ge r\mathsf{E}_{n,x} 1(\gamma \ge 1/2)\gamma. \end{aligned} \tag{22}$$

Here the complementary part of the integral $\mathsf{E}_{n,x} 1(\gamma < 1/2) \int_0^{\gamma} (1 + x + t)dH(x + t) \ge 0$ was just dropped.

Further, it will be shown that (see (21))

$$F^2 \ge r\mathsf{E}_{n,x} (1 - \gamma) 1(\inf_{0 \le t \le 1} n_t > 0). \tag{23}$$

For this aim, let us introduce by induction the sequence of stopping times,

$$\gamma^1 = \gamma, \quad \gamma^{n+1} := \inf(\gamma^n < t \le 1 : x_{\gamma^n} = 0), \quad n \ge 1 \quad (\inf(\emptyset) = 1).$$

Please note that the component x_t might only have a finite number of jumps down on any finite interval. The times of jumps on the interval $[0, 1]$ are exactly the times $\gamma^n < 1$, and possibly the last jump down on this interval may or may not occur at $t = 1$. In any case, clearly,

$$\lim_{n \to \infty} \gamma^n = 1.$$

So, we have,

$$\int_{\gamma}^{1} (1 + x_t)dH(x_t) = \sum_{n=1}^{\infty} \int_{\gamma^n}^{\gamma^{n+1}} (1 + x_t)dH(x_t),$$

where for each outcome this series is almost surely a finite sum. On each interval $[\gamma^n, \gamma^{n+1}]$ we may write down

$$\int_{\gamma^n}^{\gamma^{n+1}} (1 + x_t)dH(x_t) = \int_0^{\gamma^{n+1} - \gamma^n} (1 + x_{\gamma^n} + s)dH(x_{\gamma^n} + s)$$

$$= \int_0^{\gamma^{n+1}-\gamma^n} 1(n_{x_{\gamma^n}+s} > 0)(1 + x_{\gamma^n} + s)dH(x_{\gamma^n} + s)$$

$$= \int_0^{\gamma^{n+1}-\gamma^n} 1(n_s > 0)(1+s)dH(s) \ge r(\gamma^{n+1} - \gamma^n)1(\inf_{0\le t\le 1} n_t > 0).$$

This is by virtue of assumption (6) and because at each stopping time γ^n which is less than 1 we have $x_{\gamma^n} = 0$. If $\gamma^n \ge 1$, then both sides in the latter inequality equal zero, so that the inequality still holds true. Therefore, taking a sum over n, we obtain (23), just without $1(\inf_{0\le t\le 1} n_t > 0)$ on the left-hand side. This multiplier $1(\inf_{0\le t\le 1} n_t > 0)$ in the right-hand side guarantees that its presence in the left-hand side of (23) still leads to a valid inequality, which means that the bound (23) is justified.

It follows from (22) and (23) that

$$J^3 = F^1 + F^2 \ge r\mathsf{E}_{n,x}1(\gamma \ge 1/2)\gamma + r\mathsf{E}_{n,x}(1-\gamma)1(\inf_{0\le t\le 1} n_t > 0)$$

$$= r\mathsf{E}_{n,x}1(\gamma \ge 1/2)\gamma + r\mathsf{E}_{n,x}(1-\gamma) - r\mathsf{E}_{n,x}(1-\gamma)1(\inf_{0\le t\le 1} n_t = 0)$$

$$\ge r - r\mathsf{E}_{n,x}1(\gamma < 1/2)\gamma - r\mathsf{E}_{n,x}1(\inf_{0\le t\le 1} n_t = 0) \ge \frac{r}{2} - r\mathsf{P}_{n,x}(\inf_{0\le t\le 1} n_t = 0).$$

Due to Lemma 1, if $n > M$ then (see (13))

$$\mathsf{E}_{n,x}1(\inf_{0\le t\le 1} n_t = 0) = \mathsf{P}_{n,x}(\text{at least } n-1 \text{ jobs completed on } [0,1]) \le F(1)^{n-1} \le \varepsilon.$$

(NB: In fact, at least n jobs should be completed, the first one on $[x, x+1]$; however, we prefer to have a bound independent of x. In any case, this does not change the scheme of the proof.) Likewise, if $x > x(\varepsilon)$, then

$$\mathsf{E}_{n,x}1(\inf_{0\le t\le 1} n_t = 0) \le \mathsf{P}_{n,x}(\text{at least 1 completed job on } [0,1])$$

$$\le F(x+1) - F(x) \le \varepsilon,$$

due to the choice of $x(\varepsilon)$, see (14).

Recall that ε was chosen so that $r\varepsilon < (r/2) - (1+\Lambda)$, see (12). Hence,

$$J^3 \ge \frac{r}{2} - r\varepsilon.$$

Denote

$$(0 <)\ \Delta := \frac{r}{2} - \varepsilon - 1 - \Lambda.$$

Thus, for any $(n,x) \notin K_\varepsilon$ we obtain,

$$\mathsf{E}_{n,x}(J^3_{n,x}(1) - J^1(1) - J^2(1)) \ge \Delta. \tag{24}$$

The bound (19) follows with a constant C which may be evaluated via r, Λ, ε.

2. So, with the choice of $M = M_\varepsilon$ and $x(\varepsilon)$ according to (13) and (14), respectively, we may write,

$$1((n,x) \notin K_\varepsilon)\mathsf{E}_{n,x}L(X_1) \le 1((n,x) \notin K_\varepsilon)(L(X_0) - \Delta).$$

The event $((n,x) \notin K_\varepsilon)$ may be equivalently expressed as $(\tau > 0)$. Hence, the latter inequality may also be rewritten in the form suitable for the induction:

$$\boxed{1(0<\tau)\mathsf{E}_{n,x}L(X_1) \le 1(0<\tau)(L(X_0)-\Delta).}$$

Similarly, $((n_1,x_1) \notin K_\varepsilon)$ may be equivalently expressed as $(\tau > 1)$. Therefore, we obtain

$$1(1<\tau)\mathsf{E}_{n_1,x_1}L(X_2) \le 1(1<\tau)(L(X_1)-\Delta).$$

Hence, by taking the expectations we obtain

$$\mathsf{E}_{n,x}1(1<\tau)\mathsf{E}_{n_1,x_1}L(X_2) \le \mathsf{E}_{n,x}1(1<\tau)(L(X_1)-\Delta).$$

Similarly, by induction (in what follows the notation $n_k = n_t|_{t=k}$ is used),

$$\mathsf{E}_{n,x}1(k-1<\tau)\mathsf{E}_{n_{k-1},x_{k-1}}L(X_k) \le \mathsf{E}_{n,x}1(k-1<\tau)(L(X_{k-1})-\Delta),\ k\ge 1.$$

Due to the elementary bound $1(k-1<\tau) \ge 1(k<\tau)$, this implies,

$$\boxed{\Delta\ \mathsf{E}_{n,x}1(k-1<\tau) \le \mathsf{E}_{n,x}1(k-1<\tau)L(X_{k-1}) - \mathsf{E}_{n,x}1(k<\tau)L(X_k).}$$

Summing up and dropping the negative term on the right-hand side, we obtain

$$\Delta \sum_{k=1}^{N} \mathsf{E}_{n,x}1(k-1<\tau) \le 1((n,x)\notin K_\varepsilon)L(X_0),$$

for any $N > 0$. So, by the monotone convergence theorem,

$$\Delta\, 1((n,x)\notin K_\varepsilon) \sum_{k=1}^{\infty} \mathsf{E}_{n,x}1(k-1<\tau) \le 1((n,x)\notin K_\varepsilon)L(X_0). \tag{25}$$

By virtue of the well-known relation

$$\sum_{k=1}^{\infty} \mathsf{E}_{n,x}1(k-1<\tau) = \mathsf{E}_{n,x}\sum_{k=1}^{\infty} 1(k\le\tau) = \mathsf{E}_{n,x}\sum_{k=1}^{\tau} 1 = \mathsf{E}_{n,x}\tau,$$

for the expectation of τ the bound (25) implies the following inequality with $X_0 = (n,x)$,

$$\boxed{1((n,x)\notin K_\varepsilon)\mathsf{E}_{n,x}\tau \le \Delta^{-1}1((n,x)\notin K_\varepsilon)L(X_0).} \tag{26}$$

In particular, this bound signifies that

$$\mathsf{P}_{n,x}(\tau<\infty) = 1.$$

3. Now, once the bound for the expected value of τ is established, we are ready to explain the details of how to obtain a bound for $\mathsf{E}_{n,x}\tau_0$. The rest of the proof is devoted to this implication, with the last sentences related to the corollary about the invariant measure and convergence to it.

At τ, the process X_k attains the set $(X : X \in K_\varepsilon)$, while $X_0 \notin K_\varepsilon$; hence, both

$$n_\tau \le M \quad \& \quad x_\tau \le x(\varepsilon).$$

By definition, the random variable τ is the first integer k where simultaneously

$$n_k \le M \quad \& \quad x_k \le x(\varepsilon).$$

Therefore, at $k-1$ we have either $n_k > M$, or $x_k > x(\varepsilon)$, or both. If there are no completed jobs on $(k-1,k]$, then n_t may only increase, or, at least, stay equal on this interval, while x_t certainly increases. Therefore, $\tau = k$ may only be achieved by at least one completed job; it would mean a jump down by one of the n-component and simultaneously a jump down to zero of the x-component. Then at k we obtain $x_k \le 1$, which certainly makes it less than $x(\varepsilon)$, regardless of whether or not there were other arrivals or completed jobs on $(k-1,k]$ (recall that in addition to inequality (14) we assumed that $x(\varepsilon) \ge 2$).

Now, given $n \le M$ and $x \le x(\varepsilon)$, by virtue of lemma 2, for any $T > 0$ we have with any $x \le 1$,

$$\mathsf{P}_{n,x}(\text{no arrivals on } [0,T], n(X_T) = 0) \ge p(T) > 0, \tag{27}$$

where

$$p(T) := p^T \exp(-T\Lambda) = \tilde{F}(1)^T \exp(-T\Lambda).$$

Here T is any positive integer and recall that

$$\tilde{F}(1) = \inf_{x \le 1}(F(x+1) - F(x)).$$

Note that, of course, for non-integer values of $T > 0$ there is a likewise bound, but it looks a bit more involved, and using integer T values suffices for the proof. Recall that it was assumed in (3) that $\tilde{F}(1) > 0$, and it follows from the first line of (3) that $\tilde{F}(1) < 1$. Inequality (27) implies that

$$\inf_{x \le 1} \mathsf{P}_{n_\tau,x_\tau}(\text{no arrivals on } [0,T], n(X_T) = 0) \ge p(T) > 0. \tag{28}$$

NB:Here the standard notations for homogeneous Markov processes are used, which means that X_T after stopping time τ is, actually, the value $X_{\tau+T}$.

Please note that for any $T > 0$ the event $(n(X_{\tau+T}) = 0)$ implies that

$$\tau_0 \le \tau + T.$$

Hence, we conclude,

$$\mathsf{P}_{n_\tau,x_\tau}(\text{no arrivals on } [0,T], \tau_0 \le T) \ge p(T) > 0. \tag{29}$$

and, therefore, for any x,

$$\mathsf{P}_{n,x}(\text{no arrivals on } [\tau,\tau+T], \tau_0 \le \tau + T) \ge p(T) > 0. \tag{30}$$

4. Consider now the process X started at time τ from state (n_τ, x_τ) with $x_\tau \le 1$ and $n_\tau \le M$.

Let $T := \lceil x(\varepsilon) \rceil$ and let us stop the process either at $\tau + T$, or at

$$\chi := \inf(t \ge \tau : n_t \ge M+1, \text{ or } x_t \ge x(\varepsilon) + 1),$$

whichever happens earlier. In other words, consider the stopping time

$$\chi^1 := \chi \wedge (\tau + T).$$

The event $(\chi^1 = \tau + T)$ implies that the process $L(X_t)$ does not exceed the level $M + 2 + T$ on the interval $[\tau, \tau + T]$. On the other hand, according to the arguments of step 3 – namely, due to (29) and (30) – we have,

$$\mathsf{P}_{n_\tau,x_\tau}(\tau_0 \le \chi^1) \ge p(T) > 0. \tag{31}$$

Let

$$\chi^0 := 0, \quad \tau^1 := \tau,$$

and further, let us define two sequences of stopping times by induction,

$$\chi^{k+1} := \inf(t \ge \tau^{k+1} : n_t \ge M+1, \text{ or } x_t \ge x(\varepsilon)+1) \wedge (\tau^{k+1}+T),$$

$$\tau^{k+1} := \inf(\chi^k + i,\, i \ge 0 : n_{\chi^k+i} \le M, \,\&\, x_{\chi^k+i} \le x(\varepsilon)), \quad k \ge 0.$$

Both sequences of stopping times are monotonically increasing. Note that all integers like χ^k and τ^k here in the expressions stand for upper indices, not for power functions. Let us highlight that the stopping time τ^{k+1} equals χ^k plus some integer, but χ^k may or may be not an integer itself. All these stopping times are finite with probability one, and, moreover, due to (31) and because of the strong Markov property we have almost surely,

$$1(\tau_0 > \tau^k)\mathsf{P}_{X_{\tau^k}}(\tau_0 \le \chi^{k+1}) \ge 1(\tau_0 > \tau^k)p(T).$$

Also, almost surely

$$\lim_{k\to\infty} \tau^k \ge \tau_0, \tag{32}$$

on the event where $\tau_0 < \infty$. Indeed, suppose the opposite, that is,

$$\lim_{k\to\infty} \tau^k < \tau_0 < \infty. \tag{33}$$

Then on a finite interval of time $[0, \tau_0]$ the process n_t either crosses the interval $[M, M+1]$ and back an infinite number of times, or the process x_t an infinite number of times crosses the interval $[x(\varepsilon), x(\varepsilon)+1]$ and back; in any case, it would mean an infinite number of arrivals to $[0, \tau_0]$. Since $\Lambda < \infty$, the first option is clearly not possible. If the second option occurred, it would mean that there were an infinite number of completed jobs on $[0, \tau_0]$; this is not possible for various reasons, for example, because this would also require an infinite number of arrivals on the same interval, and this possibility we have already excluded. Thus, the assumption (33) may only happen with probability zero, and so, (32) with $\tau_0 < \infty$ holds true with probability one.

Using the strong Markov property at time χ^i (see [17], Section 4), we obtain by induction,

$$\mathsf{P}_{X_0}(\tau_0 > \chi^k) \le (1-p(T))^k, \quad k \ge 1. \tag{34}$$

5. Denote

$$d^k := \chi^k - \tau^k, \quad \delta^k := \tau^{k+1} - \chi^k.$$

Also by induction, it follows from (26) and from the elementary bound by definition

$$d^k \le T,$$

that there exists a constant C such that

$$\mathsf{E}_{X_0}(\underbrace{(\chi^{k+1} - \chi^k)}_{=:\delta^k + d^k} | X_{\chi^k}) \le C, \quad \forall k \ge 1. \tag{35}$$

Using the representation

$$\chi^{k+1} = \overbrace{\tau^1 + d^1}^{=\chi^1} + \sum_{i=2}^{k+1}(\delta^{i-1} + d^i),$$

and due to (32), we estimate,

$$\mathsf{E}_{X_0}\tau_0 = \sum_{k=0}^{\infty} \mathsf{E}_{X_0}\tau_0 1(\chi^k < \tau_0 \le \chi^{k+1}) \le \sum_{k=0}^{\infty} \mathsf{E}_{X_0}\chi^{k+1} 1(\chi^k < \tau_0) 1(\chi^{k+1} \ge \tau_0)$$

$$= \sum_{k=0}^{\infty} \mathsf{E}_{X_0}\left(\tau^1 + d^1 + \sum_{i=2}^{k+1}(\delta^{i-1} + d^i)\right) 1(\chi^k < \tau_0) 1(\chi^{k+1} \ge \tau_0). \tag{36}$$

Now we are going to estimate this sum by some geometric type series in combination with bounds (34) and (35). A small issue is that we are not able to use Hölder's, or Cauchy–Buniakowskii–Schwarz inequality here because we only possess a first moment bound for τ^k, while higher moments are not available. This minor obstacle is resolved in the next step of the proof by the following arguments using conditional expectation with respect to suitable sigma-algebras.

6. We have,

$$\mathsf{E}_{X_0}\left(\tau^1 + d^1 + \sum_{i=2}^{k+1}(\delta^{i-1} + \underbrace{d^i}_{\le T})\right) 1(\chi^k < \tau_0) 1(\chi^{k+1} \ge \tau_0)$$

$$= \mathsf{E}_{X_0}\left(\tau^1 + d^1 + \sum_{i=1}^{k}(\delta^{i-1} + \underbrace{d^i}_{\le T})\right) \prod_{j=1}^{k} 1(\chi^j < \tau_0)$$

$$+ \mathsf{E}_{X_0}(\delta^k + \underbrace{d^{k+1}}_{\le T}) \prod_{j=1}^{k} 1(\chi^j < \tau_0).$$

Let us investigate the last term. Since each random variable $1(\chi^j \le \tau_0)$ is $\mathcal{F}_{\chi^k}$-measurable for any $j \le k$, and because, due to the strong Markov property and the bound (34),

$$\mathsf{E}_{X_0} \prod_{j=1}^{k} 1(\chi^j < \tau_0) = \mathsf{E}_{X_0} 1(\chi^k < \tau_0) \le (1 - p(T))^k.$$

(It was used that $\chi^k < \tau_0$ implies $\chi^j < \tau_0$ for all $j < k$ as well.) Moreover, by virtue of the inequality (16) and since by definition all $d^k \le T$, we have for $k \ge 1$,

$$\mathsf{E}_{X_0}(\delta^k + d^{k+1}) \prod_{j=1}^{k} 1(\chi^j < \tau_0) = \mathsf{E}_{X_0} \mathsf{E}_{X_0}\left((\delta^k + d^{k+1}) \prod_{j=1}^{k} 1(\chi^j < \tau_0) | \mathcal{F}_{\chi^k}\right)$$

$$= \mathsf{E}_{X_0} \prod_{j=1}^{k} 1(\chi^j < \tau_0) \overbrace{\mathsf{E}\left((\delta^k + d^{k+1}) | \mathcal{F}_{\chi^k}\right)}^{\le C} \le C(1 - p(T))^k$$

with some finite constant C.

Let us inspect the previous term. Using that δ^{k-1} and all random variables $1(\chi^j < \tau_0)$ are $\mathcal{F}_{\tau^k}$-measurable for any $j \le k - 1$, we obtain by induction,

$$\mathsf{E}_{X_0}\left((\delta^{k-1} + \underbrace{d^k}_{\le T})\right) \prod_{j=1}^{k} 1(\chi^j < \tau_0) \le T \mathsf{E}_{X_0} \prod_{j=1}^{k} 1(\chi^j < \tau_0) + \mathsf{E}_{X_0} \delta^{k-1} \prod_{j=1}^{k} 1(\chi^j < \tau_0)$$

$$\le T(1 - p(T))^k + \mathsf{E}_{X_0} \mathsf{E}_{X_0}\left(\delta^{k-1} \prod_{j=1}^{k} 1(\chi^j < \tau_0) | \mathcal{F}_{\tau^k}\right)$$

$$= T(1-p(T))^k + \mathsf{E}_{X_0}\delta^{k-1}\prod_{j=1}^{k-1}1(\chi^j < \tau_0)\underbrace{\mathsf{E}_{X_0}\left(1(\chi^k < \tau_0)|\mathcal{F}_{\tau^k}\right)}_{\le 1-p(T)}$$

$$\le T(1-p(T))^k + (1-p(M))\mathsf{E}_{X_0}\mathsf{E}_{X_0}\left(\delta^{k-1}\prod_{j=1}^{k-1}1(\chi^j < \tau_0)|\mathcal{F}_{\chi^{k-1}}\right)$$

$$= T(1-p(T))^k + (1-p(M))\mathsf{E}_{X_0}\prod_{j=1}^{k-1}1(\chi^j < \tau_0)\underbrace{\mathsf{E}_{X_0}\left(\delta^{k-1}|\mathcal{F}_{\chi^{k-1}}\right)}_{\le C}$$

$$\le T(1-p(T))^k + C(1-p(T))^k = C(1-p(T))^k. \quad (37)$$

Also by induction, we find that a similar upper bound with the multiplier $(1-p(T))^k$ holds for each term in the sum in the right-hand side of (36), for $2 \le i < k$, and $k \ge 3$. Indeed, using the identity $1(\chi^k < \tau_0) = \prod_{j=1}^{k}1(\chi^j < \tau_0)$ we have for $2 \le i < k$,

$$\mathsf{E}_{X_0}\left((\delta^{i-1} + \underbrace{d^i}_{\le T})\right)\left(\prod_{j=1}^{k-1}1(\chi^j < \tau_0)\right)1(\chi^k < \tau_0)$$

$$= \mathsf{E}_{X_0}\mathsf{E}_{X_0}\left(1(\chi^k < \tau_0)(\delta^{i-1} + \underbrace{d^i}_{\le T})\prod_{j=1}^{k-1}1(\chi^j < \tau_0)|\mathcal{F}_{\tau^k}\right)$$

$$\le \mathsf{E}_{X_0}(\delta^{i-1} + T)\prod_{j=1}^{k-1}1(\chi^j < \tau_0)\underbrace{\mathsf{E}_{X_0}\left(1(\chi^k < \tau_0)|\mathcal{F}_{\tau^k}\right)}_{\le 1-p(T)}$$

$$\le (1-p(T))\mathsf{E}_{X_0}(\delta^{i-1} + T)\prod_{j=1}^{k-1}1(\chi^j < \tau_0).$$

Here we used that the random variable $(\delta^{i-1} + d^i)\prod_{j=1}^{k-1}1(\chi^j < \tau_0)$ is $\mathcal{F}_{\tau^k}$-measurable.

Further, by virtue of the bound (34)

$$(1-p(T))\mathsf{E}_{X_0}(\delta^{i-1} + T)\prod_{j=1}^{k-1}1(\chi^j < \tau_0)$$

$$= (1-p(T))\mathsf{E}_{X_0}\mathsf{E}_{X_0}\left((\delta^{i-1} + T)\prod_{j=1}^{k-1}1(\chi^j < \tau_0)|\mathcal{F}_{\tau^i}\right)$$

$$= (1-p(T))\mathsf{E}_{X_0}(\delta^{i-1}+T)\prod_{j=1}^{i-1}1(\chi^j<\tau_0)\underbrace{\mathsf{E}_{X_0}\left(\prod_{j=i}^{k-1}1(\chi^j<\tau_0)|\mathcal{F}_{\tau^i}\right)}_{\le(1-p(T))^{k-i}}$$

$$\le (1-p(T))^{k-i+1}\mathsf{E}_{X_0}(\delta^{i-1}+T)\prod_{j=1}^{i-1}1(\chi^j<\tau_0)$$

$$\le (1-p(T))^{k-i+1}C(1-p(T))^{i-1} = C(1-p(T))^k. \tag{38}$$

We used the bound (37) with d^k replaced by its upper bound T (as in the calculus leading to (37)) and with k replaced by i.

The first term is estimated similarly with the only change that instead of the constant C we obtain a multiplier $C(x+n+1)$ which is a function of the initial data $X_0=(n,x)$ and which makes the resulting bound non-uniform with respect to the initial data:

$$(1-p(T))\mathsf{E}_{n,x}(\tau^1+d^1)\prod_{j=1}^{k-1}1(\chi^j<\tau_0)\le C(n+x+1)(1-p(T))^k. \tag{39}$$

Overall, collecting the bounds (37)–(39), we obtain,

$$\mathsf{E}_{X_0}\left(\tau^1+d^1+\sum_{i=2}^{k+1}(\delta^{i-1}+\underbrace{d^i}_{\le T})\right)1(\chi^k<\tau_0)1(\chi^{k+1}\ge\tau_0)\le kCL(X_0)(1-p(T))^k.$$

Therefore, it follows that

$$\mathsf{E}_{X_0}\tau_0\le CL(X_0)+\sum_{k=1}^{\infty}kC(1-p(T))^k = CL(X_0)+\frac{C(1-p(T))}{p(T)^2}\le CL(X_0),$$

with some new constant C, as required.

Existence of the invariant measure for the model and the inequality (9) follows straightforwardly from the established positive recurrence (8) and in a usual way from the coupling technique, or by means of other well-known tools. Theorem 1 is proved. □

5. Two Examples

Let us provide two examples for a comparison to "local" conditions in terms of the intensity of service if the latter exists.

Example 1. *Assume that there exists the derivative function $F'(s)$ and that the hazard function $h=H'$ is no less than a constant,*

$$h(s)=H'(s)=\frac{F'(s)}{1-F(s)}\ge\mu,\quad s\ge 0. \tag{40}$$

The upper bound for $J^1_{n,x}+J^2_{n,x}$ is the same as in the proof of the theorem:

$$J^1_{n,x}+J^2_{n,x}\le 1+\Lambda.$$

To estimate $J^3_{n,x}$, in the case of either initial value n, or initial second component x, or both are large enough, by Lemma 1 we have similarly to (21) and (22) with μ in place of r,

$$J^3_{n,x}(1):=\mathsf{E}_{n,x}\int_0^1 1(n_t>0)(1+x_t)dH(x_t)$$

$$= \mathsf{E}_{n,x} \int_0^{1\wedge\gamma} \overbrace{1(n_t > 0)}^{=1}(1+x_t)\mu dt + \mathsf{E}_{n,x} \int_{1\wedge\gamma}^{1} 1(n_t > 0)(1+x_t)\mu dt$$

$$= \mathsf{E}_{n,x} \int_0^{1\wedge\gamma} (1+x+t)\mu dt + \mathsf{E}_{n,x} 1(\inf_{0\le t\le 1} n_t > 0) \underbrace{\int_{1\wedge\gamma}^{1} (1+x_t)\mu dt}_{\ge \mu(1-1\wedge\gamma)}$$

$$\underbrace{+ \mathsf{E}_{n,x} 1(\inf_{0\le t\le 1} n_t = 0) \int_{1\wedge\gamma}^{1} 1(n_t > 0)(1+x_t)\mu dt}_{\ge 0}$$

$$\ge \mathsf{E}_{n,x} \int_0^{1\wedge\gamma} \underbrace{(1+x+t)}_{\ge 1}\mu dt + \mathsf{E}_{n,x} 1(\inf_{0\le t\le 1} n_t > 0)\mu(1-1\wedge\gamma).$$

Therefore, we obtain

$$J^3_{n,x}(1) \ge \mu \mathsf{E}_{n,x}(1\wedge\gamma) + \mu \mathsf{E}_{n,x} 1(\inf_{0\le t\le 1} n_t > 0)(1-1\wedge\gamma)$$

$$= \mu \mathsf{E}_{n,x}(1\wedge\gamma) + \mu \mathsf{E}_{n,x}(1-1\wedge\gamma) - \mu \mathsf{E}_{n,x} 1(\inf_{0\le t\le 1} n_t = 0)(1-1\wedge\gamma)$$

$$\ge \mu - \mu \mathsf{E}_{n,x} 1(\inf_{0\le t\le 1} n_t = 0) \ge \mu - \mu\varepsilon,$$

where the latter inequality is due to the Lemma 1 and to the choice of the values of M and $x(\varepsilon)$, see (13) and (14).

Therefore, the condition $\mu > 1 + \Lambda$ suffices for the claims of the theorem (assuming all its other conditions are met). This should be compared with the assumption (5). The multiplier 2 in (5) may be regarded as a price for non-local, integral-type conditions, see (6) and (7). Please note, however, that assumption (40) looks clearly stronger than necessary for the bound obtained.

Example 2. *Assume that there exists the derivative function $F'(s)$ and that the hazard function $h = H'$ satisfies the condition (compare to [14])*

$$(1+s)h(s) \ge \mu, \quad s \ge 0, \tag{41}$$

with a constant μ.

Here the upper bound for $J^1_{n,x} + J^2_{n,x}$ is the same as in the proof of the theorem and as in the previous example:

$$J^1_{n,x} + J^2_{n,x} \le 1 + \Lambda.$$

To estimate $J^3_{n,x}$, in the case if either the initial value n, or the initial second component x (or both) is large enough, we have by Lemma 1, similarly to (21) and to the previous example,

$$J^3_{n,x}(1) := \mathsf{E}_{n,x} \int_0^1 1(n_t > 0)(1+x_t)dH(x_t)$$

$$= \mathsf{E}_{n,x} \int_0^{1\wedge\gamma} \overbrace{1(n_t > 0)}^{=1} \mu dt + \mathsf{E}_{n,x} \int_{1\wedge\gamma}^{1} 1(n_t > 0)\mu dt$$

$$= \mathsf{E}_{n,x}\int_0^{1\wedge\gamma}\mu dt + \mathsf{E}_{n,x}1(\inf_{0\le t\le 1} n_t > 0)\underbrace{\int_{1\wedge\gamma}^1 \mu dt}_{=\mu(1-1\wedge\gamma)}$$

$$\underbrace{+\,\mathsf{E}_{n,x}1(\inf_{0\le t\le 1} n_t = 0)\int_{1\wedge\gamma}^1 1(n_t>0)\mu dt}_{\ge 0}$$

$$\ge \mu\mathsf{E}_{n,x}(1\wedge\gamma) + \mu\mathsf{E}_{n,x}1(\inf_{0\le t\le 1} n_t > 0)(1-1\wedge\gamma)$$

$$= \mu\mathsf{E}_{n,x}(1\wedge\gamma) + \mu\mathsf{E}_{n,x}(1-1\wedge\gamma) - \mu\mathsf{E}_{n,x}1(\inf_{0\le t\le 1} n_t = 0)(1-1\wedge\gamma)$$

$$\ge \mu - \mu\mathsf{E}_{n,x}1(\inf_{0\le t\le 1} n_t = 0) \ge \mu - \mu\varepsilon,$$

again due to the definition of M and $x(\varepsilon)$, see (13) and (14). Hence, here the same condition $\mu > 1 + \Lambda$ as in the previous example suffices for the claims of the theorem (assuming all other conditions of the theorem are met). Condition (41) is clearly more relaxed than (40), but both assume the existence of the intensity of service, which is not required in Theorem 1.

6. Discussion

The result may serve as a sufficient condition for the "steady-state" property of the model $M_n/GI/1$ used as a background in [3,12]. It is plausible that the method used in this paper admits extensions to more general models. As it was said in the introduction, there is some moderate hope that it may also be applied to Erlang–Sevastyanov's type systems, which could potentially allow finding sufficient conditions for convergence rates in such systems without assuming the existence of intensity of service, thus, generalizing the results from [14].

Funding: This research was funded by the Foundation for the Advancement of Theoretical Physics and Mathematics "BASIS".

Data Availability Statement: Not applicable.

Acknowledgments: The author is grateful to two Referees for very useful remarks.

Conflicts of Interest: The author declares no conflict of interest. The funders had no role in the design of the study, in the writing of the manuscript, and in the decision to publish the results.

References

1. Borovkov, A.A. *Stochastic Processes in Queueing Theory*; Springer: New York, NY, USA, 1976. [CrossRef]
2. Morozov, E.; Steyaert, B. *Stability Analysis of Regenerative Queueing Models*; Springer: Cham, Switzerland, 2021. [CrossRef]
3. Abouee-Mehrizi, H.; Baron, O. State-dependent M/G/1 queueing systems. *Queueing Syst.* **2016**, *82*, 121–148. [CrossRef]
4. Asmussen, S. *Applied Probability and Queues*, 2nd ed.; Springer: Berlin/Heidelberg, Germany, 2003.
5. Asmussen, S.; Teugels, J.L. Convergence rates for M/G/1 queues and ruin problems with heavy tails. *J. Appl. Probab.* **1996**, *33*, 1181–1190. [CrossRef]
6. Bambos, N.; Walrand, J. On stability of state-dependent queues and acyclic queueing networks. *Adv. Appl. Probab.* **1989**, *21*, 681–701. [CrossRef]
7. Borovkov, A.A.; Boxma, O.J.; Palmowski, Z. On the Integral of the Workload Process of the Single Server Queue. *J. Appl. Probab.* **2003**, *40*, 200–225. [CrossRef]
8. Bramson, M. Stability of Queueing Networks. In *École d'Été de Probabilités de Saint-Flour XXXVI-2006, Lecture Notes in Math*; Springer: Berlin/Heidelberg, Germany, 2008; Volume 1950.

9. Fakinos, D. The Single-Server Queue with Service Depending on Queue Size and with the Preemptive-Resume Last-Come-First-Served Queue Discipline. *J. Appl. Probab.* **1987**, *24*, 758–767. [CrossRef]
10. Thorisson, H. The queue $GI/G/1$: Finite moments of the cycle variables and uniform rates of convergence. *Stoch. Proc. Appl.* **1985**, *19*, 85–99. [CrossRef]
11. Veretennikov, A.Yu. On the rate of convergence to the stationary distribution in the single-server queuing system. *Autom. Remote Control* **2013**, *74*, 1620–1629. [CrossRef]
12. Kerner, Y. The conditional distribution of the residual service time in the $M_n/G/1$ queue. *Stoch. Model.* **2008**, *24*, 364–375. [CrossRef]
13. Zverkina, G.A. On some extended Erlang–Sevastyanov queueing system and its convergence rate. *J. Math. Sci.* **2021**, *254*, 485–503. [CrossRef]
14. Veretennikov, A.Yu. On the rate of convergence for infinite server Erlang–Sevastyanov's problem. *Queueing Syst.* **2014**, *76*, 181–203. [CrossRef]
15. Veretennikov, A.Yu. An open problem about the rate of convergence in Erlang-Sevastyanov's model. *Queueing Syst.* **2022**, *100*, 357–359. [CrossRef]
16. Liptser, R.Sh.; Shiryaev, A.N. Stochastic calculus on filtered probability spaces. In *Probability Theory III, Stochastic Calculus*; Prokhorov, Yu.V., Shiryaev, A.N., Eds.; Springer: Berlin/Heidelberg, Germany, 1998; pp. 111–157.
17. Davis, M.H.A. Piecewise-Deterministic Markov Processes: A General Class of Non-Diffusion Stochastic Models. *J. R. Stat. Soc. Ser. Methodol.* **1984**, *46*, 353–388. [CrossRef]

mathematics

MDPI

Article

A Two-Server Queue with Interdependence between Arrival and Service Processes

Sindhu S [1,2], Achyutha Krishnamoorthy [3,4,*] and Dmitry Kozyrev [5,6,*]

1 Department of Mathematics, Model Engineering College, Ernakulam 682021, India; sindhus@mec.ac.in
2 Department of Mathematics, Cochin University of Science and Technology, Ernakulam 682022, India
3 Centre for Research in Mathematics, CMS College, Kottayam 686001, India
4 Department of Mathematics, Central University of Kerala, Kasargod 671316, India
5 V.A. Trapeznikov Institute of Control Sciences of Russian Academy of Sciences, 65 Profsoyuznaya Street, 117997 Moscow, Russia
6 Probability Theory and Cyber Security Department, Peoples' Friendship University of Russia (RUDN University), 6 Miklukho-Maklaya Street, 117198 Moscow, Russia
* Correspondence: krishnamoorthy@cmscollege.ac.in (A.K.); kozyrev-dv@rudn.ru (D.K.)

Abstract: In this paper, we analyse a queueing system with two servers where the arrival and service processes are interdependent. The evolution of these processes is governed by transitions on the product space of three Markov chains, which are descriptors of the arrival and service processes. The transitions in this Markov chain follow a semi-Markov rule and the exponential distribution governs the sojourn times in the states. The stability condition of the system is derived and the stationary distribution is calculated for the system in equilibrium. Several important performance measures are provided, and numerical illustrations of the model are presented.

Keywords: interdependent; Markov chain; semi-Markov process; sojourn time; heterogeneous servers

MSC: 60K25

Citation: S, S.; Krishnamoorthy, A.; Kozyrev, D. A Two-Server Queue with Interdependence between Arrival and Service Processes. *Mathematics* **2023**, *11*, 4692. https://doi.org/10.3390/math11224692

Academic Editors: Steve Drekic, Gurami Tsitsiashvili and Alexander Bochkov

Received: 31 August 2023
Revised: 9 November 2023
Accepted: 14 November 2023
Published: 18 November 2023

1. Introduction

Queueing theory is a branch of applied mathematics that studies waiting lines or queues. It finds widespread applications in various fields, including telecommunications, computer networks, manufacturing systems, and service industries. Two server queueing models are particularly significant, as they offer insights into scenarios where multiple servers handle arriving customers or tasks. While traditional queueing models assume independent arrival and service processes, in many practical situations these processes are interconnected in a way that can significantly impact system performance and behaviour. The study of queues with interdependent arrival and service processes helps in understanding how these mutual influences shape queue dynamics, waiting times, resource utilisation, and overall system efficiency.

1.1. Literature Review of Queues with Interdependence in Arrival and Service Processes

Mitchell et al. [1] investigated an M/M/1 queue in which the arrival and service processes are linked through correlated random variables, conforming to a bivariate exponential distribution. This approach sheds light on how correlations between customer service times and the time between their arrival impact the system. Courtois et al. [2] generalized the M/G/1 queuing process by assuming both the arrival and service rates to be essentially arbitrary functions of the number of customers currently using the system. In addition, they assumed that the amount of service demanded by a customer is conditioned by the queue length at the moment service is begun for that customer. Sengupta [3] investigated a single-server semi-Markovian queue with a first-come-first-served policy in which both the arrival and service processes follow a semi-Markov process. Boxma et al. [4] considered a storage model with a dual interpretation: as a distinctive queueing model featuring interdependence between service requests and subsequent inter-arrival times, and as a fluid production/inventory model operating within a two-state random environment.

Muller et al. [5] studied a queueing system characterised by partial correlation, focusing on the interplay between the quantity of work (service time) introduced by a customer and the subsequent inter-arrival time, which the authors assumed to be interdependent. The focus of the study revolved around a distinctive single-server multi-station alternating queue configuration in which the interplay between preparation and service times involves intricate auto- and cross-correlation patterns. Their investigation was carried out across two distinct scenarios. Vlasiou et al. [6] addressed a configuration involving a single-server multi-station alternating queue characterised by interrelated preparation and service times and exhibiting both auto- and cross-correlations. Adan et al. [7] presented a comprehensive investigation into a single-server queue system in which both the inter-arrival times and the service times depend on a common discrete time Markov chain. This model represents an extension of the well-known MAP/G/1 queue, incorporating interdependencies between inter-arrival and service times. Iyer et al. [8] analysed queueing models characterised by joint density of the inter-arrival and service times. These models involve a mixture of joint densities, and find natural applications in scenarios involving a single server catering to multiclass populations through a shared queue. Additionally, this approach can help to model interdependencies between inter-arrival and service times in the context of queue control models. Buchholz et al. [9] focused on a queueing system in which the inter-arrival times exhibit correlation and the service times are correlated with the inter-arrival times. They established a compelling connection between this correlated queueing model and the MMAP[K]/PH[K]/1 queue, a well-studied structure to which matrix geometric algorithms are readily applicable. Kim et al. [10] analysed the wait time distribution in an M/M/1 queue in which the inter-arrival time between the n-th and $(n+1)$-th customers along with the service time of the n-th customer are modelled as correlated random variables characterised by Downton's bivariate exponential distribution. Krishnamoorthy et al. [11] examined a queueing system with finite Markov-dependent arrival and service batch sizes, and additionally considered correlated successive inter-arrival and service time durations. Dai et al. [12] investigated a unique category of correlated queue in which the service duration of an arriving customer hinges upon the inter-arrival time separating them from the previous customer.

Moiseev et al. [13] examined a two-stage queuing system with infinite servers in which customers arrive according to a Markovian Arrival Process (MAP). In this system, the service times at the first stage and the probability of transitioning to the second stage depend on the type of request, which is determined by the state of the underlying Markov chain of MAP when the request arrives. The authors' analysis of this model utilized the multi-dimensional dynamic screening method.

1.2. Literature Review of Queues with Multiservers

The literature abounds with extensive research concerning queueing models that involve multiple servers and operate under the assumption of independent arrival and service processes. Kleinrock [14] laid the foundation for analysing queueing systems with multiple servers. The earliest works focused on fundamental models characterised by independent customer arrivals and exponential service durations. These pioneering studies offered essential insights into the behaviour of queueing systems, particularly in the context of systems featuring multiple servers. These foundational investigations laid the groundwork for subsequent advancements in the field. Cohen [15] played a crucial role in advancing this trajectory by expanding the analytical framework to accommodate batch arrivals and service times. This extension was a pivotal step, contributing to a more comprehensive understanding of the dynamics of queueing systems with multiple servers. Neuts et al. [16] addressed a notable challenge by highlighting the analytical complexity of queueing systems involving more than two heterogeneous servers. They aptly observed that in such scenarios the steady-state behaviour of the system can only be explored through algorithmic approaches. Building upon this observation, Krishna Kumar et al. [17] studied an M/M/2 queueing system characterised by heterogeneous servers.

There are a number of recent papers dealing with two-server queueing systems, including [18–28], to which interested readers may refer.

A semi-Markov process is both a Markov renewal process and a generalization of the Markov process. Unlike Markov renewal processes, which focus on the number of visits to each state, a semi-Markov process considers random variables describing the time spent in each state before transitioning. In discrete-time Markov processes, transitions occur after fixed unit time intervals; in continuous-time Markov processes, transitions follow exponential distributions for the time spent in each state. A semi-Markov process resembles a Markov process in terms of state-to-state transitions; however, it is set apart by the time spent in each state before moving to the next state being a random variable with parameter(s) that depend on both the state it is currently in and the one to be visited next.

In a semi-Markov process, changes of the occupied states take place according to a Markov chain rule; nevertheless, a random amount of time between changes is required. Suppose that a process can be in any one of N states $1, 2, 3 \ldots, N$ and that each time it enters state $i, i = 1, 2, 3, \ldots, N$ it remains there for a random amount of time before transitioning to state j with probability P_{ij}. The state j that is to be visited next is chosen at the time that state i is visited. The sojourn time in i depends on both i and j. Let the time until the transition from i to j have a distribution F_{ij}, and let $Z(t)$ denote the state at time t; then, $\{Z(t) : t \geq 0\}$ is called a semi-Markov processes.

To define a semi-Markov process, it is necessary to have a finite state space Markov chain (MC). This MC can be one-dimensional or higher, and must be finite. Associated with the MC is its initial probability vector (IPV) and one-step transition probability matrix (TPM). The semi-Markov process evolves as follows. First, choose an initial state j according to the IPV. A particle occupies this state. At this moment, the next state (say, k) to be visited by the particle is chosen; this is sampled using the one-step TPM. The sojourn time of the particle in j is distributed based on the mean time in j as μ_{jk}. At the end of this time, the particle transits to k; at this epoch, the next state (say, m) to be visited is chosen using the one-step TPM. In k, the sojourn time has a mean value μ_{km}. This process continues; the continuous time process is referred to as a semi-Markov process. The sojourn time distribution F_{ij} can be exponential or any general distribution (with the non-negative part of the real line as support). In this paper, the sojourn time distribution is taken as the exponential distribution.

We consider a two-server queueing system in which the arrival and service processes are interdependent. Numerous studies in the existing literature have investigated the interdependence among random variables and processes through various analytical methods. In this paper, we analyse the interdependence among arrival and service processes using a semi-Markov approach. This is a new approach first introduced by Achyutha Krishnamoorthy in 2021 (see [29,30]). The procedure is as follows. Suppose that n processes evolve in an interdependent fashion, as explained below. Assume that these n processes are generated through transitions in the states of n finite state space-distinct Markov chains. They may or may not have absorbing states. Considering the product space of these Markov chains, a Markov chain is imposed on this product space. This chain has an initial probability vector and a one-step transition probability matrix. The combined process evolves through transitions in the elements of the product space. We may assume that these transitions are governed by a semi-Markov rule, with the exception of the sojourn time in each stage | state (n-tuples) being exponentially distributed based on a parameter with dependence on the current state and the one to be visited next (as governed by the one-step transition probability matrix on the product space Markov chain).

In [31], the authors examined two distinct queueing models. Model I used a single-server working vacation queueing system in which the arrival and service processes were interdependent. The evolution of the arrival and service processes occurs through transitions within the product space of two separate Markov chains of orders m and n, respectively. The transitions in the product spaces of these Markov chains are governed by a semi-Markov rule, with sojourn times in the states governed by an exponential distribution with parameters depending on the state it is in and the state to be visited next, which are determined according to the Markov chain rule. During working vacation periods, service is delivered at a reduced rate. The duration of these vacation intervals follows an exponential distribution with the parameter η. In contrast, the second model is based on independent arrival and service processes following phase-type distributions with representations $(\boldsymbol{\alpha}, T)$ of order m and $(\boldsymbol{\beta}, S)$ of order n, respectively. The service time during normal operation is the phase-type distribution indicated above, whereas that during working vacation is a phase-type distribution with representation $(\boldsymbol{\beta}, \theta S), 0 < \theta < 1$.

In the present paper, we analyse a queueing system with two servers having interdependent arrival and service processes. The evolution of these processes is governed by transitions in the product space of three Markov chains. The transitions in this Markov chain follow a semi-Markov rule and the exponential distribution governs the sojourn times in the states.

1.3. Motivation

In queueing theory literature, many papers have been developed to analyse arrival and service processes which are mutually independent with the exception that service can be rendered only when the server and at least one customer to receive the service are available. A few papers have considered successive inter-arrival times and/or successive service times to be correlated, for example, Markovian or batch Markovian arrival/ service processes. Even fewer papers have considered the interdependence of arrival and service processes. For example, the inter-arrival times A_n between the $n-1$ and n-th customers and the service time S_n of the n-th customer were correlated for $n = 1, 2, \ldots$. by Sengupta [3,32]. Van Houdt [33] subsequently filled certain gaps in the work of Sengupta [3]. In all these papers, the authors made specific distributional assumptions with respect to the arrival and service times. Thus, the question arises as to whether it is necessary to specify the distributions of the inter-arrival and service times of customers. It is this question that led us to design the presented approach to multi-server queues.

Thus, in this paper we make no assumptions about the inter-arrival and service time distribution. More importantly, no such assumption can be meaningfully made in the context of modelling this interdependence. There are at most three papers [29–31] that have appeared in the stochastic modelling literature employing this new approach, and no cases involving more than one server have been considered under this framework in the extant literature. Thus, the methodology presented here is quite novel. This procedure can be extended to queues of a finite number with more than two servers. The dimension increases by one as the number of servers increases.

Markov chains provide the simplest dependence among a sequence of random variables. In the present modelling, we assume that arrival and service rendered by the two servers evolve within the product space of three Markov chains: one each for arrival, service by server 1, and service by server 2. Thus, our approach is completely new, especially in service systems with multiple servers.

1.4. Practical Applications

In the correlated inter-arrival time and service time, we can use the following example: with A_n being the sequence of independent and identically distributed inter-arrival times and S_n the sequence of service times of successive customers, the stress is on the correlation, that is, whether the correlation is positive, negative, or zero. A negative correlation indicates that if the inter-arrival time A_n is large, the corresponding service time S_n turns out to be comparatively small, and vice versa. In the positive correlation case, this indicates a "direct proportion" between the two. In our model of interdependence, quick transitions in the arrival phases indicate faster arrival. Accordingly, the servers can accelerate their service speed or, equivalently, their service rate. In the correlated arrival and service processes indicated above, the n-th customer to arrive receives service at a future time after arrival, as the server can be busy at the customer's arrival epoch. While the situation in our model is the same, the reflection of the interdependence of arrival and service is seen simultaneously. This helps to accelerate or decelerate the rate of service provided by the servers. These kinds of adjustments can be seen in the production process; if customers arrive at a fast rate, the production needs to be adjusted to ensure that the queue size does not blow up.

In large airports, when non-domestic flights flights land the passengers are required to proceed to the immigration section. A queue of such passengers is formed; at the head of the queue, the person who manages the distribution of passengers directs them to different immigration counters depending on the slots available at each counter. The processing of each passenger passes through distinct stages in a forward direction. With only slight modifications, the model discussed in the paper can be implemented to clear long queues of passengers efficiently.

Another practical application of a two-server queueing model with interdependent arrival and service processes can be found in scenarios involving distributed data processing or computing. In such scenarios, data packets or tasks arrive at a network of servers, and the rate of arrival and processing time can be influenced by the current state of the servers in the network. The arrival rate of data packets may be influenced by the current number of idle or busy servers. For example, if both servers are idle, the arrival rate may increase, as more data processing capacity is available. Conversely, if both servers are busy, the arrival rate might decrease as new data packets wait to be processed. In such a scenario, the service rate for each server can depend on its current workload.

The salient features of the model discussed in this paper are as follows:

- We introduce a new approach to analysing two-server queueing systems with interdependent arrival and service processes.
- The presented analysis does not involve the Kronecker product/sum, unlike the queueing models with independent arrival and service processes analysed using matrix geometric methods.

Notations and abbreviations used in this paper:

- $CTMC$: continuous-time Markov chain.
- $LIQBD$: level-independent quasi-birth and death.
- $\boldsymbol{e}$: column vector of 1s of appropriate order.
- QBD: quasi-birth and death.

The remainder of the paper is organized as follows: in Section 2, the model is mathematically formulated; in Section 3, we perform a steady-state analysis of the studied queuing model after establishing the stability conditions of the system; in Section 4, performance measures are computed and presented; finally, numerical illustrations of the two models are discussed in Section 5.

2. Model Description and Mathematical Formulation

Consider a queueing system with two heterogeneous servers S_1 and S_2 in which the arrival and service processes are interdependent. When a customer arrives with both servers idle, server S_1 provides service. When both servers are occupied, an arriving customer joins an infinite capacity queue. When the system becomes empty, both servers remain idle. The arrival process passes through several stages, including back and forth. An arriving customer who finds a free server enters for service immediately. The service process provided by server S_1 has n_1 stages, while that provided by server S_2 has n_2 stages. These proceed in the forward direction. The arrival process has m stages. Three Markov chains govern the arrival and service processes. Service provided by S_1 is governed by the Markov chain $\mathcal{X} = \{X_n\}$ and service provided by S_2 is governed by the Markov chain $\mathcal{Y} = \{Y_n\}$. The state space of the Markov chain $\{X_n\}$ is $\{1, 2, 3, \ldots, n_1, n_1 + 1\}$, while that of the

Markov chain $\{Y_n\}$ is $\{1,2,3,\ldots,n_2,n_2+1\}$. The arrival process is governed by a Markov chain $\mathcal{Z} = \{Z_n\}$ with state space $\{1,2,3,\ldots,m,m+1\}$. We can consider the Markov chain on the product space $\mathcal{S} = \mathcal{X} \times \mathcal{Y} \times \mathcal{Z}$ with state space $\{\{1,2,3,\ldots,n_1,n_1+1\}\times\{1,2,3,\ldots,n_2,n_2+1\} \times \{1,2,3,\ldots,m,m+1\}\}$.

The absorbing states of this Markov chain are $\{(i,j,m+1) : 1 \leq i \leq n_1; 1 \leq j \leq n_2\} \bigcup \{(i,n_2+1,k) : 1 \leq i \leq n_1; 1 \leq k \leq m\} \bigcup \{(n_1+1,j,k) : 1 \leq j \leq n_2; 1 \leq k \leq m\} \bigcup \{(n_1+1,n_2+1,k) : 1 \leq k \leq m\} \bigcup \{(n_1+1,j,m+1) : 1 \leq j \leq n_2\} \bigcup \{(i,n_2+1,m+1) : 1 \leq i \leq n_1\} \bigcup \{(n_1+1,n_2+1,m+1)\}\}$. Changes in the first coordinate due to transitions indicate service phase changes provided by server S_1, those in the second coordinate indicate service phase changes provided by server S_2, and those in the third coordinate indicate arrival phase changes. The transitions are interdependent in the sense that the sojourn time in any state (i,j,k) depends on this state as well as the one (say, (i',j',k')) to be visited next. This sojourn time is assumed to be exponentially distributed with parameter $\delta_{(i,j,k)(i',j',k')}$. Because the transitions are interdependent, within a short interval it is possible for none, one, two, or even all coordinates to change with positive probability when a transition occurs. The state space indicated above is for this general case; however, we assume here that at most one change takes place with positive probability. This assumption leads to an infinitesimal generator which is highly sparse. Let the initial probability vector of the arrival process be $\boldsymbol{\alpha} = (\overline{\boldsymbol{\alpha}}, \alpha_{m+1})$, where $\overline{\boldsymbol{\alpha}} = (\alpha_1, \alpha_2, \alpha_3, \ldots \alpha_m)$. We can further simplify the assumption that the service processes for both servers start from stage 1 and move in the forward direction only, that is, the service processes are in the order $1 \to 2 \to 3 \to 4 \ldots n_1 \to n_1+1$ and $1 \to 2 \to 3 \to 4 \ldots n_2 \to n_2+1$, respectively. The initial probability vectors of the service processes of both the servers are $(1,0,0,\ldots)$ n_1+1 and n_2+1 component vectors, respectively. Further, to ensure that arrivals do not occur too quickly, we assume that $\alpha_{m+1} = 0$.

2.1. The QBD Process

The model described in Section 2 can be studied as a Level-Independent Quasi-Birth–Death (LIQBD) process (see Latouche et al. [34]). First, we define the following notations. At time t, let:

$N(t)$ be the number of customers in the system.
$S_1(t)$ be the the phase of the service provided by server S_1.
$S_2(t)$ be the phase of the service provided by server S_2.
$A(t)$ be the phase of arrival.

Here, $\{(N(t), S_1(t), S_2(t), A(t)) : t \geq 0\}$ is an LIQBD process with state space

$\tilde{\Omega} = \{\{(0,*,*,k) : 1 \leq k \leq m\} \bigcup \{(1,i,*,k) : 1 \leq i \leq n_1; 1 \leq k \leq m\} \bigcup \{(1,*,j,k) : 1 \leq j \leq n_2; 1 \leq k \leq m\} \bigcup \{(q,i,j,k) : 1 \leq i \leq n_1; 1 \leq j \leq n_2; 1 \leq k \leq m; q \geq 2\}\}$. In the absence of customers, no service can be provided; this is indicated by the * in the position of the service coordinates (the third and fourth coordinates of the 4-tuple). If only one customer is in the system, either S_1 or S_2 is idle, when S_2 is idle it is represented by $(1,i,*,k)$ and when S_1 is idle it is represented by $(1,*,j,k)$, where $1 \leq i \leq n_1; 1 \leq j \leq n_2; 1 \leq k \leq m$.

2.2. Transitions

The transitions are described in Table 1.

Table 1. Transition rates.

From	To	Rate		Remarks
$(0,*,*,k)$	$(0,*,*,k')$	$\delta_{(*,*,k)(*,*,k')}$	$1 \le k,k' \le m,$	arrival phase change
$(0,*,*,k)$	$(1,1,*,k')$	$\alpha_{k'}\delta_{(*,*,k)(1,*,m+1)}$	$1 \le k,k' \le m$	arrival occurs
$(1,i,*,k)$	$(1,i,*,k')$	$\delta_{(i,*,k)(i,*,k')}$	$1 \le k,k' \le m; 1 \le i \le n_1$	arrival phase change
$(1,n_1,*,k)$	$(0,*,*,k)$	$\delta_{(n_1,*,k)(n_1+1,*,k)}$	$1 \le k \le m$	service completion at S_1
$(1,i,*,k)$	$(1,i+1,*,k)$	$\delta_{(i,*,k)(i+1,*,k)}$	$1 \le k \le m; 1 \le i \le n_1-1$	service phase change at S_1
$(1,i,*,k)$	$(2,i,1,k')$	$\delta_{(i,*,k)(i,1,m+1)}\alpha_{k'}$	$1 \le k,k' \le m, 1 \le i \le n_1$	arrival occurs
$(1,*,j,k)$	$(2,1,j,k')$	$\delta_{(*,j,k)(1,j,m+1)}\alpha_{k'}$	$1 \le k,k' \le m, 1 \le j \le n_2$	arrival occurs
$(1,*,j,k)$	$(1,*,j+1,k)$	$\delta_{(*,j,k)(*,j+1,k)}$	$1 \le j \le n_1-1; 1 \le k \le m;$	service phase change at S_2
$(1,*,j,k)$	$(1,*,j,k')$	$\delta_{(*,j,k)(*,j,k')}$	$1 \le j \le n_2; 1 \le k,k' \le m$	arrival phase change
$(1,*,n_2,k)$	$(0,*,*,k)$	$\delta_{(*,n_2,k)(*,n_2+1,k)}$	$1 \le k \le m$	service completion at S_2
$(2,n_1,j,k)$	$(1,*,j,k)$	$\delta_{(n_1,j,k)(n_1+1,j,k)}$	$1 \le j \le n_2; 1 \le k \le m$	service completion at S_1
$(2,i,n_2,k)$	$(1,i,*,k)$	$\delta_{(i,n_2,k)(i,n_2+1,k)}$	$1 \le i \le n_1; 1 \le k \le m$	service completion at S_2
(q,i,j,k)	$(q+1,i,j,k')$	$\alpha_{k'}\delta_{(i,j,k)(i,j,m+1)}$	$1 \le k,k' \le m; 1 \le i \le n_1; 1 \le j \le n_2; q \ge 2$	arrival occurs
(q,i,j,k)	(q,i,j,k')	$\delta_{(i,j,k)(i,j,k')}$	$1 \le k,k' \le m; 1 \le i \le n_1, 1 \le j \le n_2; q \ge 2$	arrival phase change
(q,i,j,k)	$(q,i+1,j,k)$	$\delta_{(i,j,k)(i+1,j,k)}$	$1 \le k \le m; 1 \le i \le n_1-1; 1 \le j \le n_2; q \ge 2$	service phase change at S_1
(q,i,j,k)	$(q,i,j+1,k)$	$\delta_{(i,j,k)(i,j+1,k)}$	$1 \le k \le m; 1 \le i \le n_1; 1 \le j \le n_2-1; q \ge 2$	service phase change at S_2
(q,n_1,j,k)	$(q-1,1,j,k)$	$\delta_{(n_1,j,k)(n_1+1,j,k)}$	$1 \le k \le m; 1 \le j \le n_2; q \ge 3$	service completion at S_1
(q,i,n_2,k)	$(q-1,i,1,k)$	$\delta_{(i,n_2,k)(i,n_2+1,k)}$	$1 \le k \le m; 1 \le i \le n_1; q \ge 3$	service completion at S_2

The infinitesimal generator of the CTMC is

$$Q^* = \begin{bmatrix} \mathcal{B}_1 & \mathcal{B}_0 & & & & \\ \mathcal{B}_2 & \mathcal{C}_1 & \mathcal{C}_0 & & & \\ & \mathcal{C}_2 & \mathcal{A}_1 & \mathcal{A}_0 & & \\ & & \mathcal{A}_2 & \mathcal{A}_1 & \mathcal{A}_0 & \\ & & & \mathcal{A}_2 & \mathcal{A}_1 & \mathcal{A}_0 \\ & & & \ddots & \ddots & \ddots \end{bmatrix}.$$

Here, $\mathcal{B}_1$ is a square matrix of order m which contains the transition rates within level 0, $\mathcal{B}_0$ is a matrix of order $m \times (n_1+n_2)m$ which contains transition rates from level 0 to level 1, $\mathcal{B}_2$ is a matrix of order $(n_1+n_2)m \times m$ which contains transition rates from level 1 to level 0, $\mathcal{C}_1$ is a square matrix of order $(n_1+n_2)m$ which contains the transition rates within the level 1, $\mathcal{C}_0$ is a matrix of order $(n_1+n_2)m \times n_1n_2m$ which contains transition rates from level 1 to level 2, $\mathcal{C}_2$ is a matrix of order $n_1n_2m \times (n_1+n_2)m$ which contains transition rates from level 2 to level 1, $\mathcal{A}_0$ represents transition rates from level n to level $n+1$ for $n \ge 2$, $\mathcal{A}_1$ represents transition rates within the level n for $n \ge 2$, and $\mathcal{A}_2$ represents transition rates from the level n to level $n-1$ for $n \ge 3$. All of these are square matrices of order n_1n_2m.

The matrix $\mathcal{B}_1$ is provided by

$$\begin{bmatrix} b_1 & \delta_{(*,*,1)(*,*,2)} & \delta_{(*,*,1)(*,*,3)} & \cdots\cdots & \delta_{(*,*,1)(*,*,m)} \\ \delta_{(*,*,2)(*,*,1)} & b_2 & \delta_{(*,*,2)(*,*,3)} & \cdots\cdots & \delta_{(*,*,2)(*,*,m)} \\ \delta_{(*,*,3)(*,*,1)} & \delta_{(*,*,3)(*,*,2)} & b_3 & \cdots\cdots & \delta_{(*,*,3)(*,*,m)} \\ \cdots\cdots & \cdots\cdots & \cdots\cdots & \cdots\cdots & \cdots\cdots \\ \cdots\cdots & \cdots\cdots & \cdots\cdots & \cdots\cdots & \cdots\cdots \\ \delta_{(*,*,m)(*,*,1)} & \delta_{(*,*,m)(*,*,2)} & \delta_{(*,*,m)(*,*,3)} & \cdots\cdots & b_m \end{bmatrix},$$

which is a square matrix of order m where $b_i = -[\sum_{j=1, j\neq i}^{m} \delta_{(*,*,i)(*,*,j)} + \delta_{(*,*,i)(1,*,m+1)}]; 1 \leq i \leq m$.

We define the matrix F as

$$\begin{bmatrix} \alpha_1\delta_{(*,*,1)(1,*,m+1)} & \alpha_2\delta_{(*,*,1)(1,*,m+1)} & \alpha_3\delta_{(*,*,,1)(1,*,m+1)} & \cdots\cdots & \alpha_m\delta_{(*,*,1)(1,*,m+1)} \\ \alpha_1\delta_{(*,*,2)(1,*,m+1)} & \alpha_2\delta_{(*,*,2)(1,*,m+1)} & \alpha_3\delta_{(*,*,2)(1,*,m+)} & \cdots\cdots & \alpha_m\delta_{(*,*,2)(1,*,m+1)} \\ \alpha_1\delta_{(*,*,3)(1,*,m+1)} & \alpha_2\delta_{(*,*,3)(1,*,m+1)} & \alpha_3\delta_{(*,*,3)(1,*,m+1)} & \cdots\cdots & \alpha_m\delta_{(*,*,3)(1,*,m+1)} \\ \cdots\cdots & \cdots\cdots & \cdots\cdots & \cdots\cdots & \cdots\cdots \\ \cdots\cdots & \cdots\cdots & \cdots\cdots & \cdots\cdots & \cdots\cdots \\ \alpha_1\delta_{(*,*,m)(1,*,m+1)} & \alpha_2\delta_{(*,*,m)(1,*,m+1)} & \alpha_3\delta_{(*,*,m)(1,*,m+1)} & \cdots\cdots & \alpha_m\delta_{(*,*,m)(1,*,m+1)} \end{bmatrix},$$

which is a square matrix of order m.

Then, $\mathcal{B}_0$ can be expressed as $\begin{bmatrix} F & \mathbf{0} & \mathbf{0} & \cdots\cdots & \mathbf{0} \end{bmatrix}$.

We write D_1 as

$$\begin{bmatrix} \delta_{(n_1,*,1)(n_1+1,*,1)} & & & & \\ & \delta_{(n_1,*,2)(n_1+1,*,2)} & & & \\ & & \delta_{(n_1,*,3)(n_1+1,*,3)} & & \\ & & & \ddots & \\ & & & & \delta_{(n_1,*,m)(n_1+1,*,m)} \end{bmatrix},$$

which is a square matrix of order m.

We define the matrix D_2 as

$$\begin{bmatrix} \delta_{(*,n_2,1)(*,n_2+1,1)} & & & & \\ & \delta_{(*,n_2,2)(*,n_2+1,2)} & & & \\ & & \delta_{(*,n_2,3)(*,n_2+1,3)} & & \\ & & & \ddots & \\ & & & & \delta_{(*,n_2,m)(*,n_2+1,m)} \end{bmatrix},$$

which is a square matrix of order m.

$$\text{Then, } \mathcal{B}_2 = \begin{bmatrix} \mathbf{0} \\ \mathbf{0} \\ \vdots \\ D_1 \\ \mathbf{0} \\ \mathbf{0} \\ \vdots \\ D_2 \end{bmatrix}.$$

We define

$$E_i \begin{bmatrix} e_{i1} & \delta_{(i,*,1)(i,*,2)} & \delta_{(i,*,1)(i,*,3)} & \cdots & \delta_{(i,*,1)(i,*,m)} \\ \delta_{(i,*,2)(i,*,1)} & e_{i2} & \delta_{(i,*,2)(i,*,3)} & \cdots & \delta_{(i,*,2)(i,*,m)} \\ \delta_{(i,*,3)(i,*,1)} & \delta_{(i,*,3)(i,*,2)} & e_{i3} & \cdots & \delta_{(i,*,3)(i,*,m)} \\ \cdots\cdots & \cdots\cdots & \cdots\cdots & \cdots\cdots & \cdots\cdots \\ \cdots\cdots & \cdots\cdots & \cdots\cdots & \cdots\cdots & \cdots\cdots \\ \delta_{(i,*,m)(i,*,1)} & \delta_{(i,*,m)(i,*,2)} & \delta_{(i,*,m)(i,*,3)} & \cdots & e_{im} \end{bmatrix}$$

which as a square matrix of order m, where

$$e_{iq} = -[\sum_{k=1,q\neq k}^{m} \delta_{(i,*,q)(i,*,k)} + \delta_{(i,*,q)(i,1,m+1)} + \delta_{(i,*,q)(i+1,*,q)}]; 1 \leq i \leq n_1, 1 \leq q \leq m;$$

$$G_i = \begin{bmatrix} \delta_{(i,*,1)(i+1,*,1)} & & & & \\ & \delta_{(i,*,2)(i+1,*,2)} & & & \\ & & \delta_{(i,*,3)(i+1,*,3)} & & \\ & & & \ddots & \\ & & & & \delta_{(i,*,m)(i+1,*,m)} \end{bmatrix}$$

as a square matrix of order m with $1 \leq i \leq n_1 - 1$;

$$H_j = \begin{bmatrix} h_{j1} & \delta_{(*,j,1)(*,j,2)} & \delta_{(*,j,1)(*,j,3)} & \cdots & \delta_{(*,j,1)(*,j,m)} \\ \delta_{(*,j,2)(*,j,1)} & h_{j2} & \delta_{(*,j,2)(*,j,3)} & \cdots & \delta_{(*,j,2)(*,j,m)} \\ \delta_{(*,j,3)(*,j,1)} & \delta_{(*,j,3)(*,j,2)} & h_{j3} & \cdots & \delta_{(*,j,3)(*,j,m)} \\ \cdots\cdots & \cdots\cdots & \cdots\cdots & \cdots\cdots & \cdots\cdots \\ \cdots\cdots & \cdots\cdots & \cdots\cdots & \cdots\cdots & \cdots\cdots \\ \delta_{(*,j,m)(*,j,1)} & \delta_{(*,j,m)(*,j,2)} & \delta_{(*,j,m)(*,j,3)} & \cdots & h_{jm} \end{bmatrix}$$

as a square matrix of order m, where

$$h_{jp} = -[\sum_{p=1,p\neq k}^{m} \delta_{(*,j,p)(*,j,k)} + \delta_{(*,j,p)(*,j,m+1)} + \delta_{(*,j,p)(*,j+1,p)}]; 1 \leq j \leq n_2, 1 \leq k \leq m,$$

and

$$J_j = \begin{bmatrix} \delta_{(*,i,1)(*,i+1,1)} & & & & \\ & \delta_{(*,i,2)(*,i+1,2)} & & & \\ & & \delta_{(*,i,3)(*,i+1,3)} & & \\ & & & \ddots & \\ & & & & \delta_{(*,i,m)(*,i+1,m)} \end{bmatrix}$$

as a square matrix of order m with $1 \leq j \leq n_2 - 1$.

Using these, we can write

$$\mathcal{C}_1 = \begin{bmatrix} E_1 & G_1 & & & & & & & & & \\ & E_2 & G_2 & & & & & & & & \\ & & \ddots & \ddots & & & & & & & \\ & & & E_{n_1-1} & G_{n_1-1} & & & & & & \\ & & & & E_{n_1} & \mathbf{0} & \mathbf{0} & & & & \\ & & & & \mathbf{0} & H_1 & J_1 & & & & \\ & & & & & & H_2 & J_2 & & & \\ & & & & & & & \ddots & \ddots & & \\ & & & & & & & & H_{n_2-1} & J_{n_2-1} \\ & & & & & & & & & H_{n_2} \end{bmatrix}.$$

Let

$$K_i = \begin{bmatrix} \alpha_1\delta_{(i,*,1)(i,1,m+1)} & \alpha_2\delta_{(i,*,1)(i,1,,m+1)} & \alpha_3\delta_{(i,*,1)(i,1,m+1)} & \cdots\cdots & \alpha_m\delta_{(i,*,1)(i,1,m+1)} \\ \alpha_1\delta_{(i,*,2)(i,1,m+1)} & \alpha_2\delta_{(i,*,2)(i,1,,m+1)} & \alpha_3\delta_{(i,*,2)(i,1,m+1)} & \cdots\cdots & \alpha_m\delta_{(i,*,2)(i,1,m+1)} \\ \alpha_1\delta_{(i,*,3)(i,1,m+1)} & \alpha_2\delta_{(i,*,3)(i,1,,m+1)} & \alpha_3\delta_{(i,*,3)(i,1,m+1)} & \cdots\cdots & \alpha_m\delta_{(i,*,3)(i,1,m+1)} \\ \cdots\cdots & \cdots\cdots & \cdots\cdots & \cdots\cdots & \cdots\cdots \\ \cdots\cdots & \cdots\cdots & \cdots\cdots & \cdots\cdots & \cdots\cdots \\ \alpha_1\delta_{(i,*,m)(i,1,m+1)} & \alpha_2\delta_{(i,*,m)(i,1,m+1)} & \alpha_3\delta_{(i,*,m)(i,1,m+1)} & \cdots\cdots & \alpha_m\delta_{(i,*,m)(i,1,m+1)} \end{bmatrix}$$

where i varies from 1 to n_1. K_i is a square matrix of order m.

Let

$$L_j = \begin{bmatrix} \alpha_1\delta_{(*,j,1)(1,j,m+1)} & \alpha_2\delta_{(*,j,1)(1,j,m+1)} & \alpha_3\delta_{(*,j,1)(1,j,m+1)} & \cdots\cdots & \alpha_m\delta_{(*,j,1)(1,j,m+1)} \\ \alpha_1\delta_{(*,j,2)(i,j,m+1)} & \alpha_2\delta_{(*,j,2)(1,j,,m+1)} & \alpha_3\delta_{(*,j,2)(1,j,m+1)} & \cdots\cdots & \alpha_m\delta_{(*,j,2)(1,j,m+1)} \\ \alpha_1\delta_{(*,j,3)(*,j,m+1)} & \alpha_2\delta_{(*,j,3)(*,j,m+1)} & \alpha_3\delta_{(*,j,3)(1,j,m+1)} & \cdots\cdots & \alpha_m\delta_{(*,j,3)(1,j,m+1)} \\ \cdots\cdots & \cdots\cdots & \cdots\cdots & \cdots\cdots & \cdots\cdots \\ \cdots\cdots & \cdots\cdots & \cdots\cdots & \cdots\cdots & \cdots\cdots \\ \alpha_1\delta_{(*,j,m)(1,j,m+1)} & \alpha_2\delta_{(*,j,m)(1,j,m+1)} & \alpha_3\delta_{(*,j,m)(1,j,m+1)} & \cdots\cdots & \alpha_m\delta_{(*,j,m)(1,j,m+1)} \end{bmatrix}$$

where j varies from 1 to n_2. L_j is a square matrix of order m.

$\mathcal{K}_i = \begin{bmatrix} K_i & \mathbf{0} & \mathbf{0} & \cdots\cdots & \mathbf{0} \end{bmatrix}$ is a matrix of order $m \times n_2 m$.

$$\mathcal{L} = \begin{bmatrix} L_1 & \mathbf{0} & \mathbf{0} & \cdots\cdots & \mathbf{0} \\ \mathbf{0} & L_2 & \mathbf{0} & \cdots\cdots & \mathbf{0} \\ \mathbf{0} & \mathbf{0} & L_3 & \cdots\cdots & \mathbf{0} \\ \cdots\cdots & \cdots\cdots & \cdots\cdots & \cdots\cdots & \cdots\cdots \\ \cdots\cdots & \cdots\cdots & \cdots\cdots & \cdots\cdots & \cdots\cdots \\ \mathbf{0} & \mathbf{0} & \mathbf{0} & \cdots\cdots & L_{n_2} \end{bmatrix} \text{ is a square matrix of order } n_2 m.$$

With these notations, $\mathcal{C}_0$ takes the form

$$\mathcal{C}_0 = \begin{bmatrix} \mathcal{K}_1 & \mathbf{0} & \mathbf{0} & \cdots\cdots & \mathbf{0} \\ \mathbf{0} & \mathcal{K}_2 & \mathbf{0} & \cdots\cdots & \mathbf{0} \\ \mathbf{0} & \mathbf{0} & \mathcal{K}_3 & \cdots\cdots & \mathbf{0} \\ \cdots\cdots & \cdots\cdots & \cdots\cdots & \cdots\cdots & \cdots\cdots \\ \cdots\cdots & \cdots\cdots & \cdots\cdots & \cdots\cdots & \cdots\cdots \\ \mathbf{0} & \mathbf{0} & \mathbf{0} & \cdots\cdots & \mathcal{K}_{n_1} \\ \mathcal{L} & \mathbf{0} & \mathbf{0} & \cdots\cdots & \mathbf{0} \end{bmatrix}.$$

Let

$$M_i = \begin{bmatrix} \delta_{(i,n_2,1)(i,n_2+1,1)} & & & & \\ & \delta_{(i,n_2,2)(i,n_2+1,2)} & & & \\ & & \delta_{(i,n_2,3)(i,n_2+1,3)} & & \\ & & & \ddots & \\ & & & & \delta_{(i,n_2,m)(i,n_2+1,m)} \end{bmatrix}$$

be a square matrix of order m for $1 \le i \le n_1$. Furthermore, let $P_i = \begin{bmatrix} \mathbf{0} \\ \mathbf{0} \\ \vdots \\ M_i \end{bmatrix}$ be a matrix of order $n_2 m \times m$ with $1 \le i \le n_1$.

We define

$$Q_j = \begin{bmatrix} \delta_{(n_1,j,1)(n_1+1,j,1)} & & & & \\ & \delta_{(n_1,j,2)(n_1+1,j,2)} & & & \\ & & \delta_{(n_1,j,3)(n_1+1,j,3)} & & \\ & & & \ddots & \\ & & & & \delta_{(n_1,j,m)(n_1+1,j,m)} \end{bmatrix}$$

as a square matrix of order m for $1 \le j \le n_2$.

$V = \begin{bmatrix} Q_1 & \mathbf{0} & \mathbf{0} & \cdots\cdots & \mathbf{0} \\ \mathbf{0} & Q_2 & \mathbf{0} & \cdots\cdots & \mathbf{0} \\ \mathbf{0} & \mathbf{0} & Q_3 & \cdots\cdots & \mathbf{0} \\ \cdots\cdots & \cdots\cdots & \cdots\cdots & \cdots\cdots & \cdots\cdots \\ \cdots\cdots & \cdots\cdots & \cdots\cdots & \cdots\cdots & \cdots\cdots \\ \mathbf{0} & \mathbf{0} & \mathbf{0} & \cdots\cdots & Q_{n_2} \end{bmatrix}$ is a square matrix of order $n_2 m$.

Then,

$$C_2 = \begin{bmatrix} P_1 & \mathbf{0} & \mathbf{0} & \mathbf{0} & \mathbf{0} & \cdots\cdots & \mathbf{0} & \mathbf{0} \\ \mathbf{0} & P_2 & \mathbf{0} & \mathbf{0} & \mathbf{0} & \cdots\cdots & \mathbf{0} & \mathbf{0} \\ \mathbf{0} & \mathbf{0} & P_3 & \mathbf{0} & \mathbf{0} & \cdots\cdots & \mathbf{0} & \mathbf{0} \\ \cdots\cdots & \cdots\cdots & \cdots\cdots & \cdots\cdots & \cdots\cdots & \cdots\cdots & \cdots\cdots & \cdots\cdots \\ \cdots\cdots & \cdots\cdots & \cdots\cdots & \cdots\cdots & \cdots\cdots & \cdots\cdots & \cdots\cdots & \cdots\cdots \\ \mathbf{0} & \mathbf{0} & \mathbf{0} & \mathbf{0} & \mathbf{0} & \cdots\cdots & P_{n_1} & V \end{bmatrix}.$$

We define

$$S_{ij} = \begin{bmatrix} S_{ij1} & \delta_{(i,j,1)(i,j,2)} & \delta_{(i,j,1)(i,j,3)} & \cdots\cdots & \delta_{(i,j,1)(i,j,m)} \\ \delta_{(i,j,2)(i,j,1)} & S_{ij2} & \delta_{(i,j,2)(i,j,,3)} & \cdots\cdots & \delta_{(i,j,2)(i,j,,m)} \\ \delta_{(i,j,3)(i,j,1)} & \delta_{(i,j,3)(i,j,2)} & S_{ij3} & \cdots\cdots & \delta_{(i,j,3)(i,j,m)} \\ \cdots\cdots & \cdots\cdots & \cdots\cdots & \cdots\cdots & \cdots\cdots \\ \cdots\cdots & \cdots\cdots & \cdots\cdots & \cdots\cdots & \cdots\cdots \\ \delta_{(i,j,m)(i,j,1)} & \delta_{(i,j,,m)(i,j,2)} & \delta_{(i,j,m)(i,j,3)} & \cdots\cdots & S_{ijm} \end{bmatrix}$$

as a square matrix of order m with $1 \le j \le n_2$.

In the above, $S_{ijh} = -[\sum_{k=1,h\neq k}^{m+1} \delta_{(i,j,h)(i,j,k)} + \delta_{(i,j,h)(i,j+1,h)} + \delta_{(i,j,h)(i+1,j,h)}]$ for $1 \le h \le m$ with $1 \le j \le n_2$.

Take

$$T_{ij} = \begin{bmatrix} \delta_{(i,j,1)(i,j+1,1)} & & & & \\ & \delta_{(i,j,2)(i,j+1,2)} & & & \\ & & \delta_{(i,j,3)(i,j+1,3)} & & \\ & & & \ddots & \\ & & & & \delta_{(i,j,m)(i,j+1,m)} \end{bmatrix},$$

which is a square matrix of order m with $1 \leq j \leq n_2 - 1$.

Let $N_i = \begin{bmatrix} S_{i1} & T_{i1} & & & & \\ & S_{i2} & T_{i2} & & & \\ & & S_{i3} & T_{i3} & & \\ & & & \ddots & \ddots & \\ & & & & S_{i(n_2-1} & T_{i(n_2-1)} \\ & & & & & S_{in_2} \end{bmatrix}$ be a square matrix of order mn_2 with $1 \leq i \leq n_1$.

Let $U_{ij} = \begin{bmatrix} \delta_{(i,j,1)(i+1,j,1)} & & & & \\ & \delta_{(i,j,2)(i+1,j,2)} & & & \\ & & \delta_{(i,j,3)(i+1,j,3)} & & \\ & & & \ddots & \\ & & & & \delta_{(i,j,m)(i+1,j,m)} \end{bmatrix}$ be a square matrix of order m for $1 \leq j \leq n_2$.

Let $W_i = \begin{bmatrix} U_{i1} & & & & & \\ & U_{i2} & & & & \\ & & U_{i3} & & & \\ & & & \ddots\ddots & & \\ & & & & U_{i(n_2-1)} & \\ & & & & & U_{in_2} \end{bmatrix}$ be a square matrix of order mn_2 with $1 \leq i \leq n_1 - 1$.

Using the above notations, $\mathcal{A}_1$ can be written as

$$\mathcal{A}_1 = \begin{bmatrix} N_1 & W_1 & & & & \\ & N_2 & W_2 & & & \\ & & N_3 & W_3 & & \\ & & & \ddots & \ddots & \\ & & & & N_{n_1-1} & W_{n_1-1} \\ & & & & & N_{n_1} \end{bmatrix}.$$

Let $\mathcal{D}_i = \begin{bmatrix} P_i & \mathbf{0} & \mathbf{0} & \ldots\ldots & \mathbf{0} \end{bmatrix}$ be a square matrix of order n_2m for $1 \leq i \leq n_1$. Using these, $\mathcal{A}_2$ can be expressed as

$$\mathcal{A}_2 = \begin{bmatrix} \mathcal{D}_1 & & & & & \\ & \mathcal{D}_2 & & & & \\ & & \mathcal{D}_3 & & & \\ & & & \ddots & & \\ & & & & \mathcal{D}_{n_1-1} & \\ V & & & & & \mathcal{D}_{n_1} \end{bmatrix}.$$

We define $\mathcal{F}_{ij}$ as follows:

$$\mathcal{F}_{ij} = \begin{bmatrix} \alpha_1\delta_{(i,j,1)(i,j,m+1)} & \alpha_2\delta_{(i,j,1)(i,j,m+1)} & \alpha_3\delta_{(i,j,1)(i,j,m+1)} & \cdots\cdots & \alpha_m\delta_{(i,j,1)(i,j,m+1)} \\ \alpha_1\delta_{(i,j,2)(i,j,m+1)} & \alpha_2\delta_{(i,j,2)(i,j,m+1)} & \alpha_3\delta_{(i,j,2)(i,j,m+1)} & \cdots\cdots & \alpha_m\delta_{(i,j,2)(i,j,m+1)} \\ \alpha_1\delta_{(i,j,3)(i,j,m+1)} & \alpha_2\delta_{(i,j,3)(i,j,m+1)} & \alpha_3\delta_{(i,j,3)(i,j,m+1)} & \cdots\cdots & \alpha_m\delta_{(i,j,3)(i,j,m+1)} \\ \cdots\cdots & \cdots\cdots & \cdots\cdots & \cdots\cdots & \cdots\cdots \\ \cdots\cdots & \cdots\cdots & \cdots\cdots & \cdots\cdots & \cdots\cdots \\ \alpha_1\delta_{(i,j,m)(i,j,m+1)} & \alpha_2\delta_{(i,j,m)(i,j,m+1)} & \alpha_3\delta_{(i,j,m)(i,j,m+1)} & \cdots\cdots & \alpha_m\delta_{(i,j,m)(i,j,m+1)} \end{bmatrix}$$

where j varies from 1 to n_2. $\mathcal{F}_{ij}$ is a square matrix of order m.

Then, we can write

$$\mathcal{G}_i = \begin{bmatrix} \mathcal{F}_{i1} & & & & & \\ & \mathcal{F}_{i2} & & & & \\ & & \mathcal{F}_{i3} & & & \\ & & & \ddots\ddots & & \\ & & & & \mathcal{F}_{in_2-1} & \\ & & & & & \mathcal{F}_{in_2} \end{bmatrix},$$

which is a square matrix of order mn_2 with $1 \leq i \leq n_1$.

These help us to represent $\mathcal{A}_0$ as follows:

$$\mathcal{A}_0 = \begin{bmatrix} \mathcal{G}_1 & & & & & \\ & \mathcal{G}_2 & & & & \\ & & \mathcal{G}_3 & & & \\ & & & \ddots & & \\ & & & & \mathcal{G}_{n_1-1} & \\ & & & & & \mathcal{G}_{n_1} \end{bmatrix}.$$

Next, we proceed to compute the stability condition of the system and obtain the system state probability vectors. From each of these vectors, the system occupying a specific state can be computed.

3. Steady State Analysis

In this section, we perform steady-state analysis of the queueing model by initially determining the stability condition of the system.

3.1. Stability Condition

The infinitesimal generator $\mathcal{A} = \mathcal{A}_0 + \mathcal{A}_1 + \mathcal{A}_2$ is

$$\begin{bmatrix} \mathcal{G}_1 + \mathcal{N}_1 + \mathcal{D}_1 & \mathcal{U}_1 & & & & \\ & \mathcal{G}_2 + \mathcal{N}_2 + \mathcal{D}_2 & \mathcal{U}_2 & & & \\ & & \mathcal{G}_3 + \mathcal{N}_3 + \mathcal{D}_3 & \mathcal{U}_3 & & \\ & & & \ddots\ddots & & \\ & & & & \mathcal{G}_{n_1-1} + \mathcal{N}_{n_1-1} + \mathcal{D}_{n_1-1} & \mathcal{U}_{n_1} - 1 \\ V & & & & & \mathcal{G}_{n_1} + \mathcal{N}_{n_1} + \mathcal{D}_{n_1}, \end{bmatrix}$$

which is a square matrix of order $n_1 n_2 m$

Let $\boldsymbol{\pi} = (\boldsymbol{\pi}_1, \boldsymbol{\pi}_2, \boldsymbol{\pi}_3, \ldots\ldots, \boldsymbol{\pi}_{n_1})$ denote the steady-state probability vector of the generator matrix $\mathcal{A}$.

Here, $\boldsymbol{\pi}$ is of dimension $1 \times n_1 n_2 m$ and $\boldsymbol{\pi}_r$ is of dimension $1 \times n_2 m$ for $r = 1, 2, \ldots n_1$.

The steady state probability vector $\boldsymbol{\pi}$ satisfies the equations

$$\boldsymbol{\pi}\mathcal{A} = 0, \boldsymbol{\pi}\mathbf{e} = 1. \quad (1)$$

Using Equation (1), we obtain

$$\pi_1[\mathcal{G}_1+\mathcal{N}_1+\mathcal{D}_1]+\pi_{n_1}V=\mathbf{0} \tag{2}$$

$$\pi_1\mathcal{U}_1+\pi_2[\mathcal{G}_2+\mathcal{N}_2+\mathcal{D}_2]=\mathbf{0} \tag{3}$$

$$\pi_2\mathcal{U}_2+\pi_3[\mathcal{G}_3+\mathcal{N}_3+\mathcal{D}_3]=\mathbf{0} \tag{4}$$

$$\pi_{n_1-1}\mathcal{U}_{n_1-1}+\pi_{n_1}[\mathcal{G}_{n_1}+\mathcal{N}_{n_1}+\mathcal{D}_{n_1}]=\mathbf{0} \tag{5}$$

$$\pi_1\times\mathbf{e}+\pi_2\times\mathbf{e}+\ldots.+\pi_{n_1}\times\mathbf{e}=1. \tag{6}$$

By solving the above system equations, we can find $\pi_1,\pi_2,\pi_3,\ldots\ldots\pi_{n_1}$.
The LIQBD description of the model indicates that the queueing system is stable if and only if the left drift exceeds the right drift (see Theorem 3.1.1 of Neuts [35]), that is,

$$\pi\mathcal{A}_0\mathbf{e}<\pi\mathcal{A}_2\mathbf{e}, \tag{7}$$

$$\pi\mathcal{A}_0\mathbf{e}=\pi_1\mathcal{G}_1\mathbf{e}+\pi_2\mathcal{G}_2\mathbf{e}+\pi_3\mathcal{G}_3\mathbf{e}+\ldots+\pi_{n_1}\mathcal{G}_{n_1}\mathbf{e}=\sum_{i=1}^{n_1}\pi_i\mathcal{G}_i\mathbf{e}, \tag{8}$$

$$\pi A_2\mathbf{e}=\pi_1\mathcal{D}_1\mathbf{e}+\pi_2\mathcal{D}_2\mathbf{e}+\pi_3\mathcal{D}_3\mathbf{e}+\ldots+\pi_{n_1}\mathcal{D}_{n_1}\mathbf{e}+\pi_{n_1}V\mathbf{e}=\sum_{i=1}^{n_1}\pi_i\mathcal{D}_i\mathbf{e}+\pi_{n_1}V\mathbf{e}. \tag{9}$$

Theorem 1. *The given system is stable if and only if*

$$\sum_{i=1}^{n_1}\pi_i\mathcal{G}_ie<\sum_{i=1}^{n_1}\pi_i\mathcal{D}_ie+\pi_{n_1}Ve. \tag{10}$$

3.2. The Steady State Probability Vector of Q^*

Let $\boldsymbol{x}$ be the steady state probability vector of Q^*: $\boldsymbol{x}=(\boldsymbol{x}_0,\boldsymbol{x}_1,\boldsymbol{x}_2\ldots)$, where $\boldsymbol{x}_0$ is of dimension $1\times m$, $\boldsymbol{x}_1$ is of dimension $1\times(n_1+n_2)m$, and $\boldsymbol{x}_2,\boldsymbol{x}_3,\ldots\boldsymbol{x}_{n_1}$ are of dimension $1\times n_1n_2m$.

Under the stability condition, we have $\boldsymbol{x_i}=\boldsymbol{x}_2R^{i-2}, i\geq 3$, where the matrix R is the minimal nonnegative solution to the matrix quadratic equation

$$R^2\mathcal{A}_2+R\mathcal{A}_1+\mathcal{A}_0=0$$

and the vectors $\boldsymbol{x}_0$, $\boldsymbol{x}_1$ and $\boldsymbol{x}_2$ are obtained by solving the equations

$$\boldsymbol{x}_0\mathcal{B}_1+\boldsymbol{x}_1\mathcal{B}_2=0 \tag{11}$$

$$\boldsymbol{x}_0\mathcal{B}_0+\boldsymbol{x}_1\mathcal{C}_1+\boldsymbol{x_2}\mathcal{C}_2=0 \tag{12}$$

$$\boldsymbol{x}_1\mathcal{C}_0+\boldsymbol{x}_2(\mathcal{A}_1+R\mathcal{A}_2)=0 \tag{13}$$

subject to the normalizing condition

$$\boldsymbol{x}_0\boldsymbol{e}+\boldsymbol{x}_1\boldsymbol{e}+\boldsymbol{x}_2(I-R)^{-1}\boldsymbol{e}=1. \tag{14}$$

Having obtained the system state probability vectors, we can now make use of them to compute important performance measures.

4. Performance Measures

Formulae for key performance measures are derived.

- The probability that both of the servers are idle: $p_{idle}=\boldsymbol{x}_0\mathbf{e}$.
- The probability that only Server 1 (S_1) is idle:

$$p_{1idle}=\sum_{j=1}^{n_2}\sum_{k=1}^{m}\boldsymbol{x}_{1*jk}$$

- The probability that only Server 2 (S_2) is idle:

$$p_{2idle} = \sum_{i=1}^{n_1} \sum_{k=1}^{m} \mathbf{x}_{1i*k}$$

- The probability that Server 1 is busy:

$$P_{1busy} = \sum_{q=2}^{\infty} \sum_{i=1}^{n_1} \sum_{j=1}^{n_2} \sum_{k=1}^{m} \mathbf{x}_{qijk} + \sum_{i=1}^{n_1} \sum_{k=1}^{m} \mathbf{x}_{1i*k}$$

- The probability that Server 2 is busy:

$$P_{2busy} = \sum_{q=2}^{\infty} \sum_{i=1}^{n_1} \sum_{j=1}^{n_2} \sum_{k=1}^{m} \mathbf{x}_{qijk} + \sum_{j=1}^{n_2} \sum_{k=1}^{m} \mathbf{x}_{1*jk}$$

- The probability that both the servers are busy:

$$P_{busy} = \sum_{q=2}^{\infty} \sum_{i=1}^{n_1} \sum_{j=1}^{n_2} \sum_{k=1}^{m} \mathbf{x}_{qijk}$$

- The probability of q customers being in the system: $P_q = \mathbf{x}_q \mathbf{e}$.
- The expected number of customers in the system:

$$ECS = \sum_{q=1}^{\infty} q\mathbf{x}_q \mathbf{e}$$

- The expected number of customers in the queue:

$$ECQ = \sum_{q=2}^{\infty} (q-2)\mathbf{x}_q \mathbf{e}$$

5. Numerical Illustration

In this section, we provide four numerical illustrations of our model.

We take $n_1 = 2, n_2 = 3$, and $m = 2$. The state space of the arrival process is $1,2,3$, where the transient states are 1,2 and the absorbing state is 3. The state space of the service process provided by server 1 (S_1) is $\{1,2,3\}$, where the transient states are 1,2 and the absorbing state is 3. The state space of the service process provided by server 2 (S_2) is $\{1,2,3,4\}$, where the transient states are 1,2,3 and the absorbing state is 4. Then, the state space of the Markov chain $\mathcal{X} = \{(i,j,k) : 1 \leq i \leq 3; 1 \leq j \leq 4; 1 \leq k \leq 3\}$. The absorbing states of $\mathcal{X}$ are: $\{\{(i,j,3) : 1 \leq i \leq 3; 1 \leq j \leq 4\} \bigcup \{(i,4,k) : 1 \leq i \leq 3; 1 \leq k \leq 3\} \bigcup \{(3,j,k) : 1 \leq j \leq 4; 1 \leq k \leq 3\}\} \bigcup \{(3,4,k) : 1 \leq k \leq 3\} \bigcup \{(3,j,3) : 1 \leq j \leq 4\} \bigcup \{(i,4,3) : 1 \leq i \leq 3\} \bigcup \{(3,4,3)\}$.

We assume that at most one coordinate change in a transition has a positive probability. In the absence of customers, no service can be provided, as indicated by a * symbol in the position of the service coordinates. Here, $\boldsymbol{\alpha} = (0.7, 0.3)$, $\alpha_{m+1} = 0$. The case of more than one coordinate change in a single transition (only up to three being possible in the present case, as we have the service stages provided by each server as first two coordinates and the changes in the arrival phases as the last coordinate) can be considered in a similar fashion. Consequently, we obtain a much less sparse transition rate matrix.

5.1. Illustration 1

The transition rates are taken as shown in Tables 2 and 3.

Table 2. Transition rates.

	(2,1,1)	(2,1,2)	(2,2,1)	(2,2,2)	(2,3,1)	(2,3,2)	(2,1,3)	(2,2,3)	(2,3,3)	(2,4,1)	(2,4,2)	(3,1,1)	(3,1,2)	(3,2,1)	(3,2,2)	(3,3,1)	(3,3,2)
(1,1,1)	3	0	0	0	0	0	0	0	0	0	0	0	0	0	0	0	0
(1,1,2)	0	3.1	0	0	0	0	0	0	0	0	0	0	0	0	0	0	0
(1,2,1)	0	0	3.1	0	0	0	0	0	0	0	0	0	0	0	0	0	0
(1,2,2)	0	0	0	2.3	0	0	0	0	0	0	0	0	0	0	0	0	0
(1,3,1)	0	0	0	0	2.8	0	0	0	0	0	0	0	0	0	0	0	0
(1,3,2)	0	0	0	0	0	2.3	0	0	0	0	0	0	0	0	0	0	0
(2,1,1)	−10.7	3.1	2.8	0	0	0	1.7	0	0	0	0	3.1	0	0	0	0	0
(2,1,2)	2.6	−12.3	0	3.1	0	0	3.1	0	0	0	0	0	3.5	0	0	0	0
(2,2,1)	0	0	−13.1	2.4	3.8	0	0	3.2	0	0	0	0	0	3.7	0	0	0
(2,2,2)	0	0	2.7	−11.1	0	2.2	0	3.1	0	0	0	0	0	0	3.1	0	0
(2,3,1)	0	0	0	0	−9.5	1.1	0	0	2.6	3	0	0	0	0	0	2.8	0
(2,3,2)	0	0	0	0	1.9	−12.1	0	0	2.5	0	4.2	0	0	0	0	0	3.5

Using the transition rates shown in Tables 2 and 3, we obtain the values of the performance measures as follows.

$$
\begin{aligned}
P_{idle} &= 0.0134\\
P_{1idle} &= 0.0420\\
P_{2idle} &= 0.0365\\
P_{1busy} &= 0.9445\\
P_{2busy} &= 0.9501\\
P_{busy} &= 0.9081\\
ECS &= 10.7079\\
ECQ &= 8.8133
\end{aligned}
$$

5.2. Illustration 2

The transition rates are taken as shown in Tables 4 and 5.

Using the transition rates shown in Tables 4 and 5, we obtain the values of the performance measures are as follows.

$$
\begin{aligned}
P_{idle} &= 0.0040\\
P_{1idle} &= 0.0122\\
P_{2idle} &= 0.0113\\
P_{1busy} &= 0.9838\\
P_{2busy} &= 0.9847\\
P_{busy} &= 0.9725\\
ECS &= 40.4592\\
ECQ &= 38.4907
\end{aligned}
$$

5.3. Illustration 3

The transition Rates are taken as shown in Tables 6 and 7.

Table 3. Transitions rates.

	(∗,∗,1)	(∗,∗,2)	(1,∗,1)	(1,∗,2)	(2,∗,1)	(2,∗,2)	(∗,1,1)	(∗,1,2)	(∗,2,1)	(∗,2,2)	(∗,3,1)	(∗,3,2)	(1,1,1)	(1,1,2)	(1,2,1)	(1,2,2)	(1,3,1)	(1,3,2)	(1,4,1)	(1,4,2)	(1,∗,3)	(3,∗,1)	(3,∗,2)	(∗,4,1)	(∗,4,2)	(1,1,3)	(1,2,3)	(1,3,3)	(2,1,3)
(∗,∗,1)	−4.5	1.3	0	0	0	0	0	0	0	0	0	0	0	0	0	0	0	0	0	0	3.2	0	0	0	0	0	0	0	0
(∗,∗,2)	2.4	−5.2	0	0	0	0	0	0	0	0	0	0	0	0	0	0	0	0	0	0	2.8	0	0	0	0	0	0	0	0
(1,∗,1)	0	0	−8.1	3.2	3.1	0	0	0	0	0	0	0	0	0	0	0	0	0	0	0	0	0	0	0	0	1.8	0	0	0
(1,∗,2)	0	0	2.4	−6.7	0	2.7	0	0	0	0	0	0	0	0	0	0	0	0	0	0	0	0	0	0	0	1.6	0	0	0
(2,∗,1)	0	0	0	0	−6.5	1.5	0	0	0	0	0	0	0	0	0	0	0	0	0	0	0	2.8	0	0	0	0	0	0	2.2
(2,∗,2)	0	0	0	0	2.1	−5.9	0	0	0	0	0	0	0	0	0	0	0	0	0	0	0	0	1.9	0	0	0	0	0	1.9
(∗,1,1)	0	0	0	0	0	0	−6.2	1.7	2.6	0	0	0	0	0	0	0	0	0	0	0	0	0	0	0	0	1.9	0	0	0
(∗,1,2)	0	0	0	0	0	0	2.2	−6.4	0	2.4	0	0	0	0	0	0	0	0	0	0	0	0	0	0	0	1.8	0	0	0
(∗,2,1)	0	0	0	0	0	0	0	0	−6.1	2.4	2.1	0	0	0	0	0	0	0	0	0	0	0	0	0	0	0	1.6	0	0
(∗,2,2)	0	0	0	0	0	0	0	0	2.9	−8	0	2.9	0	0	0	0	0	0	0	0	0	0	0	0	0	0	2.2	0	0
(∗,3,1)	0	0	0	0	0	0	0	0	0	0	−6.7	2.2	0	0	0	0	0	0	0	0	0	0	0	2.1	0	0	0	2.4	0
(∗,3,2)	0	0	0	0	0	0	0	0	0	0	2.1	−8.5	0	0	0	0	0	0	0	0	0	0	0	0	3.8	0	0	2.6	0
(1,1,1)	0	0	0	0	0	0	0	0	0	0	0	0	−11.2	3	2.6	0	0	0	0	0	0	0	0	0	0	2.6	0	0	0
(1,1,2)	0	0	0	0	0	0	0	0	0	0	0	0	2.8	−11.7	0	2.7	0	0	0	0	0	0	0	0	0	3.1	0	0	0
(1,2,1)	0	0	0	0	0	0	0	0	0	0	0	0	0	0	−8.9	1.5	2.2	0	0	0	0	0	0	0	0	0	2.1	0	0
(1,2,2)	0	0	0	0	0	0	0	0	0	0	0	0	0	0	3.2	−10	0	2.5	0	0	0	0	0	0	0	0	2	0	0
(1,3,1)	0	0	0	0	0	0	0	0	0	0	0	0	0	0	0	0	−10.7	1.9	4.2	0	0	0	0	0	0	0	0	1.8	0
(1,3,2)	0	0	0	0	0	0	0	0	0	0	0	0	0	0	0	0	2.5	−11.4	0	3.4	0	0	0	0	0	0	0	3.2	0

Table 4. Transition rates.

	(2,1,1)	(2,1,2)	(2,2,1)	(2,2,2)	(2,3,1)	(2,3,2)	(2,1,3)	(2,2,3)	(2,3,3)	(2,4,1)	(2,4,2)	(3,1,1)	(3,1,2)	(3,2,1)	(3,2,2)	(3,3,1)	(3,3,2)
(1,1,1)	3.2	0	0	0	0	0	0	0	0	0	0	0	0	0	0	0	0
(1,1,2)	0	3.2	0	0	0	0	0	0	0	0	0	0	0	0	0	0	0
(1,2,1)	0	0	3.3	0	0	0	0	0	0	0	0	0	0	0	0	0	0
(1,2,2)	0	0	0	2.5	0	0	0	0	0	0	0	0	0	0	0	0	0
(1,3,1)	0	0	0	0	2.7	0	0	0	0	0	0	0	0	0	0	0	0
(1,3,2)	0	0	0	0	0	2.6	0	0	0	0	0	0	0	0	0	0	0
(2,1,1)	−11.1	3.5	2.6	0	0	0	1.8	0	0	0	0	3.2	0	0	0	0	0
(2,1,2)	2.4	−11.9	0	3.3	0	0	3	0	0	0	0	0	3.2	0	0	0	0
(2,2,1)	0	0	−13.2	2.6	3.2	0	0	3.5	0	0	0	0	0	3.9	0	0	0
(2,2,2)	0	0	2.5	−12.8	0	3.4	0	3.5	0	0	0	0	0	0	3.4	0	0
(2,3,1)	0	0	0	0	−9.3	1	0	0	2.9	3.1	0	0	0	0	0	2.3	0
(2,3,2)	0	0	0	0	1.5	−12.3	0	0	2.2	0	4.9	0	0	0	0	0	3.7

Using the transition rates shown in Tables 6 and 7, we obtain the values of the performance measures as follows.

$$
\begin{aligned}
P_{idle} &= 0.0023 \\
P_{1idle} &= 0.0064 \\
P_{2idle} &= 0.0059 \\
P_{1busy} &= 0.9913 \\
P_{2busy} &= 0.9918 \\
P_{busy} &= 0.9854 \\
ECS &= 78.0583 \\
ECQ &= 76.0752
\end{aligned}
$$

5.4. Illustration 4

The transition rates are taken as shown in Tables 8 and 9.

Using the transition rates shown in Tables 8 and 9, we obtain the values of the performance measures as follows.

$$
\begin{aligned}
P_{idle} &= 0.0020 \\
P_{1idle} &= 0.0055 \\
P_{2idle} &= 0.0051 \\
P_{1busy} &= 0.9925 \\
P_{2busy} &= 0.9929 \\
P_{busy} &= 0.9874 \\
ECS &= 89.0997 \\
ECQ &= 87.1143
\end{aligned}
$$

From the above numerical illustrations, it can be concluded that when the transition rates increase, the values of ECS, ECQ, P_{1busy}, P_{2busy}, and P_{busy} increase, while the values of P_{idle}, P_{1idle}, and P_{2idle} decrease as the transition rates increases.

Table 5. Transition rates.

	(∗,∗,1)	(∗,∗,2)	(1,∗,1)	(1,∗,2)	(2,∗,1)	(2,∗,2)	(∗,1,1)	(∗,1,2)	(∗,2,1)	(∗,2,2)	(∗,3,1)	(∗,3,2)	(1,1,1)	(1,1,2)	(1,2,1)	(1,2,2)	(1,3,1)	(1,3,2)	(1,4,1)	(1,4,2)	(1,∗,3)	(3,∗,1)	(3,∗,2)	(∗,4,1)	(∗,4,2)	(1,1,3)	(1,2,3)	(1,3,3)	(2,1,3)
(∗,∗,1)	−4.6	1.5	0	0	0	0	0	0	0	0	0	0	0	0	0	0	0	0	0	0	3.1	0	0	0	0	0	0	0	0
(∗,∗,2)	2	−4.5	0	0	0	0	0	0	0	0	0	0	0	0	0	0	0	0	0	0	2.5	0	0	0	0	0	0	0	0
(1,∗,1)	0	0	−8.9	3.6	3.4	0	0	0	0	0	0	0	0	0	0	0	0	0	0	0	0	0	0	0	0	1.9	0	0	0
(1,∗,2)	0	0	2.9	−7.3	0	2.9	0	0	0	0	0	0	0	0	0	0	0	0	0	0	0	0	0	0	0	1.5	0	0	0
(2,∗,1)	0	0	0	0	−6.1	2	0	0	0	0	0	0	0	0	0	0	0	0	0	0	0	2	0	0	0	0	0	0	2.1
(2,∗,2)	0	0	0	0	2.8	−5.9	0	0	0	0	0	0	0	0	0	0	0	0	0	0	0	0	1.5	0	0	0	0	0	1.6
(∗,1,1)	0	0	0	0	0	0	−6.1	1.9	2.1	0	0	0	0	0	0	0	0	0	0	0	0	0	0	0	0	2.1	0	0	0
(∗,1,2)	0	0	0	0	0	0	2.7	−6.8	0	2.6	0	0	0	0	0	0	0	0	0	0	0	0	0	0	0	1.5	0	0	0
(∗,2,1)	0	0	0	0	0	0	0	0	−6.2	2.5	2.3	0	0	0	0	0	0	0	0	0	0	0	0	0	0	0	1.4	0	0
(∗,2,2)	0	0	0	0	0	0	0	0	2.7	−7.4	0	2.6	0	0	0	0	0	0	0	0	0	0	0	0	0	0	2.1	0	0
(∗,3,1)	0	0	0	0	0	0	0	0	0	0	−7	2.4	0	0	0	0	0	0	0	0	0	0	0	2	0	0	0	2.6	0
(∗,3,2)	0	0	0	0	0	0	0	0	0	0	1.9	−8.6	0	0	0	0	0	0	0	0	0	0	0	0	3.9	0	0	2.8	0
(1,1,1)	0	0	0	0	0	0	0	0	0	0	0	0	−10.2	3.1	2.3	0	0	0	0	0	0	0	0	0	0	1.6	0	0	0
(1,1,2)	0	0	0	0	0	0	0	0	0	0	0	0	2.5	−11.1	0	2.9	0	0	0	0	0	0	0	0	0	2.5	0	0	0
(1,2,1)	0	0	0	0	0	0	0	0	0	0	0	0	0	0	−10.5	1.7	3.2	0	0	0	0	0	0	0	0	0	2.3	0	0
(1,2,2)	0	0	0	0	0	0	0	0	0	0	0	0	0	0	2.2	−8.5	0	2.1	0	0	0	0	0	0	0	0	1.7	0	0
(1,3,1)	0	0	0	0	0	0	0	0	0	0	0	0	0	0	0	0	−12.1	2.1	4.5	0	0	0	0	0	0	0	0	2.8	0
(1,3,2)	0	0	0	0	0	0	0	0	0	0	0	0	0	0	0	0	2.8	−11.6	0	3.2	0	0	0	0	0	0	0	3	0

Table 6. Transition rates.

	(2,1,1)	(2,1,2)	(2,2,1)	(2,2,2)	(2,3,1)	(2,3,2)	(2,1,3)	(2,2,3)	(2,3,3)	(2,4,1)	(2,4,2)	(3,1,1)	(3,1,2)	(3,2,1)	(3,2,2)	(3,3,1)	(3,3,2)
(1,1,1)	3.2	0	0	0	0	0	0	0	0	0	0	0	0	0	0	0	0
(1,1,2)	0	3.7	0	0	0	0	0	0	0	0	0	0	0	0	0	0	0
(1,2,1)	0	0	3.5	0	0	0	0	0	0	0	0	0	0	0	0	0	0
(1,2,2)	0	0	0	2.7	0	0	0	0	0	0	0	0	0	0	0	0	0
(1,3,1)	0	0	0	0	2.9	0	0	0	0	0	0	0	0	0	0	0	0
(1,3,2)	0	0	0	0	0	2.9	0	0	0	0	0	0	0	0	0	0	0
(2,1,1)	−12.2	3.8	2.9	0	0	0	2.1	0	0	0	0	3.4	0	0	0	0	0
(2,1,2)	2.7	−12.8	0	3.3	0	0	3.2	0	0	0	0	0	3.6	0	0	0	0
(2,2,1)	0	0	−15.4	2.9	3.8	0	0	3.8	0	0	0	0	0	4.9	0	0	0
(2,2,2)	0	0	2.8	−12.6	0	3	0	3.6	0	0	0	0	0	0	3.2	0	0
(2,3,1)	0	0	0	0	−11.2	1.3	0	0	3.1	3.9	0	0	0	0	0	2.9	0
(2,3,2)	0	0	0	0	2	−13.9	0	0	2.6	0	5.4	0	0	0	0	0	3.9

Table 7. Transitions rates.

	(∗,∗,1)	(∗,∗,2)	(1,∗,1)	(1,∗,2)	(2,∗,1)	(2,∗,2)	(∗,1,1)	(∗,1,2)	(∗,2,1)	(∗,2,2)	(∗,3,1)	(∗,3,2)	(1,1,1)	(1,1,2)	(1,2,1)	(1,2,2)	(1,3,1)	(1,3,2)	(1,4,1)	(1,4,2)	(1,∗,3)	(3,∗,1)	(3,∗,2)	(∗,4,1)	(∗,4,2)	(1,1,3)	(1,2,3)	(1,3,3)	(2,1,3)
(∗,∗,1)	−5.1	1.8	0	0	0	0	0	0	0	0	0	0	0	0	0	0	0	0	0	0	3.3	0	0	0	0	0	0	0	0
(∗,∗,2)	2.4	−5	0	0	0	0	0	0	0	0	0	0	0	0	0	0	0	0	0	0	2.6	0	0	0	0	0	0	0	0
(1,∗,1)	0	0	−9.2	3.8	3.5	0	0	0	0	0	0	0	0	0	0	0	0	0	0	0	0	0	0	0	0	1.9	0	0	0
(1,∗,2)	0	0	3.1	−7.9	0	3.2	0	0	0	0	0	0	0	0	0	0	0	0	0	0	0	0	0	0	0	1.6	0	0	0
(2,∗,1)	0	0	0	0	−8	2.3	0	0	0	0	0	0	0	0	0	0	0	0	0	0	0	3.5	0	0	0	0	0	0	2.2
(2,∗,2)	0	0	0	0	3	−6.6	0	0	0	0	0	0	0	0	0	0	0	0	0	0	0	0	1.9	0	0	0	0	0	1.7
(∗,1,1)	0	0	0	0	0	0	−6.8	2.2	2.4	0	0	0	0	0	0	0	0	0	0	0	0	0	0	0	0	2.2	0	0	0
(∗,1,2)	0	0	0	0	0	0	2.9	−7.2	0	2.7	0	0	0	0	0	0	0	0	0	0	0	0	0	0	0	1.6	0	0	0
(∗,2,1)	0	0	0	0	0	0	0	0	−6.6	2.6	2.4	0	0	0	0	0	0	0	0	0	0	0	0	0	0	0	1.6	0	0
(∗,2,2)	0	0	0	0	0	0	0	0	2.9	−7.9	0	2.8	0	0	0	0	0	0	0	0	0	0	0	0	0	0	2.2	0	0
(∗,3,1)	0	0	0	0	0	0	0	0	0	0	−7.7	2.6	0	0	0	0	0	0	0	0	0	0	0	2.2	0	0	0	2.9	0
(∗,3,2)	0	0	0	0	0	0	0	0	0	0	2.2	−9.4	0	0	0	0	0	0	0	0	0	0	0	0	4.1	0	0	3.1	0
(1,1,1)	0	0	0	0	0	0	0	0	0	0	0	0	−10.7	3.4	2.4	0	0	0	0	0	0	0	0	0	0	1.7	0	0	0
(1,1,2)	0	0	0	0	0	0	0	0	0	0	0	0	2.7	−12.3	0	3.1	0	0	0	0	0	0	0	0	0	2.8	0	0	0
(1,2,1)	0	0	0	0	0	0	0	0	0	0	0	0	0	0	−11.6	1.9	3.4	0	0	0	0	0	0	0	0	0	2.8	0	0
(1,2,2)	0	0	0	0	0	0	0	0	0	0	0	0	0	0	2.5	−9.5	0	2.4	0	0	0	0	0	0	0	0	1.9	0	0
(1,3,1)	0	0	0	0	0	0	0	0	0	0	0	0	0	0	0	0	−13.1	2.4	4.7	0	0	0	0	0	0	0	0	3.1	0
(1,3,2)	0	0	0	0	0	0	0	0	0	0	0	0	0	0	0	0	3	−12.7	0	3.5	0	0	0	0	0	0	0	3.3	0

Table 8. Transition rates.

	(2, 1, 1)	(2, 1, 2)	(2, 2, 1)	(2, 2, 2)	(2, 3, 1)	(2, 3, 2)	(2, 1, 3)	(2, 2, 3)	(2, 3, 3)	(2, 4, 1)	(2, 4, 2)	(3, 1, 1)	(3, 1, 2)	(3, 2, 1)	(3, 2, 2)	(3, 3, 1)	(3, 3, 2)
(1, 1, 1)	3.3	0	0	0	0	0	0	0	0	0	0	0	0	0	0	0	0
(1, 1, 2)	0	3.8	0	0	0	0	0	0	0	0	0	0	0	0	0	0	0
(1, 2, 1)	0	0	3.6	0	0	0	0	0	0	0	0	0	0	0	0	0	0
(1, 2, 2)	0	0	0	2.8	0	0	0	0	0	0	0	0	0	0	0	0	0
(1, 3, 1)	0	0	0	0	3	0	0	0	0	0	0	0	0	0	0	0	0
(1, 3, 2)	0	0	0	0	0	3	0	0	0	0	0	0	0	0	0	0	0
(2, 1, 1)	−12.6	3.9	3	0	0	0	2.2	0	0	0	0	3.5	0	0	0	0	0
(2, 1, 2)	2.8	−13.2	0	3.4	0	0	3.3	0	0	0	0	0	3.7	0	0	0	0
(2, 2, 1)	0	0	−15.8	3	3.9	0	0	3.9	0	0	0	0	0	5	0	0	0
(2, 2, 2)	0	0	3	−13.4	0	3.2	0	3.8	0	0	0	0	0	0	3.4	0	0
(2, 3, 1)	0	0	0	0	−12.1	1.4	0	0	3.2	4.5	0	0	0	0	0	3	0
(2, 3, 2)	0	0	0	0	2.1	−14.7	0	0	2.7	0	5.9	0	0	0	0	0	4

Table 9. Transitions rates.

	(∗,∗,1)	(∗,∗,2)	(1,∗,1)	(1,∗,2)	(2,∗,1)	(2,∗,2)	(∗,1,1)	(∗,1,2)	(∗,2,1)	(∗,2,2)	(∗,3,1)	(∗,3,2)	(1,1,1)	(1,1,2)	(1,2,1)	(1,2,2)	(1,3,1)	(1,3,2)	(1,4,1)	(1,4,2)	(1,∗,3)	(3,∗,1)	(3,∗,2)	(∗,4,1)	(∗,4,2)	(1,1,3)	(1,2,3)	(1,3,3)	(2,1,3)
(∗,∗,1)	−5.3	1.9	0	0	0	0	0	0	0	0	0	0	0	0	0	0	0	0	0	0	3.4	0	0	0	0	0	0	0	0
(∗,∗,2)	2.5	−5.2	0	0	0	0	0	0	0	0	0	0	0	0	0	0	0	0	0	0	2.7	0	0	0	0	0	0	0	0
(1,∗,1)	0	0	−9.5	3.9	3.6	0	0	0	0	0	0	0	0	0	0	0	0	0	0	0	0	0	0	0	0	2	0	0	0
(1,∗,2)	0	0	3.2	−8.2	0	3.3	0	0	0	0	0	0	0	0	0	0	0	0	0	0	0	0	0	0	0	1.7	0	0	0
(2,∗,1)	0	0	0	0	−8.3	2.4	0	0	0	0	0	0	0	0	0	0	0	0	0	0	0	3.6	0	0	0	0	0	0	2.3
(2,∗,2)	0	0	0	0	3.1	−6.9	0	0	0	0	0	0	0	0	0	0	0	0	0	0	0	0	2	0	0	0	0	0	1.8
(∗,1,1)	0	0	0	0	0	0	−7.1	2.3	2.5	0	0	0	0	0	0	0	0	0	0	0	0	0	0	0	0	2.3	0	0	0
(∗,1,2)	0	0	0	0	0	0	3	−7.5	0	2.8	0	0	0	0	0	0	0	0	0	0	0	0	0	0	0	1.7	0	0	0
(∗,2,1)	0	0	0	0	0	0	0	0	−6.9	2.7	2.5	0	0	0	0	0	0	0	0	0	0	0	0	0	0	0	1.7	0	0
(∗,2,2)	0	0	0	0	0	0	0	0	3	−8.2	0	2.9	0	0	0	0	0	0	0	0	0	0	0	0	0	0	2.3	0	0
(∗,3,1)	0	0	0	0	0	0	0	0	0	0	−8	2.7	0	0	0	0	0	0	0	0	0	0	0	2.3	0	0	0	3	0
(∗,3,2)	0	0	0	0	0	0	0	0	0	0	2.3	−9.7	0	0	0	0	0	0	0	0	0	0	0	0	4.2	0	0	3.2	0
(1,1,1)	0	0	0	0	0	0	0	0	0	0	0	0	−11.1	3.5	2.5	0	0	0	0	0	0	0	0	0	0	1.8	0	0	0
(1,1,2)	0	0	0	0	0	0	0	0	0	0	0	0	2.8	−12.7	0	3.2	0	0	0	0	0	0	0	0	0	2.9	0	0	0
(1,2,1)	0	0	0	0	0	0	0	0	0	0	0	0	0	0	−12	2	3.5	0	0	0	0	0	0	0	0	0	2.9	0	0
(1,2,2)	0	0	0	0	0	0	0	0	0	0	0	0	0	0	2.6	−9.9	0	2.5	0	0	0	0	0	0	0	0	2	0	0
(1,3,1)	0	0	0	0	0	0	0	0	0	0	0	0	0	0	0	0	−13.5	2.5	4.8	0	0	0	0	0	0	0	0	3.2	0
(1,3,2)	0	0	0	0	0	0	0	0	0	0	0	0	0	0	0	0	3.1	−13.1	0	3.6	0	0	0	0	0	0	0	3.4	0

6. Conclusions

In this paper, we have analysed a queuing system in which the arrival process and services handled by two servers are interdependent. The evolution of these processes is governed by transitions occurring in the product space of three Markov chains according to a semi-Markov rule, with sojourn times in states following the exponential distribution. Using the matrix geometric method, we have analysed the model and computed various performance measures. Additionally, we have included numerical illustrations of the model.

We propose extending the model presented here to a more general case. In this paper, we have assumed that upon the arrival of a customer in the system, the starting phase of the next arrival is chosen according to the initial probability vector. It is possible to extend this to the case in which the next arrival phase is the phase in which transition occurs consequent to the arrival of the customer.

In this paper, we have discussed service by both servers in stages while moving in the forward direction. This could be extended to the case in which the service processes are in stages and can make several forward and backward jumps. In addition, it would interesting to study how the system evolves with a larger number of servers. In future work, we propose analysing a PH/two-server interdependent service model and an MAP/two-server interdependent service model and comparing them numerically.

Author Contributions: Conceptualization, S.S. and A.K.; methodology, S.S. and A.K.; validation, S.S. and A.K.; formal analysis, A.K.; writing—original draft preparation, S.S. and A.K.; writing—review and editing, S.S., A.K. and D.K.; supervision, A.K.; project administration, A.K.; funding acquisition, D.K. All authors have read and agreed to the published version of the manuscript.

Funding: This paper has been supported by the RUDN University Strategic Academic Leadership Program (recipient D.K., funding acquisition; writing—original draft preparation). The reported study was funded by Russian Science Foundation, project number 22-49-02023 (recipient D.K., writing—review and editing).

Institutional Review Board Statement: Not applicable.

Informed Consent Statement: Not applicable.

Data Availability Statement: No publicly available database is used in this paper. So there is no link available. The given numerical works are based on parameter values which do not breach the stability condition.

Conflicts of Interest: The authors declare no conflict of interest. The funders had no role in the design of the study, in the collection, analysis, or interpretation of data, in the writing of the manuscript, or in the decision to publish the results.

References

1. Mitchell, C.R.; Paulson, A.S. M/M/1 queues with interdependent arrival and service processes. *Nav. Res. Logist. Q.* **1979**, *26*, 47–56. [CrossRef]
2. Courtois, P.J.; Georges, J. On a single-server finite queuing model with state-dependent arrival and service processes. *Oper. Res.* **1971**, *19*, 424–435. [CrossRef]
3. Sengupta, B. The semi-Markovian queue: Theory and applications. *Stoch. Model.* **1990**, *6*, 383–413. [CrossRef]
4. Boxma, O.J.; Perry, D. A queueing model with dependence between service and interarrival times. *Eur. J. Oper. Res.* **2001**, *128*, 611–624. [CrossRef]
5. Müller, A. On the waiting times in queues with dependency between interarrival and service times. *Oper. Res. Lett.* **2000**, *26*, 43–47. [CrossRef]
6. Vlasiou, M.; Adan, I.J.; Boxma, O.J. A two-station queue with dependent preparation and service times. *Eur. J. Oper. Res.* **2009**, *195*, 104–116. [CrossRef]
7. Adan, I.J.B.F.; Kulkarni, V.G. Single-server queue with Markov-dependent inter-arrival and service times. *Queueing Syst.* **2003**, *45*, 113–134. [CrossRef]
8. Iyer, S.K.; Manjunath, D. Queues with dependency between interarrival and service times using mixtures of bivariates. *Stoch. Model.* **2006**, *22*, 3–20. [CrossRef]

9. Buchholz, P.; Kriege, J. Fitting correlated arrival and service times and related queueing performance. *Queueing Syst.* **2017**, *85*, 337–359. [CrossRef]
10. Kim, B.; Kim, J. The waiting time distribution for a correlated queue with exponential interarrival and service times. *Oper. Res. Lett.* **2018**, *46*, 268–271. [CrossRef]
11. Krishnamoorthy, A.; Joshua, A.N. A $BMAP/BMSP/1$ queue with Markov dependent arrival and Markov dependent service batches. *J. Ind. Manag. Optim.* **2021**, *17*, 2925–2941. [CrossRef]
12. Dai, W.; Hu, J.Q. Correlated queues with service times depending on inter-arrival times. *Queueing Syst.* **2022**, *100*, 41–60. . [CrossRef]
13. Moiseev, A.; Shklennik, M.; Polin, E. Infinite-server queueing tandem with Markovian arrival process and service depending on its state. *Ann. Oper. Res.* **2023**, *326*, 261–269. [CrossRef]
14. Kleinrock, L. *Queueing Systems: Theory*; John Wiley: Hoboken, NJ, USA, 1975; Volume 1.
15. Cohen, J.W. Batch arrivals in queueing systems. *Queueing Syst.* **1982**, *1*, 25–52.
16. Neuts, M.F.; Takahashi, Y. *Asymptotic Behavior of the Stationary Distributions in the GI/PH/c Queue with Heterogeneous Servers*; Applied Mathematics Institute Technical Report; University of Delaware: Newark, NJ, USA, 1980; Volume 57B, 30p.
17. Kumar, B.K.; Madheswari, S.P. An M/M/2 queueing system with heterogeneous servers and multiple vacations. *Math. Comput. Model.* **2005**, *41*, 1415–1429. [CrossRef]
18. Krishnamoorthy, A.; Divya, V. A Two-Server Queueing System with Processing of Service Items by a Server. In *Applied Probability and Stochastic Processes*; Springer Nature: Singapore, 2020; pp. 307–333.
19. Ayyappan, G.; Archana, G. Analysis of MAP/PH1, PH2/2 Queueing Model with Working Breakdown, Repairs, Optional Service, and Balking. *Appl. Appl. Math. Int. J. (AAM)* **2023**, *18*, 1.
20. Bouchentouf, A.A.; Messabihi, A. Heterogeneous two-server queueing system with reverse balking and reneging. *Opsearch* **2018**, *55*, 251–267. [CrossRef]
21. Klimenok, V.; Dudin, A.; Vishnevsky, V. Priority multi-server queueing system with heterogeneous customers. *Mathematics* **2020**, *8*, 1501. [CrossRef]
22. Krishnamoorthy, A.; Sreenivasan, C. An M/M/2 queueing system with heterogeneous servers including one with working vacation. *Int. J. Stoch. Anal.* **2012**, *2012*, 145867. [CrossRef]
23. Lin, B.; Lin, Y.; Bhatnagar, R. Optimal policy for controlling two-server queueing systems with jockeying. *J. Syst. Eng. Electron.* **2022**, *33*, 144–155. [CrossRef]
24. Samouylov, K.; Dudina, O.; Dudin, A. Analysis of Multi-Server Queueing System with Flexible Priorities. *Mathematics* **2023**, *11*, 1040. [CrossRef]
25. Saravanan, V.; Poongothai, V.; Godhandaraman, P. Admission control policy of a two heterogeneous server finite capacity retrial queueing system with maintenance activity. *OPSEARCH* **2023**, 1–24. . [CrossRef]
26. Tian, N.; Zhang, Z.G. A two threshold vacation policy in multiserver queueing systems. *Eur. J. Oper. Res.* **2006**, *168*, 153–163. [CrossRef]
27. Yue, D.; Yue, W.; Yu, J.; Tian, R. A heterogeneous two-server queuing system with balking and server breakdowns. In Proceedings of the Eight International Symposium on Operations Research and its Applications (ISORA'09), Zhangjiajie, China, 20–22 September 2009.
28. Zhao, C.; Wang, Z. The impact of line-sitting on a two-server queueing system. *Eur. J. Oper. Res.* **2023**, *308*, 782–800. [CrossRef]
29. Krishnamoorthy, A. Analysis of Interdependent Processes: A Semi-Markov Approach. *Bull. Kerala Math. Assoc.* **2020**, *17*, 49–68.
30. Krishnamoorthy, A. Analysis of Reliability of Interdependent Serial, Parallel and The General k-out-of-n: G System: A New Approach. *J. Indian Soc. Probab. Stat.* **2022**, 483–496. [CrossRef]
31. Sindhu, S.; Krishnamoorthy, A.; Kozyrev, D. On Queues with Working Vacation and Interdependence between Arrival and Service Processes. *Mathematics* **2023**, *11*, 2280. [CrossRef]
32. Sengupta, B. Markov processes whose steady state distribution is matrix-exponential with an application to the GI/PH/1 queue. *Adv. Appl. Probab.* **1989**, *21*, 159–180. [CrossRef]
33. Van Houdt, B. A matrix geometric representation for the queue length distribution of multitype semi-Markovian queues. *Perform. Eval.* **2012**, *69*, 299–314. [CrossRef]
34. Latouche, G.; Ramaswami, V. *Introduction to Matrix Analytic Methods in Stochastic Modeling*; Society for Industrial and Applied Mathematics: Philadelphia, PA, USA, 1999.
35. Neuts, M.F. *Matrix-Geometric Solutions in Stochastic Models: An Algorithmic Approach*; Courier Corporation: Chelmsford, MA, USA, 1994.

mathematics

MDPI

Article

The k-out-of-n:G System Viewed as a Multi-Server Queue

Achyutha Krishnamoorthy [1,*,†], Anu Nuthan Joshua [2,*] and Ambily P. Mathew [1]

1 Department of Mathematics, CMS College Kottayam, Kottayam 686001, India
2 Department of Mathematics, Union Christian College, Aluva 683102, India
* Correspondence: krishnamoorthy@cmscollege.ac.in (A.K.); anunuthanjoshua@uccollege.edu.in (A.N.J.)
† Current address: Department of Mathematics, Central University of Kerala, Kasargod 671316, India.

Abstract: This paper extends the k-out-of-n:G reliability system to a multi-server queue. We study a multi-server reliability-queuing model with the N-policy of repair. The queuing system considered here has n servers, each of which has identically and exponentially distributed service times with parameter μ. Servers are subject to breakdown at an exponential rate γ. The repair process follows the N-policy of repair. Although these servers work independently of each other, service can be provided only when k functional servers are available in the system. We study the model in the steady state, using the matrix analytic method. We evaluate some associated performance measures and provide graphical/numerical illustrations. We consider an optimization problem, and the results of the study are presented.

Keywords: reliability; k-out of n system; N-policy; HAP system; breakdown

MSC: 60K25; 60K10; 60K20; 90B22; 90B25

Citation: Krishnamoorthy, A.; Joshua, A.N.; Mathew, A.P. The k-out-of-n:G System Viewed as a Multi-Server Queue. *Mathematics* **2024**, *12*, 210. https://doi.org/10.3390/math12020210

Academic Editors: Gurami Tsitsiashvili, Alexander Bochkov and Alexander Zeifman

Received: 11 November 2023
Revised: 8 December 2023
Accepted: 2 January 2024
Published: 8 January 2024

1. Introduction

This paper extends the reliability system k-out-of-n:G to a multi-server queue (n parallel server, with server failure and N-policy of repair). As in the k-out-of-n:G system, at least k servers should be operational in order for the whole system to work (system can provide service). At the moment that the number of operational servers is reduced to $k-1$, the service is disrupted. As a result, the system can start operation only after one working server is added to the system. This model can be applied directly to the high-altitude platform (HAP) system used for smooth telecommunication when all other facilities fail, due to natural calamities. The system collapses when the number of operational servers is reduced to $k-1$, irrespective of the number of customers in the system. As with parallel (at least one operational component should be UP for the system to function)/serial (all components should be operational for the system to be functional) systems in the reliability context, we encounter a similar situation in the queuing model discussed in this paper. Thus, this paper extends the k-out-of-n reliability system to the queuing system and simultaneously to the classical multi-server queuing system. In the proposed queuing model, an element of dependency on the number of operational servers arises. Thus, it differs from the classical multi-server queuing models in that at least k ($k=1$, for the classical multi-server system with server failures and their repair) servers should be operational to provide a service. This phenomenon can be interpreted as the system becoming overloaded when the number of operational servers reduces to $k-1$.

The mathematical foundation of reliability was laid by the pioneering work of Barlow and Proschan (1965) [1]. The element of interest is the reliability of the machine in a given interval of time or failure-free operation up to time t. Naturally, one investigates the distribution of the number of failed components at any random time when the machine is in operation. The k-out-of-n system is classified as COLD, WARM or HOT, depending on whether the operational components are also subject to deterioration when the system is down. In the COLD system, no operational components fail when the system is DOWN (not operational); for the WARM system, the operational components fail at a reduced rate when the system is DOWN compared to that when it is UP (operational); and in the case of a HOT system, the failure rate of components is the same, irrespective of whether the system is UP or DOWN. As such, for k-out-of-n systems, the notion of "providing service" was not in vogue until Krishnamoorthy, Sathian and Viswanath (2016) [2] introduced it. They considered separately two distinct cases of providing service. In one of them, the system serves (repairs) failed machines in an organization. The k-out-of-n system is considered as a single server subject to failure. Therefore, at most, one failed machine undergoes repair by the server at a time. The repair of failed components is according to some specific policy. In this case, the system state remains finite because it is assumed that there are only a limited number of machines in the organization. In the second model, the authors assumed that when the server (the k-out-of-n system) is idle, it provides a service to external customers. However, priority is given to the repair of internal customers (for example, failed machines in the organization). For these two cases, the authors investigated continuous-time server reliability.

The N-policy was introduced by Yadin and Naor in [3]. The N-policy of repair was introduced by Krishnamoorthy et al. (2002) [4] and is as follows: when the number of operational components comes down to level $N(k \leq N \leq n)$, a repair facility starts repairing failed units, one at a time. This repair process continues until all failed components are brought back to an operational state. It is assumed that the components repaired are as good as new. This assumption is essential for mathematical tractability. There is another means of keeping the system reliability high—placing an order for $n-k+1$ components when the number of operational components goes down to N. It takes a certain amount of time, deterministic or random, for the order to materialize. Sometimes, this may take place only after the system becomes non-operational. The higher the N value, the higher the probability of the system remaining UP until the order materialization. We can further improve the system reliability by combining the N-policy of repair and the order placement for new components. The order can be placed above the level at which the number of operating components drops to N or we can place the order at this level or even below this level. This policy can be further modified for the cancellation of a placed order if the number of operational components reaches the maximum n through the repair process before the materialization of the order placed. Thus, various extensions are possible. N-policy queuing systems with server breakdowns (both working and non-working breakdowns) are studied extensively in the queuing literature (for example, [5–7]).

Other than the N-policy of repair, for policies such as D (the accumulated work load reaching or crossing the level D (D is a continuous random variable)) and T, the time until the repair facility is activated after the system completely returns to a fully functional state (all components are in an operational state due to repair), are used. Additional solutions are different combinations of these policies. In the queuing literature, the N-policy of repair is the best- and the T-policy is the worst-performing. Similarly, comparisons among the combinations of policies can also be performed.

Highlights of This Work

- The present work is the first to consider the k-out-of-n system as a multi-server queue.
- Unlike the classical multi-server queue, in the present work, at least k servers should be operational to provide a service. The multi-server case can be deduced from the present work by assuming that $k = 1$.
- Two birth and death processes encountered in the analysis are (i) the customer's arrival and service processes and (ii) the accumulation of servers that break while providing a service; until this number reaches $n - N$ (pure death process), the repair process starts from this point of time and, with this, we have a birth and death process. Note that we assume all random variables (inter-arrival times, service time, repair

time of failed servers) involved to be exponentially distributed. The combined process turns out to be a birth and death process.

- The system under consideration reduces to the classical $M/M/\infty$ queue if we assume that the servers (system components) do not fail and that $k = 1$. Then, we take the limit as $n \to \infty$.

The remaining part of this paper is as follows. In Section 2, the description of the problem is given. In Section 3, the mathematical modeling of the problem and its analysis are presented. The stability condition is derived in Section 4 and the steady state probabilities found. Section 5 outlines certain distributions and performance measures of the system's behavior. Numerical and graphical illustrations providing insights into the working of the system are included in Section 6. An optimization problem is considered in Section 7. Concluding remarks are provided in Section 8.

2. Model Description

The model considered here is one that extends the k-out-of-n:G reliability system to a multi-server queue, especially to a multi-server queue with n servers/units working in parallel.

We consider a multi-server queueing system where customers arrive according to a Poisson process with parameter λ. The service facility consists of n servers/units, each of which can provide service to individual customers. We are considering a k-out-of-n:G system All the n parallel servers have identically and exponentially distributed service times with parameter μ. Although these servers work independently, service will take place only if at least k servers are operational or in working condition. These servers are susceptible to breakdown. Breakdown occurs at exponentially distributed time intervals with rate γ. Repair will take place according to the N-policy of repair as described in the previous section. When the number of operational components comes down to the level $N(k \leq N < n)$, a repair facility starts repairing failed units one at a time. The repair process continues until all failed components are restored to an operational state. It is assumed that the time taken to repair each unit is exponentially distributed with parameter δ. The system considered here is COLD, i.e., when the system fails due to the lack of at least k operational units, the units that are operational do not deteriorate until the system restarts again, with the failed units replaced with new ones. If we are considering a k-out-of-$n:F$ COLD system, system failure occurs when any k of the n units fail. Here, the model under consideration is a k-out-of-n:G COLD system.

We formulate the problem mathematically as follows.

3. Mathematical Formulation

Define, for $t \geq 0$,

$N_1(t)$: the number of customers in the system at time t;

$N_2(t)$: the number of operational units in the system at time t;

$R(t)$: the status of the server that repairs the failed components at time t.

$$R(t) = \begin{cases} 0 & if\ the\ repair\ is\ OFF \\ 1 & if\ the\ repair\ is\ ON \end{cases}$$

Then, $\{(N_1(t), N_2(t), R(t))|t \geq 0\}$ is a regular irreducible $CTMC$ on state space $\Omega = \{(n_1, n_2, 1) : n_1 \geq 0; k-1 \leq n_2 \leq n-1\} \bigcup \{(n_1, n_2, 0) : n_1 \geq 0; N+1 \leq n_2 \leq n\}$.

The generator matrix for this process when the states are arranged lexicographically is of the form

$$Q = \begin{bmatrix} A_1^0 & A_0 & & & & & & \\ A_2^1 & A_1^1 & A_0 & & & & & \\ & A_2^2 & A_1^2 & A_0 & & & & \\ & & \ddots & \ddots & & & & \\ & & & A_2^{n-1} & A_1^{n-1} & A_0 & & \\ & & & & A_2 & A_1 & A_0 & \\ & & & & & \ddots & \ddots & \ddots \end{bmatrix}$$

A_1^i contains transitions within level i for $0 \le i \le n-1$. A_2^i (which are diagonal matrices) contain transitions from level i to $i-1$ for $1 \le i \le n-1$. A_2 (diagonal matrix) contains transitions from i to $i-1$ and A_1 within level i. A_0 (diagonal matrix) contains transitions from i to $i+1, \forall i$. All the matrices are square matrices of dimension $d = 2n - (N+k) + 1$.

$$A_0 = \begin{bmatrix} 0 & \\ & \lambda I_{d-1} \end{bmatrix}$$

$$A_{0_{(n_1,n_2,n_3)}}^{(n_1+1,n_2,n_3)} = \lambda$$

For $1 \le i \le k$,

$$A_2^i = \begin{bmatrix} 0 & \\ & i\mu I_{d-1} \end{bmatrix}$$

$$A_{2(n_1,n_2,n_3)}^{i\,(n_1-1,n_2,n_3)} = n_1\mu$$

For $k+1 \le i \le N$,

$$A_2^i = \begin{bmatrix} 0 & & & & & \\ & k\mu & & & & \\ & & (k+1)\mu & & & \\ & & & \ddots & & \\ & & & & (i-1)\mu & \\ & & & & & i\mu I \end{bmatrix}$$

I is an identity matrix of order $d - (i-k+1)$.

$$A_{2(n_1,n_2,n_3)}^{i\,(n_1-1,n_2,n_3)} = \begin{cases} n_2\mu & if\ k \le n_2 \le i-1 \\ i\mu & if\ i \le n_2 \le n \end{cases}$$

For $N+1 \le i \le n-1$,

$$A_2^i = \begin{bmatrix} 0 & & & & & & & & \\ & k\mu & & & & & & & \\ & & (k+1)\mu & & & & & & \\ & & & \ddots & & & & & \\ & & & & N\mu & & & & \\ & & & & & I_2 \otimes (N+1)\mu & & & \\ & & & & & & \ddots & & \\ & & & & & & & I_2 \otimes (i-1)\mu & \\ & & & & & & & & i\mu I \end{bmatrix}$$

$$A^{i\,(n_1-1,n_2,n_3)}_{2(n_1,n_2,n_3)} = \begin{cases} n_2\mu & if\, k \le n_2 \le i-1 \\ i\mu & if\, i \le n_2 \le n \end{cases}$$

$$A_2 = \begin{bmatrix} 0 & & & & & & & & \\ & k\mu & & & & & & & \\ & & (k+1)\mu & & & & & & \\ & & & \ddots & & & & & \\ & & & & N\mu & & & & \\ & & & & & I_2 \otimes (N+1)\mu & & & \\ & & & & & & \ddots & & \\ & & & & & & & I_2 \otimes (n-1)\mu & \\ & & & & & & & & n\mu \end{bmatrix}$$

$$A_{2\,(n_1,n_2,n_3)}^{(n_1-1,n_2,n_3)} = n_2\mu$$

To easily represent matrix A_1^i, the states $\{(n_1,n_2,1) : k-1 \le n_2 \le N\}$ of order $N-k+2$ are grouped together and given subscript 1, the states $\{(n_1,n_2,r) : N+1 \le n_2 \le n-1, r = 0,1\}$ of order $2(n-1-N)$ are grouped together and given subscript 2, and the state $\{(n_1,n_2,0) : n_2 = n\}$ is given subscript 3.

$$A_1^0 = L$$

$$A_1^i = L - A_2^i$$

$$L = \begin{bmatrix} L_{11} & L_{12} & \mathbf{0} \\ L_{21} & L_{22} & L_{23} \\ \mathbf{0} & L_{32} & L_{33} \end{bmatrix}$$

$$L_{11} = \gamma E^- + \delta E^+ - \delta I_{N-k+2} - (\lambda+\gamma)\begin{bmatrix} 0 & \\ & I_{N-k+1} \end{bmatrix}$$

where E^- and E^+ (refer to page 19) are square matrices of dimension $N-k+2$.

$$L_{12} = E_{N-k+21} \otimes [0\,\delta]$$

$$L_{21} = E'_{N-k+21} \otimes [\gamma\,\gamma]'$$

E_{N-k+21} is a matrix (refer to page 19) of order $(N-k+2) \times (n-1) - N$.

$$L_{22} = E^+ \otimes \begin{bmatrix} 0 & 0 \\ 0 & \delta \end{bmatrix} + E^- \otimes \begin{bmatrix} \gamma & 0 \\ 0 & \gamma \end{bmatrix} - I_{n-1-N} \otimes \begin{bmatrix} 0 & 0 \\ 0 & \delta \end{bmatrix} - (\lambda+\gamma)I_{2(n-1-N)}$$

where E^- and E^+ (refer to page 19) are square matrices of dimension $(n-1)-N$.

$$L_{23} = e_{n-1-N} \otimes [0\,\delta]'$$

$$L_{32} = e'_{n-1-N} \otimes [\gamma\,0],$$

e_{n-1-N} is a column vector with 1 in the $(n-1-N)$ position.

$$L_{33} = -(\lambda+\gamma)$$

$$A_1 = L - A_2.$$

The transitions in $A_1^0 = L$ excluding diagonal entries are given as

$$A_{1(n_1,n_2,n_3)}^{0(n_1,m_2,m_3)} = \begin{cases} \delta & if n_3 = 1, m_3 = 1, m_2 = n_2 + 1, k-1 \leq n_2 \leq n-2 \\ \delta & if n_3 = 1, m_3 = 0, n_2 \leq n-1, m_2 = n \\ \gamma & if n_3 = m_3, m_2 = n_2 - 1, n_2 \neq N+1 \\ \gamma & if n_3 = 0, m_3 = 1, m_2 = N, n_2 = N+1 \end{cases}$$

The transitions from one state to another are given below:

- Transitions due to the arrival of a customer to the system:
  $(n_1, n_2, r) \to (n_1 + 1, n_2, r)$ with rate λ when $n_2 \neq k-1$.

- Transitions due to the service completion of a customer:
  $(n_1, n_2, r) \to (n_1 - 1, n_2, r)$ with rate
  $n_1\mu$ when $1 \leq n_1 \leq k$;
  $n_2\mu$ when $k+1 \leq n_1 \leq n, n_2 \leq n_1$;
  $n_1\mu$ when $k+1 \leq n_1 \leq n, n_2 > n_1$;

- Transitions due to the breakdown of an operational unit in the k-out-of-N system:
  $(n_1, n_2, r) \to (n_1, n_2 - 1, r)$ with rate γ when $n_2 \geq k, n_2 \neq N+1, r = 0, 1$.
  $(n_1, N+1, 0) \to (n_1, N, 1)$ with rate γ.

- Transitions due to the repair of an operational unit in the k-out-of-N system:
  $(n_1, n_2, 1) \to (n_1, n_2 + 1, 1)$ with rate δ when $k-1 \leq n_2 \leq n-1$.
  $(n_1, n-1, 1) \to (n_1, n, 0)$ with rate δ.

4. Stability Analysis

4.1. Stability Condition

Let $A = A_2 + A_1 + A_0$.

$$A = \begin{bmatrix} -\delta & \delta & & & & & & & & & \\ \gamma & -\gamma-\delta & \delta & & & & & & & & \\ & \gamma & -\gamma-\delta & \delta & & & & & & & \\ & & \ddots & \ddots & & & & & & & \\ & & & \gamma & -\gamma-\delta & \delta & & & & & \\ & & & & \gamma & -\gamma-\delta & B_1 & & & & \\ & & & & & B_2 & B_3 & B_4 & & & \\ & & & & & & B_5 & B_3 & B_4 & & \\ & & & & & & & \ddots & \ddots & \ddots & \\ & & & & & & & B_5 & B_3 & B_4 & \\ & & & & & & & & B_5 & B_3 & B_6 \\ & & & & & & & & & B_7 & -\gamma \end{bmatrix}$$

where

$$B_1 = \begin{bmatrix} 0 & \delta \end{bmatrix}, B_2 = \begin{bmatrix} \gamma \\ \gamma \end{bmatrix}, B_3 = \begin{bmatrix} -\gamma & 0 \\ 0 & -\gamma-\delta \end{bmatrix}, B_4 = \begin{bmatrix} 0 & 0 \\ 0 & \delta \end{bmatrix},$$

$$B_5 = \begin{bmatrix} \gamma & 0 \\ 0 & \gamma \end{bmatrix}, B_6 = \begin{bmatrix} 0 \\ \delta \end{bmatrix}, B_7 = \begin{bmatrix} \gamma & 0 \end{bmatrix}$$

Let $\boldsymbol{\pi} = (\pi_{k-1}, \pi_k, \pi_{k+1}...\pi_{N-1}, \pi_N, \boldsymbol{\pi_{N+1}}, ...\boldsymbol{\pi_{n-1}}, \pi_n)$ be the steady-state probability vector of the infinitesimal generator matrix A.
In other words,

$$\boldsymbol{\pi} A = \mathbf{0}; \boldsymbol{\pi}\mathbf{e} = 1. \tag{1}$$

Define

$$\mathcal{U}_{n-1} = \frac{1}{\gamma} B_6$$

$$\mathcal{U}_i = \begin{cases} (B_3 + B_7 \mathcal{U}_{n-1})^{-1}(-B_4) & for\ i = n-2 \\ (B_3 + B_5 \mathcal{U}_{i+1})^{-1}(-B_4) & for\ N-1 \leq i \leq n-3 \\ (B_3 + B_5 \mathcal{U}_{N+1})^{-1}(-B_1) & for\ i = N \end{cases}$$

For $k \leq i \leq N$,

$$\pi_i = (\frac{\delta}{\gamma})^{i-(k-1)} \pi_{k-1}$$

For $N+1 \leq i \leq n-1$,

$$\boldsymbol{\pi_i} = \boldsymbol{\pi_{i-1}} \mathcal{U}_{i-1}$$

$$\pi_n = \boldsymbol{\pi_{n-1}} \mathcal{U}_{n-1}$$

From the normalizing condition $\boldsymbol{\pi}\mathbf{e} = 1$, we have

$$\pi_{k-1} [\sum_{i=0}^{k} (\frac{\delta}{\gamma})^i + (\frac{\delta}{\gamma})^k \sum_{l=N}^{n-1} \prod_{j=N}^{l} \mathcal{U}_j \mathbf{e}] = 1. \tag{2}$$

The infinitesimal generator of this Markov chain indicates that it is a level-independent quasi-birth–death process. Thus, this queuing system is stable if and only if $\boldsymbol{\pi} A_0 \mathbf{e} < \boldsymbol{\pi} A_2 \mathbf{e}$; see Neuts [8]. The stability condition is given as follows:

Theorem 1. *The given system is stable if and only if*

$$\lambda < \sum_{i=0}^{N-k} (k+i)\mu (\frac{\delta}{\gamma})^{i+1} \pi_{k-1} + \sum_{l=N+1}^{n} l\mu \prod_{j=N}^{l} U_j e (\frac{\delta}{\gamma})^k \pi_{k-1} \tag{3}$$

4.2. Steady-State Probability Vector

Assuming the stability of the system, we proceed to find the steady-state probability of the system states.

Let $\boldsymbol{x}$ be the steady-state probability vector of $\mathcal{Q}$, i.e., $\boldsymbol{x}$ satisfies $\boldsymbol{x}\mathcal{Q} = \mathbf{0}$ and $\boldsymbol{x}\mathbf{e} = 1$. We partition this vector as

$$\boldsymbol{x} = (\boldsymbol{x}_0, \boldsymbol{x}_1, \boldsymbol{x}_2 \ldots),$$

where $\boldsymbol{x}_i = x_{(i,n_2,r)}$ are of dimension $d = 2n - (N+k) + 1$.

$$\boldsymbol{x}_{n+i} = \boldsymbol{x}_n R^i, i \geq 1$$

where the matrix R is the minimal non-negative solution to the matrix quadratic equation

$$R^2 A_2 + R A_1 + A_0 = 0$$

and the vectors $\boldsymbol{x}_0, \boldsymbol{x}_1, \cdots, \boldsymbol{x}_n \ldots$ are obtained by solving the equations

$$\boldsymbol{x}_0 A_1^0 + \boldsymbol{x}_1 A_2^1 = 0 \tag{4}$$

$$\boldsymbol{x}_{i-1} A_0 + \boldsymbol{x}_i A_1^i + \boldsymbol{x}_2 A_2^i = 0, \text{ for } i \leq i \leq n-2 \tag{5}$$

$$\boldsymbol{x}_{n-2} A_0 + \boldsymbol{x}_{n-1} A_1^{n-1} + \boldsymbol{x}_n A_2 = 0 \tag{6}$$

$$\boldsymbol{x}_{n-1} A_0 + \boldsymbol{x}_n (A_1 + R A_2) = 0 \tag{7}$$

subject to the normalizing condition

$$\sum_{i=0}^{n-1} \mathbf{x}_i \mathbf{e} + \mathbf{x}_n (I-R)^{-1} \mathbf{e} = 1 \tag{8}$$

4.3. Special Case

If we consider a system in which $k = 1$ and assume that the servers do not break down, the above queuing model reduces to the classical $M/M/n$ queuing system. In this case, $\{N_1(t); t \geq 0\}$ forms a $CTMC$ on state space $\{0, 1, 2, 3, ...\}$ The infinitesimal generator matrix reduces to the form

$$Q' = \begin{bmatrix} -\lambda & \lambda & & & & & \\ \mu & -(\lambda+\mu) & \lambda & & & & \\ & 2\mu & -(\lambda+2\mu) & \lambda & & & \\ & & \ddots & \ddots & \ddots & & \\ & & & n\mu & -(\lambda+n\mu) & \lambda & \\ & & & & \ddots & \ddots & \ddots \end{bmatrix}$$

The above system is stable if and only if $\lambda < n\mu$. For further details, refer to [9].

5. System Characteristics

5.1. Distribution of Server Idle Times

Some servers are not operational when the number of customers in the system is less than the number of working servers. Here, we compute the distribution of server idle times.

Consider the Markov Chain $\{(N_1(t), N_2(t), R(t)) | t \geq 0\}$ on state space, $\Omega_1 = \{(n_1, n_2, 1) : 0 \leq n_1 \leq n-1; n_1 < n_2 : k-1 \leq n_2 \leq n-1\} \bigcup \{(n_1, n_2, 0) : 0 \leq n_1 \leq n-1; n_1 < n_2 : N+1 \leq n_2 \leq n\} \bigcup \{*\}$. Here, $\{*\}$ denotes the absorbing state indicating that all servers are busy. The time to absorption of this $CTMC$ to $\{*\}$ is the time for which the $n_2 - n_1$ operational units are idle. The infinitesimal generator matrix of this Markov chain is of the form

$$Q_1 = \begin{bmatrix} T_1 & T_1^0 \\ 0 & 0 \end{bmatrix}$$

$$Q_1 = \begin{bmatrix} A_1^0 & A_0 & & & & & & \\ A_2^1 & A_1^1 & A_0 & & & & & \\ & \ddots & \ddots & & & & & \\ & & A_2^{k-2} & A_1^{k-2} & A_0 & & & \\ & & & C_2^{k-1} & C_1^{k-1} & C_0^{k-1} & & \\ & & & & \ddots & \ddots & \ddots & \\ & & & & & C_2^{n-2} & C_1^{n-2} & C_0^{n-2} \\ & & & & & & C_2^{n-1} & C_1^{n-1} \end{bmatrix}$$

For $k-1 \leq i \leq N-1$,

$$C_2^i = \begin{bmatrix} \mathbf{0} & i\mu I_{2n-i-1-N} \end{bmatrix}$$

$$C_0^i = \begin{bmatrix} \mathbf{0} \\ \lambda I_{2n-i-2-N} \end{bmatrix}$$

For $N \leq i \leq n-2$,

$$C_2^i = \begin{bmatrix} \mathbf{0} & i\mu I_{2(n-i)-1} \end{bmatrix}$$

$$C_0^i = \begin{bmatrix} \mathbf{0} \\ \lambda I_{2(n-i)-3} \end{bmatrix}$$

$$C_2^{n-1} = \begin{bmatrix} 0 & 0 & (n-1)\mu \end{bmatrix}$$
$$C_1^{n-1} = \begin{bmatrix} -(n-1)\mu \end{bmatrix}$$

The matrices $C_1^i; k-1 \leq i \leq N-1$ are square matrices of order $2n-(i+1+N)$ obtained by deleting $i-1$ rows and columns from $A_1^i : k-1 \leq i \leq N-1$. $C_1^i; N \leq i \leq n-1$ are square matrices of order $2n-2-i; N \leq i \leq n-1$ obtained by deleting $N-1+(i-N)$ rows and columns from $A_1^i; N \leq i \leq n-1$.

The matrix T_1^0 is a column matrix $[\mathbf{0},..\mathbf{0}, T_1^{k-1},..T_1^{N-1}, T_1^N,..T_1^{n-1}]$. $T_1^i; k-1 \leq i \leq N-1$ is a column matrix with the first entry $\lambda+\gamma$. $T_1^i; N \leq i \leq n-2$ is a column matrix with the first two entries $\lambda+\gamma$. $T_1^{n-1} = [\lambda+\gamma]$

Theorem 2. *Let U be the random variable designating the server idle time. Then, $P(U > t) = \boldsymbol{\alpha}.e^{\mathcal{Q}_1 t}\mathbf{e}$. This distribution is also of the PH type with representation $PH(\boldsymbol{\alpha}, \mathcal{Q}_1)$, where $\boldsymbol{\alpha} = \frac{1}{d}(x_{(0,k-1,1)}, ..., x_{(0,N,1)} ... x_{(0,n,0)} ..., x_{(k-1,k,1)} .. x_{(k-1,n,0)}, ... x_{(n-2,n-1,0)}, .., x_{(n-2,n,0)}, x_{(n-1,n,0)})$, and d is the normalizing constant.*

5.2. Distribution of First Passage Time from an Inoperative State to n-Operational Server State

Under the assumption that a sufficiently large number of customers are present in the system, we compute the distribution of time taken for the system to pass from a state in which there are $k-1$ operational servers to a state in which there are n operational servers.

Consider the Markov chain $\{(N_2(t), 1) | t \geq 0\}$ on state space $\Omega_2 = \{(n_2, 1) : k-1 \leq n_2 \leq n-1\} \bigcup \{*\}$. Here, $\{*\}$ denotes the absorbing state indicating that all servers are working. The time to absorption of this $CTMC$ to $\{*\}$ is the first passage time from state $(k-1, 1)$ to the state $\{*\}$. The infinitesimal generator matrix of this Markov chain, when the states are arranged in ascending order of n_2, is of the form

$$\mathcal{Q}_2 = \left[\begin{array}{ccccc|c} -\delta & \delta & & & & 0 \\ \gamma & -(\delta+\gamma) & \delta & & & 0 \\ & \ddots & \ddots & \ddots & & \\ & & \gamma & -(\delta+\gamma) & \delta & \\ & & & \gamma & -(\delta+\gamma) & \delta \\ \hline 0 & 0 & 0 & 0 & 0 & 0 \end{array}\right]$$

Theorem 3. *Let V be the random variable denoting the first passage time from an inoperative state to an n-operational server state, under the assumption of a large number of customers in the system. Then, V has a PH distribution with representation $PH(\boldsymbol{\beta}, \mathcal{Q}_2)$, where $\boldsymbol{\beta} = (1, 0, ..., 0)$.*

5.3. Distribution of First Passage Time from an n-Operational Server State to an Inoperative State

Here, we also compute the distribution of time taken for the system to pass from a state in which there are n operational servers to a state in which there are $k-1$ operational servers, under the assumption that a sufficiently large number of customers are present in the system.

Consider the Markov chain $\{(N_2(t), r) | t \geq 0\}$ on state space $\Omega_3 = \{(n_2, 1) : k \leq n_2 \leq n-1\} \bigcup \{(n_2, 0) : N+1 \leq n_2 \leq n\} \bigcup \{*\}$. Here, $\{*\}$ denotes the absorbing state indicating that only $k-1$ servers are working/operational. The time to absorption of this $CTMC$ to $\{*\}$ is the first passage time from state $(n, 0)$ to the state $\{*\}$. When the states are arranged in decreasing order of n_2, the infinitesimal generator matrix of this Markov chain is of the form

$$\mathcal{Q}_3 = \begin{bmatrix} S_1 & S_1^0 \\ \mathbf{0} & 0 \end{bmatrix}$$

where

$$S_1 = \begin{bmatrix} S_{11} & S_{12} & \mathbf{0} \\ S_{21} & S_{22} & S_{23} \\ \mathbf{0} & S_{32} & S_{33} \end{bmatrix}$$

$$S_{11} = -\gamma$$

$$S_{12} = \gamma$$

$$S_{21} = e_1 \otimes [0, \delta]',$$

is e_1 is a column matrix of order $n-1-N$.

$$S_{22} = I_{(n-1-N)} \otimes \begin{bmatrix} -\gamma & \\ & -(\gamma+\delta) \end{bmatrix} + E^- \otimes \begin{bmatrix} 0 & 0 \\ 0 & \delta \end{bmatrix} + E^+ \otimes \begin{bmatrix} \gamma & 0 \\ 0 & \gamma \end{bmatrix}$$

where E^- and E^+ are square matrices (refer to page 19) of dimension $n-1-N$.

$$S_{23} = E_{(n-1)-N1} \otimes [\gamma\ \gamma]'$$

$E_{(n-1)-N1}$ is a matrix (refer to page 19) of order $(n-1) - N \times (N-k)$.

$$S_{32} = E_{1(n-1)-N} \otimes [0\ \delta]$$

$E_{1(n-1)-N}$ is a matrix of order $(N-k) \times (n-1) - N$.

$$S_{33} = -(\lambda+\gamma) I_{N-k} + \delta E^- + \gamma E^+$$

where E^- and E^+ are square matrices (refer to page 19) of dimension $N-k$.

$$S_1^0 = \gamma e_{2n-(N+k)}$$

Theorem 4. *Let W be the random variable denoting the first passage time from the n-operational server state to an inoperative state, under the assumption of a large number of customers in the system. Then, W has a PH distribution with representation $PH(\chi, Q_3)$, where $\chi = (1, 0, ..., 0)$.*

5.4. Distribution of Number of Times That the System Becomes Inoperative before Reaching the n Server State

Under the assumption of a large number of customers in the system, we compute the distribution of the number of times that the system hits the $k-1$ server state before reaching the n server state. Let Δ denote the absorbing state indicating the number of working servers hitting n. Then, $(M(t), N_2(t))$, (where $M(t)$ is the number of times that the system hits the $k-1$ server state before reaching the n server state) is the Markov chain on state space $\Omega_4 = \{\Delta\} \bigcup \{(m, n_2) : m \geq 0, k-1 \leq n_2 \leq n-1\}$. The infinitesimal generator is of the form

$$Q_4 = \begin{bmatrix} \mathbf{0} & \mathbf{0} & \mathbf{0} & \mathbf{0} & \mathbf{0} & \cdots \\ \delta e_{n-k} & M_0 & M_1 & \mathbf{0} & \mathbf{0} & \cdots \\ \delta e_{n-k} & \mathbf{0} & M_0 & M_1 & \mathbf{0} & \cdots \\ \delta e_{n-k} & \mathbf{0} & \mathbf{0} & M_0 & M_1 & \\ \vdots & & & \ddots & \ddots & \end{bmatrix}$$

where

$$M_0 = \begin{bmatrix} -\delta & \delta e'_{n-k} \\ \mathbf{0} & \gamma E^- + \delta E^+ - (\gamma+\delta) I_{n-k} \end{bmatrix}$$

$$M_1 = \gamma E_{21}$$

E_{21} is a matrix of order $n-k-1$ with 1 in the $(2,1)$ position and E^-, E^+ are matrices (refer to page 19) of order $n-k$.

Let y_m denote the probability that the system hits the $k-1$ server state m times before reaching the n server state.

$$y_0 = -\delta\psi M_0^{-1}\mathbf{e}$$

$$y_m = (-1)^m \delta\psi (M_1 M_0^{-1})^m M_0^{-1}\mathbf{e}$$

where ψ is the initial probability vector $\psi = \frac{1}{d_1}(\pi_{k-1}, ..\pi_N, ...\pi_{(N+1,1)}...\pi_{(n-1,1)})$, and d_1 is the normalizing constant.

Theorem 5. *The expected number of times that the system hits the $k-1$ server state before reaching the n server state is $\sum m y_m$.*

5.5. Other Performance Measures

- Fraction of time for which the system is under repair:

  $F_{Repair} = \sum_{n_1=0}^{\infty} \sum_{n_2=k-1}^{n-1} x_{(n_1,n_2,1)}$

- Fraction of time for which the servers are idle:

  $F_{Idle} = \sum_{n_1=0}^{\infty} x_{(n_1,k-1,1)} + \sum_{n_1=0,n_1\leq n_2}^{\infty} \sum_{n_2=k-1}^{n} \sum_{r=0,1} x_{(n_1,n_2,r)}$

- System reliability, i.e., the probability that at least k servers are operational:

  $S = \sum_{n_1=0}^{\infty} \sum_{n_2\geq k}^{n} \sum_{r=0,1} x_{(n_1,n_2,r)} = 1 - \sum_{n_1=0}^{\infty} x_{(n_1,k-1,1)}$

- Average number of customers in the system:

  $E_{System} = \sum_{n_1=0}^{\infty} n_1 x_{n_1}$

- Average number of failed servers in the system:

  $E_{FS} = \sum_{n_2=k-1}^{n-1} (n-n_2) \sum_{n_1=0}^{\infty} \sum_{r=0,1} x_{(n_1,n_2,r)}$

- Average number of servers that are idle:

  $E_{IS} = \sum_{n_1=0}^{\infty} (k-1) x_{(n_1,k-1,1)} + \sum_{n_1=0,n_1\leq n_2}^{\infty} \sum_{n_2=k-1}^{n} \sum_{r=0,1} (n_2-n_1) x_{(n_1,n_2,r)}$

6. Numerical Illustrations

In this section, we give some numerical examples that show the effect of the level N of the N-policy and the repair rate δ on certain performance measures. For this, we consider a 5-out-of-20 : G system.

6.1. Effect of Level N on the Performance of the System

In this numerical example, the following parameters are kept fixed with values as given below:

$$\lambda = 10; \mu = 4; \gamma = 2; \delta = 1$$

From Table 1 and Figures 1–4, the following conclusions can be made:

- The fraction of time for which the servers are under repair is the minimum for a value $N = 7$. To minimize F_{Repair}, we need to initiate the repair process when the number of non-operational servers is 13.

- The fraction of time for which the servers are idle can be minimized if we choose $N = 8$. It plays a crucial role in enhancing the cost-effectiveness of the model.
- The system reliability is maximum when $N = 7$. We can achieve a more reliable system if we take into consideration the results in Figure 3.
- The expected number of customers found waiting in the system is the minimum if we choose $N = 8$. The value of the expected number of customers is low as the servers are working in parallel.

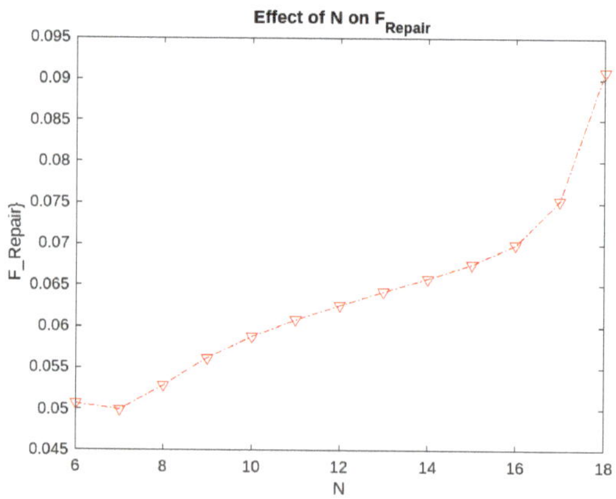

Figure 1. Effect of parameter N on the fraction of time for which the servers are under repair.

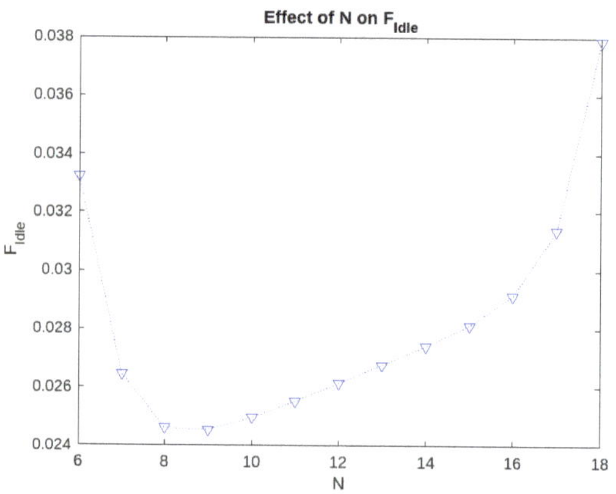

Figure 2. Effect of the level N on the fraction of time for which the servers are idle.

Table 1. Effect of N on F_{Repair},F_{Idle}, S and E_{repair}.

N	F_{Repair}	F_{Idle}	S	E_{System}
6	0.0507	0.0332	0.9691	0.0651
7	0.0499	0.0265	0.9726	0.0530
8	0.0528	0.0246	0.9722	0.0505
9	0.0562	0.0245	0.9708	0.0510
10	0.0588	0.0250	0.9696	0.0523
11	0.0608	0.0255	0.9685	0.0537
12	0.0626	0.0261	0.9676	0.0551

Table 1. *Cont.*

N	F_{Repair}	F_{Idle}	S	E_{System}
13	0.0642	0.0268	0.9667	0.0564
14	0.0658	0.0274	0.9658	0.0578
15	0.0676	0.0281	0.9649	0.0594
16	0.0700	0.0291	0.9636	0.0615
17	0.0753	0.0314	0.9608	0.0662
18	0.0909	0.0379	0.9527	0.0799

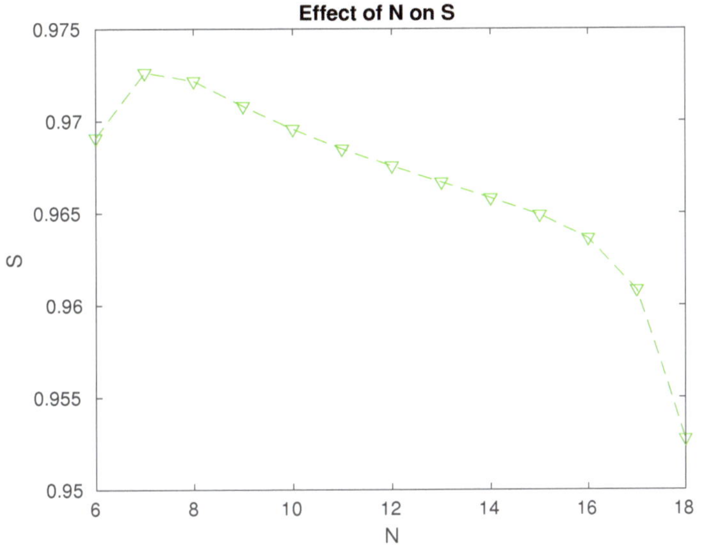

Figure 3. Effect of the level N on the reliability of the system.

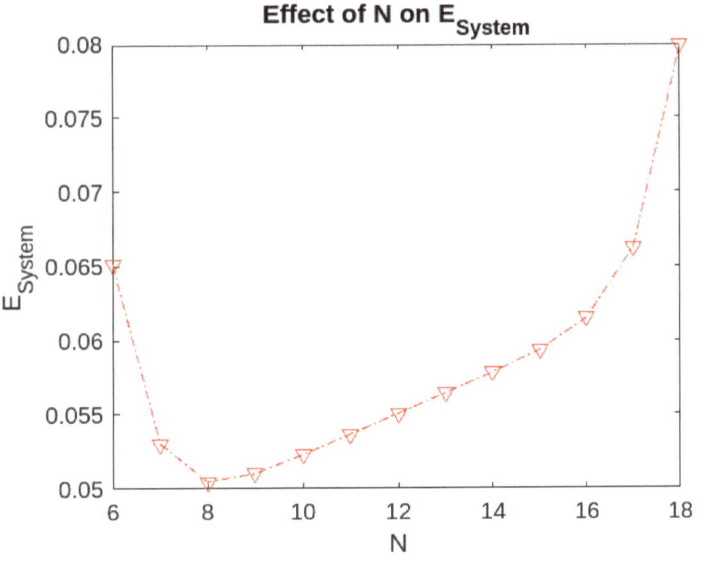

Figure 4. Effect of the level N on E_{System}.

The above example illustrates that the analysis can guide us in properly managing the repair policy with specific objectives.

6.2. *Effect of the Repair Rate δ on the Performance of the System*

In this section, we study the effect of the repair rate δ on the performance of the system. In this example, the following parameters are kept fixed with values as given below:

$$\lambda = 35; \mu = 26; \gamma = 3; N = 6$$

From Table 2 and Figures 5–8, the following conclusions can be made:

- The fraction of time for which the servers are under repair is the minimum for a value $\delta = 4$. We can decide to what extent we need to enhance the repair facilities or resources based on the data in Figure 5.
- The fraction of time for which the servers are idle is maximum for $\delta = 2$. The data in Figure 6 play a crucial role in specifically designing the facilities so that the idle time of the servers can be effectively utilized and the entire system can be managed accordingly.
- The system reliability increases with increasing values of δ as expected. The rate at which this increase occurs helps us to decide on the optimum value to be considered when compared to the resources or efforts involved in enhancing the repair facilities; see Figure 7.
- The expected number of customers found waiting in the system is maximum for $\delta = 2$.

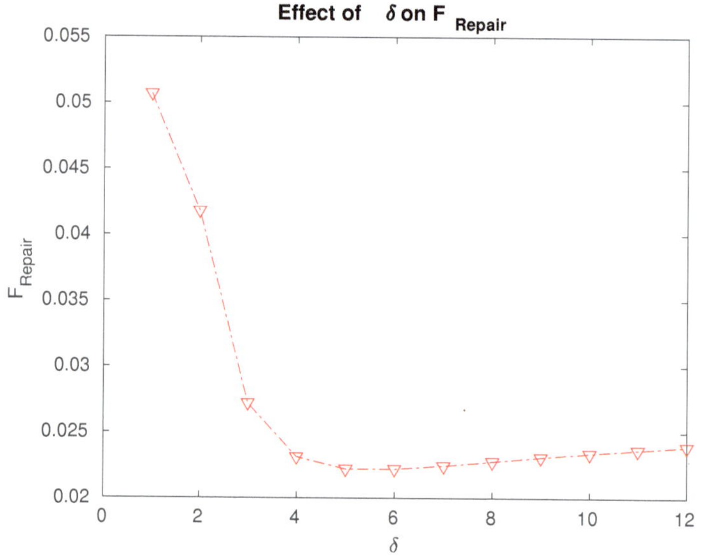

Figure 5. Effect of the repair rate δ on the fraction of time for which the servers are under repair.

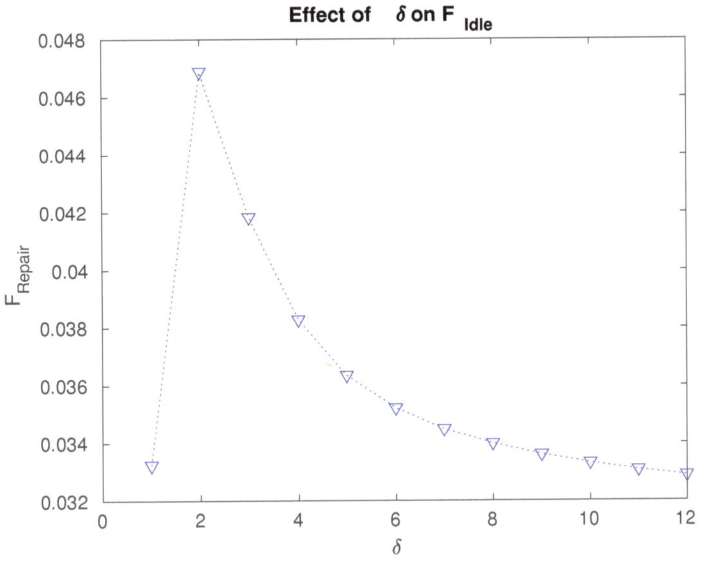

Figure 6. Effect of the repair rate δ on the fraction of time for which the servers are idle.

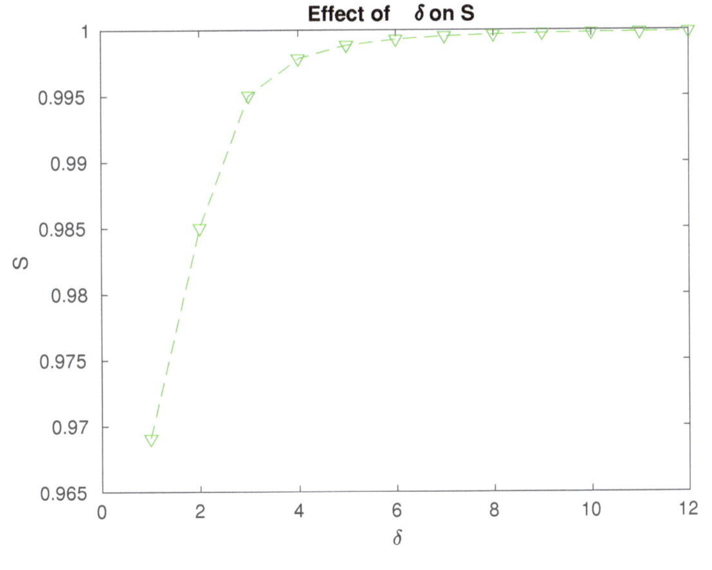

Figure 7. Effect of the repair rate δ on the reliability of the system.

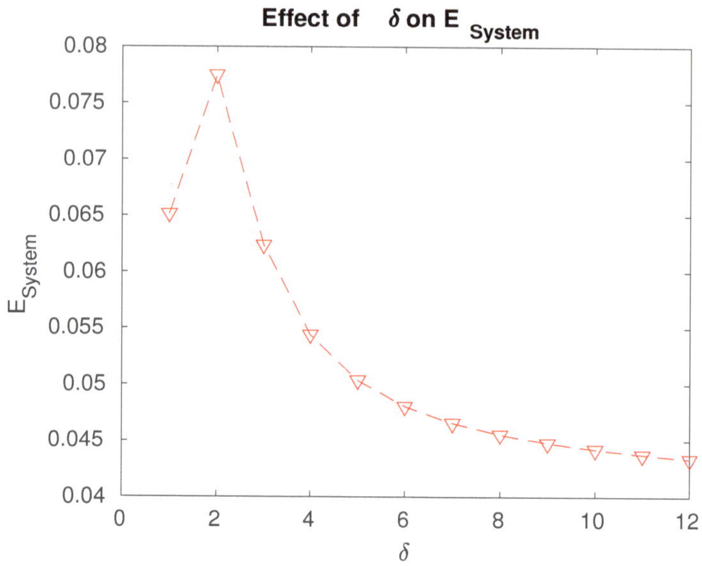

Figure 8. Effect of δ on E_{System}.

Table 2. Effect of repair rate δ on F_{Repair}, F_{Idle}, S and E_{repair}.

δ	F_{Repair}	F_{Idle}	S	E_{System}
1	0.0507	0.0332	0.9691	0.0651
2	0.0418	0.0469	0.9850	0.0775
3	0.0272	0.0418	0.9950	0.0624
4	0.0232	0.0383	0.9979	0.0544
5	0.0223	0.0364	0.9989	0.0504
6	0.0222	0.0352	0.9993	0.0481
7	0.0225	0.0345	0.9995	0.0466
8	0.0228	0.0340	0.9997	0.0456
9	0.0232	0.0336	0.9998	0.0448
10	0.0235	0.0333	0.9998	0.0442
11	0.0237	0.0331	0.9999	0.0438
12	0.0240	0.0329	0.9999	0.0434

7. Cost Analysis and Optimization Problem

Cost analysis plays an important role in decision making or in formulating policies related to the working of any system that we encounter in everyday life. In this section, we propose a cost function related to the system under consideration. We consider an optimization problem to find the value of the level N of the N-policy of the repair. With the help of numerical examples as well as graphical illustrations, we show that an optimal value N exists so that the total expected cost is the minimum. We also study the effect of the repair rate δ on the the total expected cost.

To determine the optimal level of N at which the repair facility can start working and to determine the effectiveness of enhancing the repair rate, we proceed as follows. For the cost analysis, we define the following costs.

- C_1: Establishment cost or setup cost.
- C_2: Holding cost per customer per unit time.
- C_3: Unit time cost to run the repair mechanism.
- C_4: Unit time cost incurred due to the idleness of the servers.
- C_5: unit time revenue received from the busy servers.
- C_6: Unit time revenue received when at least k servers are operational.

The expected total cost is

$$TC = C_1 + C_2 * E_{System} + C_3 * F_{Repair} + C_4 * F_{Idle} - C_5 * (1 - F_{Idle}) - C_6 * S$$

We fix the following values:

$$C_1 = 1000, C_2 = 75, C_3 = 100, C_4 = 200, C_5 = 500, C_6 = 160$$

The values of other parameters are the same as in Section 6.

- From Table 3 and Figure 9, we see that the optimal value of the level N is $N = 8$, for which the total expected cost is the minimum. Thus, it is optimal to wait until 12 servers are non-operational before starting the repair.
- From Table 3 and Figure 10, it can be seen that the expected total cost is the maximum when the repair rate is $\delta = 2$. This means that if we spend more to increase the rate at which the repair is performed from 1 to 2 or more, it will not be reflected in the total expected cost. Thus, a decision regarding the facilities to be arranged to ensure a specific repair rate can be made using this cost analysis. In this specific numerical example, it is enough to ensure a repair rate of at most $\delta = 1$.
- By analyzing the problem numerically, we can decide on the optimal values of the level N of the $N-$policy and also the optimal repair rate δ to be maintained.

Table 3. Effect of N and δ on expected total cost.

N	TC	δ	TC
6	528.2877	1	528.2877
7	523.1107	2	540.4254
8	521.8479	3	535.9006
9	521.9576	4	532.8554
10	522.4002	5	531.2721
11	522.9301	6	530.3771
12	523.4913	7	529.8193
13	524.0646	8	529.4412
14	524.6497	9	529.1666
15	525.2904	10	528.9562
16	526.1948	11	528.7882
17	528.1975	12	528.6496
18	534.0302		

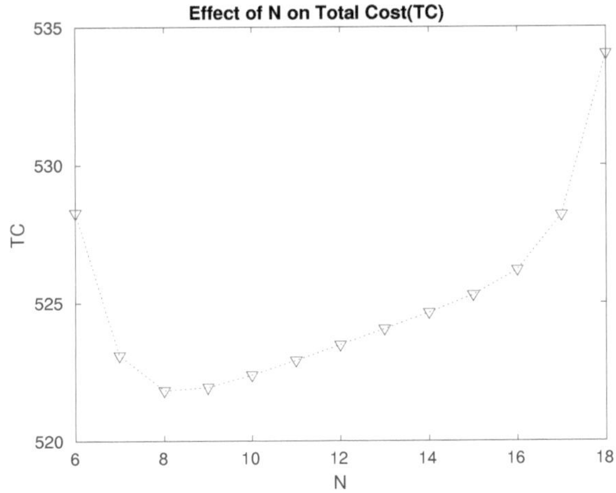

Figure 9. Effect of the changes in level N on the expected total cost.

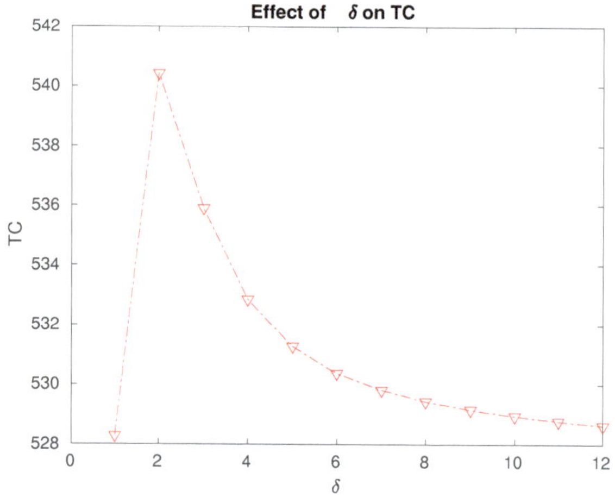

Figure 10. Effect of the repair rate δ on the expected total cost.

8. Conclusions

This paper studies a multi-server queuing system with the N-policy of repair, by viewing it as a k-out-of-n:G system. The steady-state distributions of various system states are computed. The distribution of server idle times is analyzed. Under the assumption of a sufficiently large number of customers in the system, the distribution of the first passage times from an inoperative state to an n server state and vice versa has been found. The assumption of a sufficiently large number of customers in the system is made to avoid complications that may arise due to future arrivals. Other system performance measures such as system reliability, the expected number of failed servers, etc., are computed.

The effect of an increase in the level N and the repair rate δ on various performance measures, when other parameters are kept fixed, is studied numerically and graphically. An increase in the repair rate significantly increases the system reliability and decreases the expected number of customers in the system. It reduces the idleness of servers and reduces the fraction of time for which the servers are under repair. On the other hand, an increase in N increases the fraction of time for which the servers are under repair. However, as N increases, the idleness of servers and the expected number in the system first decrease and then increase, while the system reliability first increases and then decreases. The optimal value of N depends on the parameters of the system. The examples illustrate how an optimal policy could be derived for a multi-server queuing system, keeping in mind certain specific objectives. A cost function has been constructed and the results of the cost analysis are presented.

There can be several extensions to the problem considered in this paper. An extension of the present work to one in which consecutive k-out-of-n systems provide a service to customers, either linearly or in a circular fashion, is proposed for future research. A similar study of the k-out-of-$n : F$ system is underway. Moreover, the k-out-of-n system could be analyzed under both HOT and WARM conditions.

Author Contributions: Conceptualization, A.K.; Methodology, A.N.J.; Validation, A.P.M.; Formal analysis, A.N.J.; Data curation, A.P.M.; Writing—original draft, A.K., A.N.J. and A.P.M.; Writing—review & editing, A.K., A.N.J. and A.P.M.; Supervision, A.K. All authors have read and agreed to the published version of the manuscript.

Funding: Dr. Anu Nuthan Joshua and Dr. Ambily P. Mathew received support from DST-RSF research project number 22-49-02023 (RSF) and research project number 64800 (DST) for the preparation of this publication. https://search.crossref.org/funding.

Data Availability Statement: Data are contained within the article.

Conflicts of Interest: The authors declare no conflicts of interest.

Notations and Abbreviations

The following abbreviations are used in this manuscript:

$CTMC$	Continuous-time Markov chain
QBD process	Quasi-birth–death process
$\mathbf{e}$	All one vector with appropriate dimension
I_n	Identity matrix of order $n \times n$
E^-	Square matrix of appropriate size with all zero entries except the entries $(E^-)_{(j,j-1)}$, which are equal to 1
E^+	Square matrix of appropriate size with all zero entries except the entries $(E^+)_{(j,j+1)}$, which are equal to 1
E_{ij}	Matrix of appropriate size with all zero entries except the entry (i, j), which is equal to 1
e_i	Column vector of appropriate dimension with all zero entries except the entry i, which is equal to 1
A'	Transpose of matrix A
$\mathbf{0}$	Matrix whose entries are 0, of appropriate size
$A \otimes B$	Kronecker product; if $A = [a_{ij}]$ is a matrix of order $m \times n$ and if B is a matrix of order $p \times q$, then $A \otimes B = [a_{ij}B]$ will denote a matrix of order $mp \times nq$

References

1. Barlow, R.E.; Proschan, F. *Mathematical Theory of Reliability*; John Wiley and Sons: New York, NY, USA, 1965.
2. Krishnamoorthy, A.; Sathian, M.K.; Narayanan Viswanath, C. Reliability of a k-out-of-n System with Repair by a Single Server Extending Service to External Customers with Pre-emption. *Electron. J. Reliab. Theory Appl.* **2016**, *11*, 61–93.
3. Yadin, M.; Naor, P. Queueing system with a removable service station *Oper. Res. Q.* **1963**, *14*, 393–405. [CrossRef]
4. Krishnamoorthy, A.; Ushakumari, P.V.; Lakshmy, B. k-out-of-n-system with repair: The N-policy. *Asia-Pac. J. Oper. Res.* **2002**, *19*, 47–61.
5. Yen, T.-C.; Wang, K.-H.; Chen, J.-Y. Optimization Analysis of the N-Policy $M/G/1$ Queue with Working Breakdowns. *Symmetry* **2020**, *12*, 583. [CrossRef]
6. Vemuri, V.K.; Boppana, V.S.N.H.P.; Kotagiri, C.; Bethapudi, R.T. Optimal strategy analysis of an N-policy two-phase $M^X/M/1$ queueing system with server startup and breakdowns. *Opsearch* **2011**, *48*, 109–122. [CrossRef]
7. Singh, C.J.; Jain, M.; Kumar, B. Analysis of queue with two phases of service and m phases of repair for server breakdown under N-policy. *Int. J. Serv. Oper. Manag.* **2013**, *16*, 373–406 [CrossRef]
8. Neuts, M.F. *Matrix—Geometric Solutions in Stochastic Models: An Algorithmic Approach*; The Johns Hopkins University Press: Baltimore, MD, USA, 1981.
9. Gross, D.; Harris, C. *Fundamentals of Queuing Theory*, 3rd ed.; John Wiley: Chichester, UK, 1988.

mathematics
MDPI

Article

Robust Transceiver Design for Correlated MIMO Interference Channels in the Presence of CSI Errors under General Power Constraints

Jae-Mo Kang [1] and Dong-Woo Lim [2,*]

[1] Department of Artificial Intelligence, Kyungpook National University, Daegu 41566, Republic of Korea; jmkang@knu.ac.kr
[2] Department of Information & Communication Engineering, Changwon National University, Changwon 51140, Republic of Korea
* Correspondence: dwlim@changwon.ac.kr

Abstract: In this paper, we consider a new design problem of optimizing a linear transceiver for correlated multiple-input multiple-output (MIMO) interference channels in the presence of channel state information (CSI) errors, which is a more realistic and practical scenario than those considered in the previous studies on uncorrelated MIMO interference channels. By taking CSI errors into account, the optimization problem is initially formulated to minimize the average mean square error (MSE) under the general power constraints. Since the objective function is not jointly convex in precoders and receive filters, we split the original problem into two convex subproblems, and then linear precoders and receive filters are obtained by solving two subproblems iteratively. It is shown that the proposed algorithm is guaranteed to converge to a local minimum. The numerical results show that the proposed algorithm can significantly reduce the sensitivity to CSI errors compared with the existing robust schemes in the correlated MIMO interference channel.

Keywords: interference channel; MIMO; stochastic robustness; MSE minimization; CSI errors; general power constraints

MSC: 94A05

Citation: Kang, J.-M.; Lim, D.-W. Robust Transceiver Design for Correlated MIMO Interference Channels in the Presence of CSI Errors under General Power Constraints. *Mathematics* **2024**, *12*, 801. https://doi.org/10.3390/math12060801

Academic Editors: Gurami Tsitsiashvili and Alexander Bochkov

Received: 16 January 2024
Revised: 26 February 2024
Accepted: 5 March 2024
Published: 8 March 2024

1. Introduction

Inter-cell interference management is an important issue in upcoming wireless communications standards such as 3GPP LTE-Advanced and IEEE 802.16m due to universal frequency reuse [1,2]. One promising way to mitigate the inter-cell interference is through cooperation among base stations, e.g., coordinated multi-point transmission (CoMP) in 3GPP LTE-advanced as well as in 4G/5G/6G. In this paper, we consider a multi-cell scenario where base stations and mobile users are equipped with multiple antennas and model it as a multiple-input multiple-output (MIMO) interference channel. An interference channel is a network that models simultaneously communicating transmitter–receiver pairs using the same resources such as time and frequency [3]. In interference channels, much attention has been focused on information-theoretic studies such as interference alignment (IA), because the exact capacity characterization of the interference channels still remains unknown except for some special cases [3].

Recently, linear transceiver designs for the MIMO interference channel have been studied in [4–7] (as well as for MIMO multiuser systems in [8] and reconfigurable intelligent surface (RIS)-assisted MIMO systems in [9]) based on various criteria with the assumption of the perfect knowledge of the channel state information (CSI) at each node. To combat the assumption of perfect CSI, mean square error (MSE)-based robust transceiver design algorithms were developed in [7] while taking CSI errors into account. The results in [4–7] have been developed under the implicit assumption of independent and identically distributed (i.i.d.) or spatially uncorrelated channel elements.

1.1. Motivations

In real environments, however, the channel elements are not only erroneous due to inaccurate channel estimation or quantization of estimated CSI [10,11], but are also correlated due to little scattering and insufficient antenna spacing, e.g., the 3GPP spatial channel model (SCM). In this case, the performance of the works [4–6] is severely sensitive to both CSI errors and channel correlation. In addition, the existing results in [7] cannot be applicable to correlated MIMO interference channels, since they cannot appropriately deal with interference leakage through the correlated channels and their corresponding CSI errors. Although joint robust transceiver design methods with imperfect CSI were developed in [12,13] for single-user point-to-point correlated MIMO channels, these methods also cannot be directly extended to the correlated MIMO interference channels due to the interference issue.

Accordingly, in [14–22], effective robust transceiver designs have been developed for the MIMO interference channel with imperfect CSI. Despite these advantages in designs, the techniques in [14–22] have limitations or face technical obstacles in the following aspects. In [14–19], the CSI errors were assumed to be spatially uncorrelated. Also, in [20], the CSI errors were assumed to be correlated only at the transmitter sides, not at both the transmitter and receiver sides. In [21], the CSI errors were assumed to be deterministic (rather than random), which however may not adequately capture the stochastic nature of the uncertainty. In [22], a robust transceiver design scheme considering the (full) spatial correlation at both the transmitter and receiver sides was developed for the MIMO interference channel. However, [22] considered only the total power constraints at the transmitters, which clearly neglects the per-antenna power constraints, and thus hinders the practical applicability in distributed MIMO scenarios where multiple single-antenna devices collaborate to form a virtual antenna array under their strict individual power constraints (although the total power constraints and per-antenna power constraints have been investigated separately in the literature, such as in [23], these constraints have not been dealt with in an integrated or mixed manner as in our work).

1.2. Contributions

The main purpose of this study is to develop a novel robust transceiver design by considering a more realistic and practical scenario than the previous studies. The main contributions of this paper are as follows:

- We propose to design a new linear transceiver to minimize the average sum-MSE for the correlated MIMO interference channel under the CSI errors considering the general power constraints that include the total power constraints and per-antenna power constraints (as well as their mixtures) as special cases. Our design approach is based on the existing stochastic robust technique [12].
- Since the optimization problem is not jointly convex in precoders and receive filters, it is difficult to find the jointly optimal solution. In order to resolve this difficulty, we leverage the alternating optimization method. More specifically, we propose to divide the original problem into two subproblems that are convex in precoders and receive filters, respectively. Then, precoders and receive filters can be obtained by solving two subproblems iteratively.
- We present extensive numerical results, which demonstrate the superior performance and effectiveness of the proposed scheme compared to the conventional schemes.

1.3. Organization

The remainder of this paper is organized as follows. Section 2 describes the system model. Section 3 presents the problem formulation and linear transceiver design design. Section 4 presents the numerical results and Section 5 concludes this paper.

1.4. Notations

$(\cdot)^{\mathsf{T}}$, $(\cdot)^{\mathsf{H}}$, and $(\cdot)^{1/2}$ denote the transpose, conjugate transpose, and matrix square root, respectively. $\text{vec}(\cdot)$ is the vectorization operator and $\text{Tr}(\cdot)$ is the trace of a matrix. Also, $\otimes$ and $\text{vec}(\cdot)$ represent the Kronecker product and vectorization operator, respectively. $\mathbb{E}_x(\cdot)$ denotes the expectation of a random variable x. A circular symmetric Gaussian random vector $\mathbf{z}$ with mean $\bar{\mathbf{z}}$ and covariance matrix $\mathbf{C}_\mathbf{z}$ is denoted as $\mathbf{z} \sim CN(\bar{\mathbf{z}}, \mathbf{C}_\mathbf{z})$.

2. System Model

We consider the K-user MIMO interference channel as shown in Figure 1, in which there are K transmitter–receiver pairs and K transmitters that simultaneously transmit independent data streams to their intended receivers. The kth transmitter and receiver are equipped with M_k antennas and N_k antennas, respectively. The transmit signal vector at the kth transmitter is denoted as $\mathbf{x}_k = \mathbf{F}_k\mathbf{s}_k$, where $\mathbf{F}_k \in \mathbb{C}^{M_k \times d_k}$ is the precoding matrix at the kth transmitter and $\mathbf{s}_k \in \mathbb{C}^{d_k \times 1}$ is the symbol vector transmitted by the kth transmitter, which satisfies $\mathbb{E}[\mathbf{s}_k\mathbf{s}_k^{\mathsf{H}}] = \mathbf{I}_{d_k}$ and $\mathbb{E}[\mathbf{s}_k\mathbf{s}_j^{\mathsf{H}}] = \mathbf{0}_{d_k \times d_j}, \forall j \neq k$.

Also, $d_k \leq \min(M_k, N_k)$ denotes the number of transmitted data streams at the kth transmitter. The received signal at the kth receiver is given by the following:

$$\mathbf{y}_k = \mathbf{H}_{kk}\mathbf{F}_k\mathbf{s}_k + \sum_{j=1, j\neq k}^{K} \mathbf{H}_{kj}\mathbf{F}_j\mathbf{s}_j + \mathbf{w}_k, \quad k = 1, \cdots, K \tag{1}$$

where $\mathbf{H}_{kj} \in \mathbb{C}^{N_k \times M_j}$ denotes the channel matrix between the j-th transmitter and kth receiver, and it is assumed to be frequency-flat. $\mathbf{w}_k \in \mathbb{C}^{N_k \times 1}$ is the additive Gaussian noise at the k-th receiver whose elements are i.i.d. $CN(0, \sigma_k^2)$. In this paper, we assume that the elements of the channel matrices $\{\mathbf{H}_{kj}\}_{k,j=1}^{K}$ are correlated and adopt the Kronecker channel model, which is widely used in MIMO systems due to its simplicity and analytical tractability [24] (p. 90). Using this model, the channel matrices can be expressed as follows:

$$\mathbf{H}_{kj} = \mathbf{R}_k^{1/2}\mathbf{H}_{kj}^{(w)}\mathbf{T}_j^{1/2}, \quad \forall k, j \tag{2}$$

where $\mathbf{H}_{kj}^{(w)} \in \mathbb{C}^{N_k \times M_j}$ is the spatially white channel matrix whose elements are i.i.d. $CN(0, 1)$. $\mathbf{T}_j \in \mathbb{C}^{M_j \times M_j}$ and $\mathbf{R}_k \in \mathbb{C}^{N_k \times N_k}$ represent the transmit correlation at the jth transmitter and receive correlation at the kth receiver, respectively. Although the channel matrices can vary block to block due to mobile location, the correlation matrices can be assumed to be constant over a large number of blocks since the scattering environments change slowly. For this reason, they can be estimated reliably at each receiver and fed back to the transmitters through reliable feedback links [25]. Hence, we reasonably assume that the channel correlation matrices are constant and available at each node, and that the CSI errors are mainly due to the imperfect channel estimation.

We define the CSI error for each link of the interference channel as follows:

$$\mathbf{\Delta}_{kj} = \mathbf{H}_{kj} - \hat{\mathbf{H}}_{kj}, \quad \forall k, j \tag{3}$$

where $\hat{\mathbf{H}}_{kj}$ is an estimate of $\mathbf{H}_{kj}$. For the Kronecker channel model, the CSI error can be modeled as $\text{vec}\left(\mathbf{\Delta}_{kj}\right) \sim CN\left(\mathbf{0}_{M_jN_k \times 1}, \sigma_{e,kj}^2(\mathbf{\Sigma}_j \otimes \mathbf{\Psi}_k)\right)$ [13,26], and it is said to have a matrix variate complex Gaussian distribution, which can be written as follows [27]:

$$\mathbf{\Delta}_{kj} \sim CN_{N_k,M_j}\left(\mathbf{0}_{N_k \times M_j}, \sigma_{e,kj}^2(\mathbf{\Sigma}_j \otimes \mathbf{\Psi}_k)\right) \tag{4}$$

where $\sigma_{e,kj}^2$ is the channel estimation error variance, and $\mathbf{\Sigma}_j \in \mathbb{C}^{M_j \times M_j}$ and $\mathbf{\Psi}_k \in \mathbb{C}^{N_k \times N_k}$ represent the column and row covariance of $\mathbf{\Delta}_{kj}$, respectively. It is assumed that the CSI error is independent of the transmitted symbol and noise. If the minimum MSE (MMSE) channel estimator is used to estimate the spatially white channel $\mathbf{H}_{kj}^{(w)}$, the matrices $\mathbf{\Sigma}_j$

and $\mathbf{\Psi}_k$ can be represented by $\mathbf{\Sigma}_j = \mathbf{T}_j$ and $\mathbf{\Psi}_k = \mathbf{R}_k$, respectively, and $\sigma^2_{e,kj}$ is given by $\sigma^2_{e,kj} = 1 - \sigma^2_{\hat{h},kj}$, where $\sigma^2_{\hat{h},kj}$ denotes the variance of each element of $\hat{\mathbf{H}}^{(w)}_{kj}$ [26]. On the other hand, if the correlated channel $\mathbf{H}_{kj}$ is estimated using the MMSE channel estimator, $\mathbf{\Sigma}_j$ and $\mathbf{\Psi}_k$ are given by $\mathbf{\Sigma}_j = \mathbf{T}_j$ and $\mathbf{\Psi}_k = \left(\mathbf{I}_{N_k} + \sigma^2_{e,kj}\mathbf{R}_k^{-1}\right)^{-1}$, and $\sigma^2_{e,kj}$ is given by $\sigma^2_{e,kj} = \sigma_k^2 \mathrm{Tr}(\mathbf{T}_j)/p_j^{(tr)}$, where $p_j^{(tr)}$ is the training signal power at the jth transmitter [13].

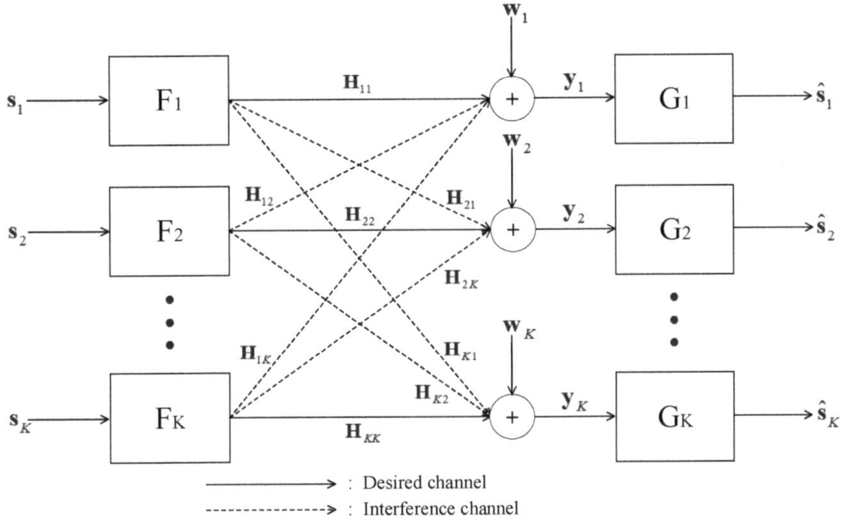

Figure 1. K-user MIMO interference channel.

3. Problem Formulation and Linear Transceiver Design

3.1. General Power Constraints

We impose the general power constraints at each transmitter, which are described by

$$\mathbb{E}\left[\left\|\mathbf{Q}_{l,k}^{1/2}\mathbf{x}_k\right\|^2\right] = \mathrm{Tr}\left(\mathbf{F}_k^{\mathsf{H}}\mathbf{Q}_{l,k}\mathbf{F}_k\right) \leq p_{l,k}, \quad l = 1, \cdots, L_k, \tag{5}$$

where $\mathbf{Q}_{l,k}$'s and $p_{l,k}$'s are weighting matrices and power budgets, respectively. Also, L_k denotes the number of constraints, which determines the type of power constraint together with the weighting matrices $\mathbf{Q}_{l,k}$'s. The general power constraints in (5) include the total power constraint, per-antenna power constraints, and a mixture between total and per-antenna power constraints as special cases via a proper selection of $\mathbf{Q}_{l,k}$'s [28,29]. For example, when $L_k = 1$ and $\mathbf{Q}_{1,k} = \mathbf{I}_{M_k}$, it corresponds to the total power constraint with $\mathrm{Tr}\left(\mathbf{F}_k^{\mathsf{H}}\mathbf{F}_k\right) \leq p_{1,k}$, where $p_{1,k}$ corresponds to the total power budget. On the other hand, when $L_k = M_k$ and $\mathbf{Q}_{l,k} = \mathbf{e}_l\mathbf{e}_l^{\mathsf{T}}, l = 1, \cdots, M_k$, where $\mathbf{e}_l = [0, \cdots, 0, 1, 0, \cdots, 0]^{\mathsf{T}}$ denotes the lth unit vector (i.e., all zeros except for the lth entry being unity), the general power constraints reduce to the per-antenna power constraints as $\mathrm{Tr}([\mathbf{F}_k\mathbf{F}_k^{\mathsf{H}}]_{l,l}) \leq p_{l,k}$, $l = 1, \cdots, M_k$, with $p_{l,k}$ corresponding to the power budget of the lth transmit antenna of the kth transmitter.

3.2. Problem Formulation

For given CSI errors with known distributions, one of the most popular approaches to design the transceiver is to optimize the average performance of the system. This approach is known as the stochastic robust technique [12,13]. By exploiting this technique, in this section, we will design a set of precoders and receive filters to minimize the *average* sum-MSE, i.e., the total discrepancy between the transmit signals and their estimates, over CSI errors under the general power constraints in (5). We select the sum-MSE minimization as an optimization criterion for the following reasons. First, an MSE-based design can improve the performance of the system in all SNR regions unlike the zero-forcing and matched filtering. Second, in the case of a transmission of control signals, it is better to minimize the total MSE for improvement of the transmission reliability. Finally, it is known that the solution of the weighted sum-MSE minimization problem maximizes the sum-rate if weight matrices are properly chosen [30].

Let $\hat{\mathbf{s}}_k = \mathbf{G}_k^{\mathsf{H}}\mathbf{y}_k$ represent the estimate of $\mathbf{s}_k$ when a linear receive filter $\mathbf{G}_k \in \mathbb{C}^{N_k \times d_k}$ is used. In order to derive an expression of the average sum-MSE over CSI errors, denoted by

$$J = \mathbb{E}_{\{\Delta_{kj}\}_{k,j=1}^{K}}\left\{\sum_{k=1}^{K}\mathbb{E}_{\mathbf{s}_k,\mathbf{w}_k}\left[\|\mathbf{s}_k-\hat{\mathbf{s}}_k\|^2\right]\right\}, \tag{6}$$

we state the following lemma.

Lemma 1 ([27]). *Let* $\mathbf{X} \in \mathbb{C}^{m\times n}$ *be a random matrix and* $\mathbf{X} \sim CN_{m,n}(\bar{\mathbf{X}}, \mathbf{A}\otimes\mathbf{B})$*; then,*

$$\begin{aligned}\mathbb{E}\left[\mathbf{X}^{\mathsf{H}}\mathbf{M}\mathbf{X}\right] &= \bar{\mathbf{X}}^{\mathsf{H}}\mathbf{M}\bar{\mathbf{X}} + Tr(\mathbf{B}\mathbf{M})\mathbf{A}^{\mathsf{T}}\\ \mathbb{E}\left[\mathbf{X}\mathbf{M}\mathbf{X}^{\mathsf{H}}\right] &= \bar{\mathbf{X}}\mathbf{M}\bar{\mathbf{X}}^{\mathsf{H}} + Tr\left(\mathbf{M}\mathbf{A}^{\mathsf{T}}\right)\mathbf{B}\end{aligned}$$

where $\mathbf{M} \in \mathbb{C}^{m\times m}$ *is a deterministic matrix, and* $\mathbf{A} \in \mathbb{C}^{n\times n}$ *and* $\mathbf{B} \in \mathbb{C}^{m\times m}$ *are positive semi-definite matrices.*

Using Lemma 1, the average sum-MSE is expressed as follows:

$$\begin{aligned}J\left(\{\mathbf{F}_k\}_{k=1}^K, \{\mathbf{G}_k\}_{k=1}^K\right) &= \mathbb{E}_{\{\Delta_{kj}\}_{k,j=1}^{K}}\left[\sum_{k=1}^{K}\mathbb{E}_{\mathbf{s}_k,\mathbf{w}_k}\left[\|\mathbf{s}_k-\hat{\mathbf{s}}_k\|^2\right]\right]\\ &= \sum_{k=1}^{K}\mathrm{Tr}\left[\mathbf{G}_k^{\mathsf{H}}\left(\sum_{j=1}^{K}\hat{\mathbf{H}}_{kj}\mathbf{F}_j\mathbf{F}_j^{\mathsf{H}}\hat{\mathbf{H}}_{kj}^{\mathsf{H}} + \sigma_{e,kj}^2\mathrm{Tr}(\mathbf{F}_j^{\mathsf{H}}\boldsymbol{\Sigma}_j^{\mathsf{T}}\mathbf{F}_j)\boldsymbol{\Psi}_k\right)\mathbf{G}_k + \sigma_k^2\mathbf{G}_k^{\mathsf{H}}\mathbf{G}_k + \mathbf{I}_{d_k}\right]\\ &\quad - 2\mathrm{Re}\left[\mathrm{Tr}(\mathbf{G}_k^{\mathsf{H}}\hat{\mathbf{H}}_{kk}\mathbf{F}_k)\right].\end{aligned} \tag{7}$$

Our goal is to design a set of precoders and receive filters that minimize the average sum-MSE under the general power constraints at each transmitter. The optimization problem can be formulated as follows:

$$\begin{aligned}&\underset{\{\mathbf{F}_k\}_{k=1}^K,\{\mathbf{G}_k\}_{k=1}^K}{\text{minimize}} \quad J\left(\{\mathbf{F}_k\}_{k=1}^K, \{\mathbf{G}_k\}_{k=1}^K\right)\\ &\text{subject to} \quad \mathrm{Tr}\left(\mathbf{F}_k^{\mathsf{H}}\mathbf{Q}_{l,k}\mathbf{F}_k\right) \le p_{l,k}, \quad \forall l,k.\end{aligned} \tag{8}$$

Since the optimization problem in (8) is not jointly convex in $\{\mathbf{F}_k\}_{k=1}^K$ and $\{\mathbf{G}_k\}_{k=1}^K$, it is difficult to find $\{\mathbf{F}_k\}_{k=1}^K$ and $\{\mathbf{G}_k\}_{k=1}^K$ jointly. However, it is convex in $\{\mathbf{F}_k\}_{k=1}^K$ for a given $\{\mathbf{G}_k\}_{k=1}^K$, and in $\{\mathbf{G}_k\}_{k=1}^K$ for a given $\{\mathbf{F}_k\}_{k=1}^K$ (in general, combining two convex problems does not necessarily guarantee that the resulting problem will remain convex. In our case, although the problem in (8) is convex in either $\{\mathbf{F}_k\}_{k=1}^K$ or $\{\mathbf{G}_k\}_{k=1}^K$, it is not jointly convex over both $\{\mathbf{F}_k\}_{k=1}^K$ and $\{\mathbf{G}_k\}_{k=1}^K$ due to the coupling between $\{\mathbf{F}_k\}_{k=1}^K$ and $\{\mathbf{G}_k\}_{k=1}^K$).

Based on this observation, we divide the original problem (8) into two convex sub-problems that can be solved iteratively via the alternating minimization method [5].

3.3. Receive Filter Design

For a given $\{\mathbf{F}_k\}_{k=1}^K$, the optimization problem in (8) becomes an unconstrained convex optimization problem with respect to $\{\mathbf{G}_k\}_{k=1}^K$, and hence the optimal receive filters can be obtained by the optimality conditions, which are given by $\frac{\partial J}{\partial \mathbf{G}_k^*} = \mathbf{0}, \forall k$. From the optimality conditions, the optimal receive filters for a given $\{\mathbf{F}_k\}_{k=1}^K$ can be found in the form of the well-known Wiener filter as follows:

$$\mathbf{G}_k = \left(\mathbf{\Gamma}_k + \sigma_k^2 \mathbf{I}_{N_k}\right)^{-1} \hat{\mathbf{H}}_{kk}\mathbf{F}_k, \quad \forall k \tag{9}$$

where $\mathbf{\Gamma}_k = \sum_{j=1}^K \hat{\mathbf{H}}_{kj}\mathbf{F}_j\mathbf{F}_j^{\mathsf{H}}\hat{\mathbf{H}}_{kj}^{\mathsf{H}} + \sum_{j=1}^K \sigma_{e,kj}^2 \mathrm{Tr}\left(\mathbf{F}_j^{\mathsf{H}}\mathbf{\Sigma}_j^{\mathsf{T}}\mathbf{F}_j\right)\mathbf{\Psi}_k$.

3.4. Precoder Design

When $\{\mathbf{G}_k\}_{k=1}^K$ are given, the optimization problem in (8) is a constrained convex optimization problem with respect to $\{\mathbf{F}_k\}_{k=1}^K$, because the objective function is a quadratic function and the general power constraints are convex with respect to $\{\mathbf{F}_k\}_{k=1}^K$. To be specific, with some mathematical manipulations, the average sum-MSE in (7) can be rewritten in terms of $\{\mathbf{F}_k\}_{k=1}^K$ as follows:

$$J = \sum_{k=1}^K \left[\mathsf{vec}(\mathbf{F}_k)^{\mathsf{H}}\mathbf{A}_k\mathsf{vec}(\mathbf{F}_k) - 2\mathrm{Re}\left\{\mathbf{b}_k^{\mathsf{H}}\mathsf{vec}(\mathbf{F}_k)\right\}\right] + const \tag{10}$$

where:

$$\mathbf{A}_k = \sum_{j=1}^K \mathbf{I}_{d_j} \otimes \left(\hat{\mathbf{H}}_{kj}^{\mathsf{H}}\mathbf{G}_j\mathbf{G}_j^{\mathsf{H}}\hat{\mathbf{H}}_{kj} + \sigma_{e,jk}^2 \mathrm{Tr}(\mathbf{G}_j^{\mathsf{H}}\mathbf{\Psi}_j^{\mathsf{T}}\mathbf{G}_j)\mathbf{\Sigma}_k^{\mathsf{T}}\right), \tag{11}$$

$$\mathbf{b}_k = \mathsf{vec}(\mathbf{G}_j^{\mathsf{H}}\hat{\mathbf{H}}_{kj}). \tag{12}$$

Similarly, the general power constraints in (5) can be equivalently written as follows:

$$\mathsf{vec}(\mathbf{F}_k)^{\mathsf{H}}\mathbf{C}_{l,k}\mathsf{vec}(\mathbf{F}_k) \le p_{l,k}, \quad \forall l,k \tag{13}$$

where:

$$\mathbf{C}_{l,k} = \mathbf{I}_{d_k} \otimes \mathbf{Q}_{l,k}. \tag{14}$$

Using the above results and introducing the slack variables $\{t_k\}_{k=1}^K$ (corresponding to the upper bound of the objective value), the precoder design problem can be recast into the following second-order cone programming (SOCP):

$$\begin{aligned} \underset{\{\mathbf{F}_k\}_{k=1}^K, \{t_k\}_{k=1}^K}{\text{minimize}} \quad & \sum_{k=1}^K t_k \\ \text{subject to} \quad & \left\|\mathbf{A}_k^{1/2}\mathsf{vec}(\mathbf{F}_k)\right\| \le \sqrt{t_k + 2\mathrm{Re}\{\mathbf{b}_k^H\mathsf{vec}(\mathbf{F}_k)\}}, \quad \forall k, \\ & \left\|\mathbf{C}_{l,k}^{1/2}\mathsf{vec}(\mathbf{F}_k)\right\| \le \sqrt{p_{l,k}}, \quad \forall l,k. \end{aligned} \tag{15}$$

Since the above SOCP problem is convex, it can be solved efficiently via convex optimization techniques such as the interior point method [31].

For a special case when only the total power constraints are imposed (i.e., $\mathrm{Tr}\left(\mathbf{F}_k^{\mathsf{H}}\mathbf{F}_k\right) \le p_{1,k}, \forall k$), the optimal precoders can be found in a closed form to satisfy the Karush–Khun–

Tucker (KKT) conditions [31]. Specifically, from the KKT conditions, the optimal precoders for a given $\{\mathbf{G}_k\}_{k=1}^K$ are expressed as follows:

$$\mathbf{F}_k = \left(\mathbf{\Phi}_k + \lambda_k \mathbf{I}_{M_k}\right)^{-1} \hat{\mathbf{H}}_{kk}^{\mathsf{H}} \mathbf{G}_k, \quad \forall k \tag{16}$$

where $\mathbf{\Phi}_k = \sum_{j=1}^K \hat{\mathbf{H}}_{jk}^{\mathsf{H}} \mathbf{G}_j \mathbf{G}_j^{\mathsf{H}} \hat{\mathbf{H}}_{jk} + \sum_{j=1}^K \sigma_{e,jk}^2 \mathrm{Tr}\left(\mathbf{G}_j^{\mathsf{H}} \mathbf{\Psi}_j \mathbf{G}_j\right) \mathbf{\Sigma}_k^{\mathsf{T}}$. This coincides with the existing result in [22] (Equation (6)). Note, that $\lambda_k \geq 0$ denotes the Lagrangian multiplier associated with the total power constraint at the kth transmitter, which is chosen to satisfy $\mathrm{Tr}\left(\mathbf{F}_k^{\mathsf{H}} \mathbf{F}_k\right) \leq p_{1,k}$. Since the total transmit power $\mathrm{Tr}\left(\mathbf{F}_k^{\mathsf{H}} \mathbf{F}_k\right)$ is a monotonically decreasing function of λ_k, the optimal value of λ_k can be found through a one-dimensional (1-D) search such as the bisection method [7].

3.5. Proposed Algorithm

In this subsection, based on the results (9) and (16) obtained in the previous subsections, we propose a robust transceiver design algorithm. The proposed algorithm is shown in Algorithm 1 and summarized as follows. First, the algorithm is initialized by an appropriate set of precoders that satisfies the power constraints. Then, $\{\mathbf{G}_k\}_{k=1}^K$ and $\{\mathbf{F}_k\}_{k=1}^K$ are iteratively updated based on (9) and (16), respectively, until the stopping criterion is met.

It is clear that the updated linear receive filters $\{\mathbf{G}_k(t+1)\}$ yield the minimum average sum-MSE for a given $\{\mathbf{F}_k(t)\}$, and hence $J(\{\mathbf{F}_k(t)\}, \{\mathbf{G}_k(t+1)\}) \leq J(\{\mathbf{F}_k(t)\}, \{\mathbf{G}_k(t)\})$. Also, since $\{\mathbf{F}_k(t+1)\}$ minimizes the average sum-MSE for a given $\{\mathbf{G}_k(t+1)\}$, we have $J(\{\mathbf{F}_k(t+1)\}, \{\mathbf{G}_k(t+1)\}) \leq J(\{\mathbf{F}_k(t)\}, \{\mathbf{G}_k(t+1)\})$. That is, the proposed algorithm monotonically decreases the average sum-MSE that is lower-bounded by zero. Thus, according to the monotonic convergence theorem [32], the proposed algorithm is guaranteed to converge to a local optimum. If initial points are properly selected, the proposed algorithm may converge to the global optimum. Several methods for choosing appropriate initial points were given in [7].

Algorithm 1 Proposed Robust Transceiver Design for MIMO Interference Channel with Generalized Power Constraints

1: Set the iteration number $n = 0$. Initialize $\mathbf{F}_j(0), \forall j$.
2: Set $n \leftarrow n + 1$. Update $\mathbf{G}_k(n), \forall k$, as in (9).
3: Update $\mathbf{F}_j(n), \forall j$ by solving the SOCP problem (15).
4: Repeat Steps 2 and 3 until the algorithm converges or works for a predetermined number of iterations.

4. Results and Discussion

4.1. Simulation Setup

In the simulations, we considered a three-user MIMO interference channel with $M_1 = M_2 = M_3 = 4$, $N_1 = N_2 = N_3 = 4$, and $d_1 = d_2 = d_3 = 2$. Also, the QPSK modulation was used for each data stream. The exponential model was used for both the transmit and receive correlation matrices, and their entries were given by $[\mathbf{T}_k]_{m,n} = \rho_T^{|m-n|}$, $\forall m, n \in \{1, \cdots, M_k\}$ and $[\mathbf{R}_k]_{m,n} = \rho_R^{|m-n|}$, $\forall m, n \in \{1, \cdots, N_k\}$, where $|\rho_T| \leq 1$ and $|\rho_R| \leq 1$ are the correlation coefficients at the transmitter and receiver, respectively [33]. Each transmitter has the same power constraint ($p_1 = p_2 = p_3$) and the noise variance at each receiver is the same ($\sigma_1^2 = \sigma_2^2 = \sigma_3^2$). We define the SNR as $\sum_{k=1}^3 (p_k/\sigma_k^2)$. We use the erroneous CSI model considered in [26]. In this case, $\mathbf{\Sigma}_k = \mathbf{T}_k$, $\mathbf{\Psi}_k = \mathbf{R}_k$, $\forall k$, and the channel estimate of each link can be expressed as $\hat{\mathbf{H}}_{kj} = \mathbf{R}_k^{1/2} \hat{\mathbf{H}}_{kj}^{(w)} \mathbf{T}_j^{1/2}, \forall k, j$. Also, we set $L_k = 1$ and $\mathbf{Q}_1 = \mathbf{I}_{M_k}, \forall k$.

4.2. Performance Comparisons

The BER performance of the proposed algorithm is compared with the following conventional schemes: a perfect CSI case, a nonrobust scheme, the robust scheme in [7]

(denoted by *Baseline scheme I*), which corresponds to $\mathbf{T}_k = \mathbf{I}_{M_k}$ and the $\mathbf{R}_k = \mathbf{I}_{N_k}$ case in the proposed algorithm, the robust scheme in [20] (denoted by *Baseline scheme II*), and a conventional robust least squares technique taking the channel perturbation into account [34]. The results were averaged over 10,000 independent trials. Figure 2 shows the average BER performance for the weekly correlated MIMO interference channel ($\rho_T = \rho_R = 0.2$) when $\sigma^2_{e,kj} = 0.05$ and $\sigma^2_{e,kj} = 0.01, \forall k, j$. The average BER performance for the strongly correlated MIMO interference channel ($\rho_T = \rho_R = 0.8$) is shown in Figure 3 for when $\sigma^2_{e,kj} = 0.05$ and $\sigma^2_{e,kj} = 0.01, \forall k, j$. From these figures, we can see that the proposed algorithm provided better BER performance and significantly reduced the sensitivity to CSI errors compared with the conventional robust schemes in the correlated MIMO interference channel. When $\sigma^2_{kj} = 0.01$, the proposed scheme degraded by roughly 1–3 dB compared to the perfect CSI case.

The convergence behavior of the proposed algorithm is depicted in Figure 4 for SNR = 5 dB and 15 dB with two different initialization methods. One is the singular matrices initialization method and the other is the random initialization method [7]. In both cases, we used the values $\sigma^2_{e,kj} = 0.01$ and $\sigma^2_{e,kj} = 0.05, \forall k, j$. From the figure, we can observe that the proposed algorithm converged within a few iterations.

As shown in Figures 5 and 6, we set $\sigma^2_{e,kj} = 0.05$ and $\sigma^2_{e,kj} = 0.01, \forall k, j$, respectively, and compared the average BER performance of the proposed and conventional schemes. From Figures 5 and 6, it can be observed that the proposed scheme still significantly outperformed the other schemes.

In Figure 7, the total achievable rate of the proposed scheme is compared with that of the distributed interference alignment (IA) scheme in [4] and those of the robust schemes in [7,20] when $\sigma^2_{e,kj} = 0.01, \forall k, j$. From the figure, we can observe that the proposed scheme yielded a higher total achievable rate than the schemes in [4,7,20], and it achieved almost the same degrees of freedom as those of the IA scheme roughly for the 0 dB to 20 dB SNR range. As the SNR increased beyond 20 dB, the performance of the proposed scheme degraded, but it still outperformed the nonrobust IA scheme and the robust schemes in [7,20] in terms of the total achievable rate.

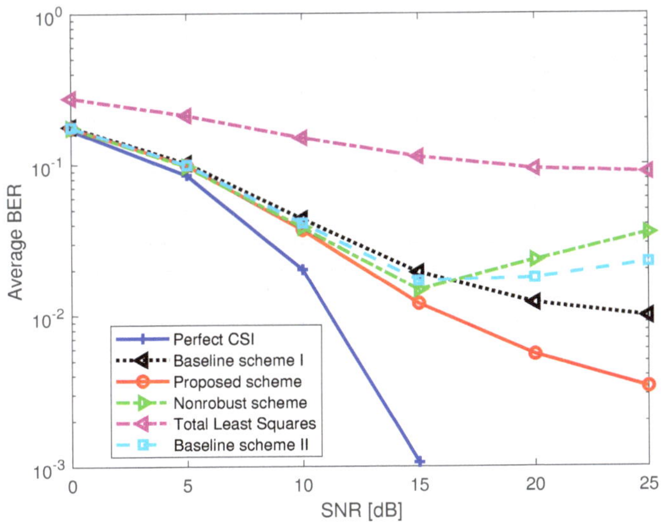

(a)

Figure 2. *Cont.*

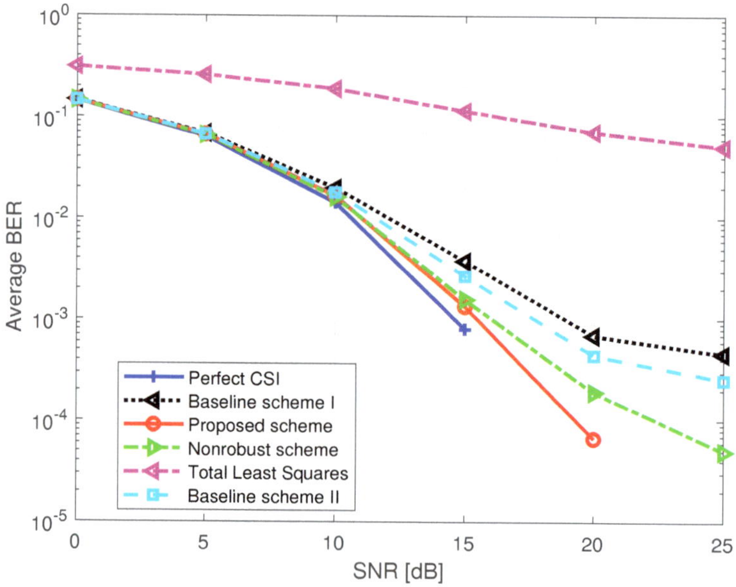

(b)

Figure 2. The average BER performance for the weakly correlated channel ($\rho_T = \rho_R = 0.2$). (**a**) $\sigma^2_{e,kj} = 0.05, \forall k, j$. (**b**) $\sigma^2_{e,kj} = 0.01, \forall k, j$.

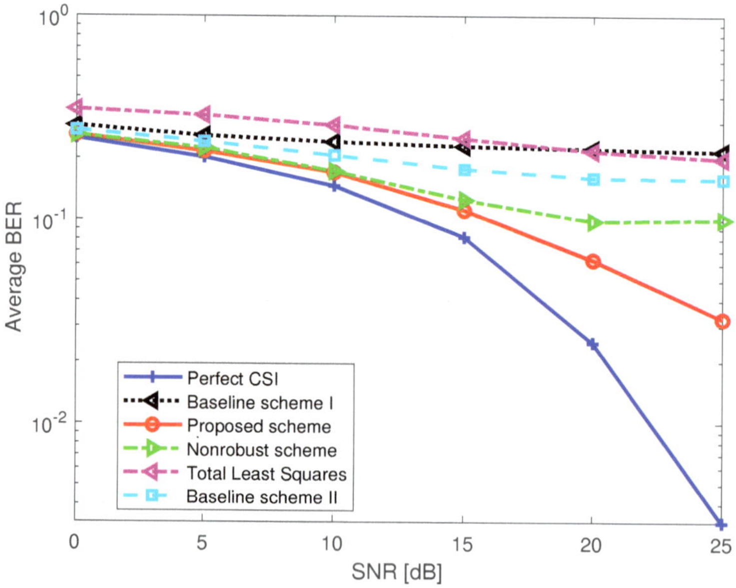

(a)

Figure 3. *Cont.*

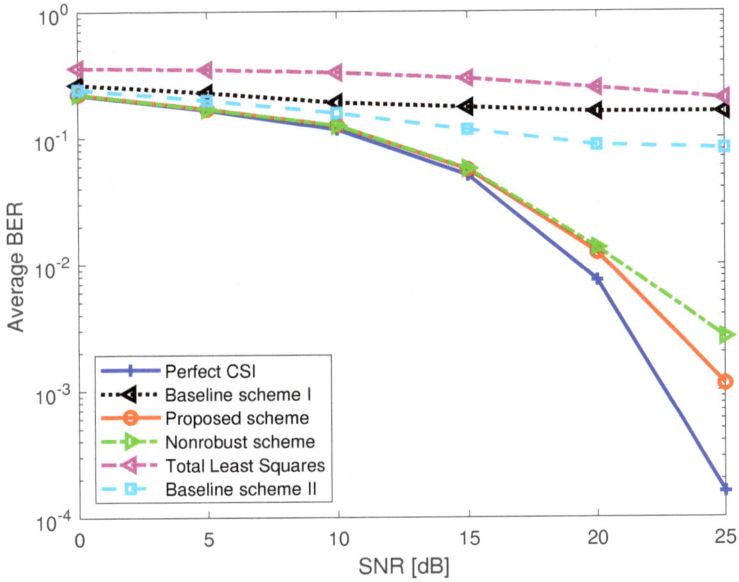

(b)

Figure 3. The average BER performance for the strongly correlated channel ($\rho_T = \rho_R = 0.8$). (**a**) $\sigma^2_{e,kj} = 0.05, \forall k, j$. (**b**) $\sigma^2_{e,kj} = 0.01, \forall k, j$.

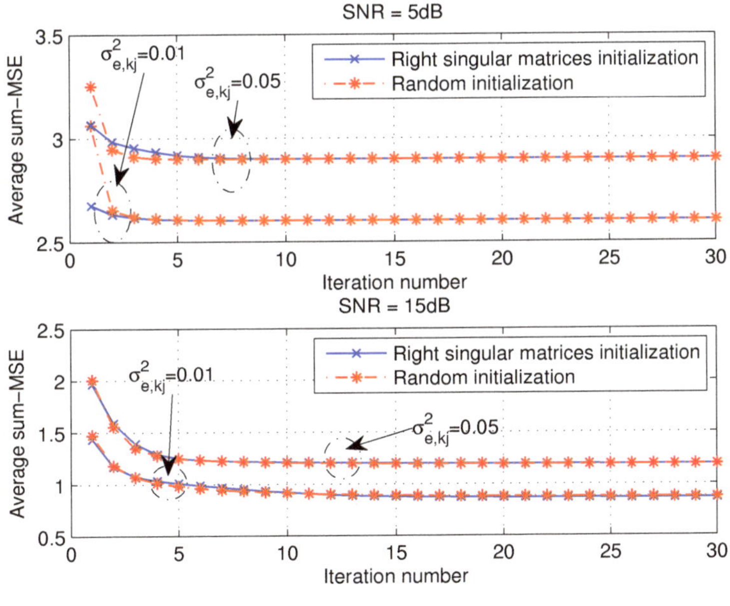

Figure 4. Convergence behavior of the proposed algorithm with two different initialization methods.

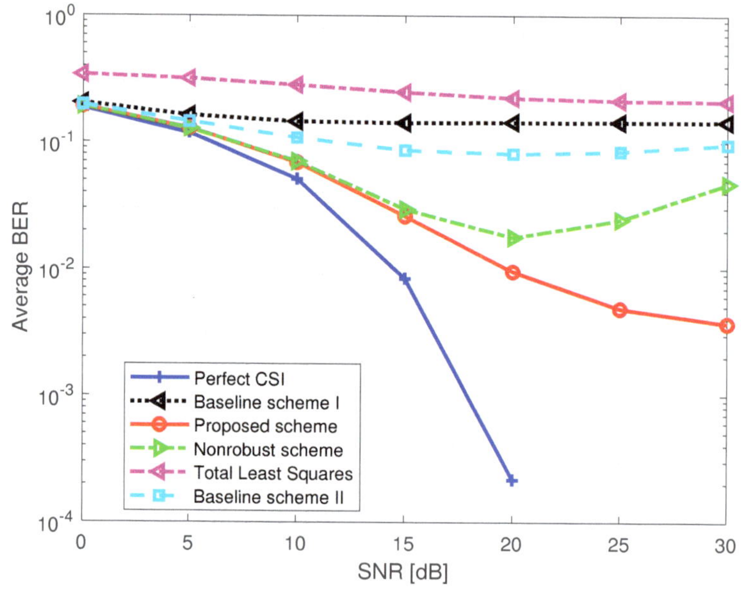

Figure 5. Average BER performance comparison when $\sigma_{e,kj}^2 = 0.05, \forall k, j$.

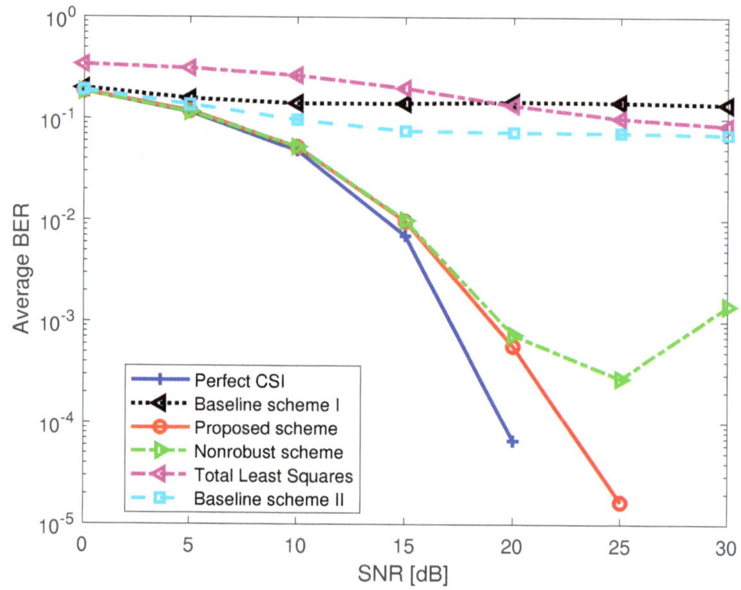

Figure 6. Average BER performance comparison when $\sigma_{e,kj}^2 = 0.01, \forall k, j$.

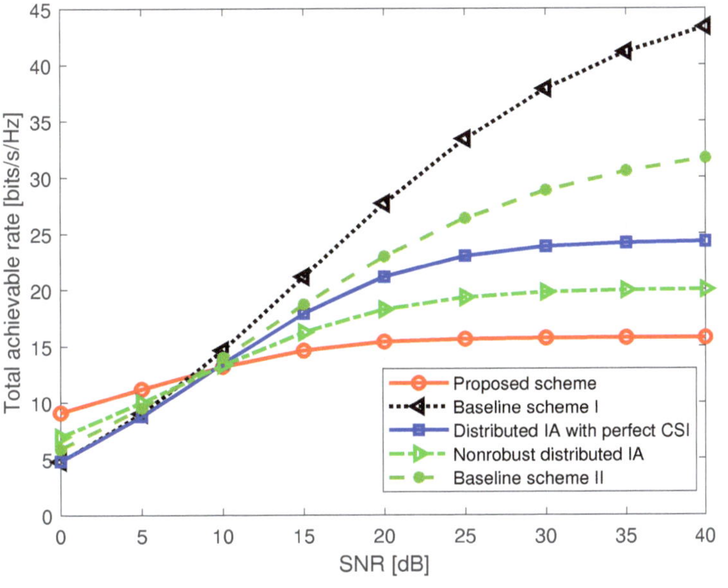

Figure 7. Comparison of the total achievable rate when $\sigma^2_{e,kj} = 0.01, \forall k, j$.

4.3. Complexity Analysis

In each iteration of Algorithm 1, the update of the receive filters $\{\mathbf{G}_k\}_{k=1}^K$ requires a computational complexity of $\mathcal{O}(\sum_{k=1}^K N_k^3)$ and the update of the precoders $\{\mathbf{F}_k\}_{k=1}^K$ requires a computational complexity of $\mathcal{O}((\sum_{k=1}^K M_k d_k + L_k)^{3.5})$. Thus, the overall complexity of the proposed scheme is estimated as $\mathcal{O}\left(N_{\text{iter}}\left(\sum_{k=1}^K N_k^3 + \left(\sum_{k=1}^K M_k d_k + L_k\right)^{3.5}\right)\right)$, where N_{iter} denotes the number of iterations. Similarly, the computational complexity of the nonrobust scheme or the robust scheme in [4] is estimated as $\mathcal{O}\left(N_{\text{iter}} \sum_{k=1}^K (M_k^3 + N_k^3)\right)$.

4.4. Discussion

Overall, from the numerical results along with the analysis on the computational complexities, it can be concluded that the proposed scheme yields a comparable performance to IA with perfect CSI and considerably surpasses the other schemes in terms of the BER and data rate at the cost of slightly higher computational complexity than the other schemes, thereby rendering it highly useful in practice. The conventional schemes suffer from poor sensitivities to the correlated CSI errors, and thus they should be of limited applicability in practice.

5. Conclusions

In this paper, we have investigated the robust transceiver design for the correlated MIMO interference channel in the presence of CSI errors, considering the general transmit power constraints. We have proposed an alternating optimization-based iterative algorithm minimizing the average sum-MSE with respect to the CSI errors, and it has been shown that the proposed algorithm is guaranteed to converge to a local optimum. Through the simulation results, the proposed algorithm has been demonstrated to better cope with the CSI errors compared to the conventional robust schemes in the correlated MIMO interference channel.

As an interesting and important focus of future research, it is deserved to study the robust transceiver design in a more general scenario, e.g., in multi-cell or multi-group multicast scenarios.

Author Contributions: Conceptualization, J.-M.K.; supervision, D.-W.L. All authors have read and agreed to the published version of the manuscript.

Funding: This work was supported by a Korea Agency for Infrastructure Technology Advancement (KAIA) grant funded by the Ministry of Land, Infrastructure and Transport (Grant RS-2023-00251002), and in part by First Research Fund Program In Life at Changwon National University in 2023.

Data Availability Statement: No new data were created or analyzed in this study. Data sharing is not applicable to this article.

Conflicts of Interest: The authors declare no conflicts of interest.

Abbreviations

List of mathematical symbols used in this paper:

Symbol	Definition
K	Number of transmitter–receiver pairs
M_k	Number of antennas at the kth transmitter
N_k	Number of antennas at the kth receiver
$\mathbf{x}_k$	Transmit signal of the kth transmitter
$\mathbf{F}_k$	Precoding matrix of the kth transmitter
$\mathbf{s}_k$	Transmit symbol (or data) of the kth transmitter
d_k	Number of data streams of the kth transmitter
$\mathbf{y}_k$	Received signal at the kth receiver
$\mathbf{w}_k$	Received noise at the kth receiver
$\mathbf{H}_{kj}$	MIMO channel matrix between the jth transmitter and the kth receiver
$\mathbf{T}_k$	Transmit correlation matrix at the kth transmitter
$\mathbf{R}_k$	Receive correlation matrix at the kth transmitter
$\hat{\mathbf{H}}_{kj}$	Estimate of $\mathbf{H}_{kj}$
$\boldsymbol{\Delta}_{kj}$	Error between $\mathbf{H}_{kj}$ and $\hat{\mathbf{H}}_{kj}$
$\boldsymbol{\Sigma}_j$	Column covariance matrix of $\boldsymbol{\Delta}_{kj}$
$\boldsymbol{\Psi}_k$	Row covariance matrix of $\boldsymbol{\Delta}_{kj}$
$\mathbf{Q}_{l,k}$	Weighting matrix for the lth power constraint at the kth transmitter
$p_{l,k}$	Power budget for the lth power constraint at the kth transmitter
$\mathbf{G}_k$	Receiver filter at the kth receiver
$\hat{\mathbf{s}}_k$	Estimate of $\mathbf{s}_k$ at the kth receiver
$J(\cdot)$	Average sum-MSE

References

1. 3rd Generation Partnership Project (3GPP). *Requirements for Further Advancements for for Evolved Universal Terrestrial Radio Access (E-UTRA)*; Physical Layer Aspects; Technical Reports; ETSI: Valbonne, France, 2008.
2. *IEEE 802.16m-07/002*; Draft IEEE 802.16m Requirements; Broadband Wireless Access Working Group, WirelessMAN: New York, NY, USA, 2010.
3. Cadambe, V.; Jafar, S.A. Interference alignment and the degrees of freedom of the K user interference channel. *IEEE Trans. Inf. Theory* **2008**, *54*, 3425–3441. [CrossRef]
4. Gomadam, K.; Cadambe, V.R.; Jafar, S.A. A distributed numerical approach to interference alignment and applications to wireless interference networks. *IEEE Trans. Inf. Theory* **2011**, *57*, 3309–3322. [CrossRef]
5. Peters, S.W.; Heath, R.W., Jr. Interference alignment via alternating minimization. In Proceedings of the 2009 IEEE International Conference on Acoustics, Speech and Signal Processing, IEEE ICASSP'09, Taipei, Taiwan, 19–24 April 2009.
6. Peters, S.W.; Heath, R.W., Jr. Cooperative algorithms for MIMO interference channels. *IEEE Trans. Veh. Technol.* **2011**, *60*, 206–218. [CrossRef]
7. Shen, H.; Li, B.; Tao, M.; Wang, X. MSE-based transceiver designs for the MIMO interference channel. *IEEE Trans. Wirel. Commun.* **2010**, *11*, 3480–3489. [CrossRef]
8. Muthu, P.S.B.; Ponnusamy, K. Design of linear precoder for correlated multiuser MIMO system with imperfect CSI. *AEU-Int. J. Electron. Commun.* **2017**, *74*, 55–62. [CrossRef]
9. Kang, J.-M.; Yun, S.; Kim, I.M.; Jung, H. MSE-based joint transceiver and passive beamforming designs for intelligent reflecting surface-aided MIMO systems. *IEEE Trans. Wireless Commun.* **2021**, *11*, 622–626. [CrossRef]
10. Wang, J.; Jafar, S.A. Robust sum-GDoF of symmetric 2 × 2 × 2 weak interference channel with heterogeneous hops. *IEEE J. Sel. Areas Commun.* **2023**, *41*, 1309–1319. [CrossRef]

11. Zhang, X.; Vaezi, M. Deep autoencoder-based Z-interference channels with perfect and imperfect CSI. *IEEE Trans. Commun.* **2024**, *72*, 861–873. [CrossRef]
12. Zhang, X.; Palomar, D.P.; Ottersten, B. Statistically robust design of linear MIMO transceivers. *IEEE Trans. Signal Process.* **2008**, *56*, 3678–3689. [CrossRef]
13. Ding, M.; Blostein, S.D. MIMO minimum total MSE transceiver design with imperfect CSI at both ends. *IEEE Trans. Signal Process.* **2009**, *57*, 1141–1150. [CrossRef]
14. Chen, C.E.; Chung, W.H. An iterative minmax per-stream MSE transceiver design for MIMO interference channel. *IEEE Wirel. Commun.* **2012**, *1*, 229–232. [CrossRef]
15. Zhang, Q.; He, C.; Jiang, L.; Li, J. Robust per-stream MSE based transceiver design for MIMO interference channel. In Proceedings of the 2013 IEEE Global Communications Conference (GLOBECOM), Atlanta, GA, USA, 9–13 December 2013; pp. 3990–3995.
16. Li, C.; Liu, F.; He, C.; Jiang, L. Sum-MSE minimization for MIMO interference channels under CSI mismatch. In Proceedings of the International Conference on Ubiquitous and Future Networks, Sapporo, Japan, 7–10 July 2015; pp. 588–590.
17. Teodoro, S.; Silva, A.; Dinis, R.; Castanheira, D.; Gameiro, A. New robust iterative minimum mean squared error-based interference alignment algorithm. *J. Eng.* **2014**, *2*, 49–50. [CrossRef]
18. Zhang, Q.; He, C.; Jiang, L. Per-stream MSE based linear transceiver design for MIMO interference channels with CSI error. *IEEE Trans. Commun.* **2015**, *63*, 1676–1689. [CrossRef]
19. Rahman, M.J.; Lampe, L. Robust MSE-based transceiver optimization for downlink cellular interference alignment. In Proceedings of the 2015 IEEE International Conference on Communications (ICC), London, UK, 8–12 June 2015; pp. 4624–4629.
20. Li, L.; Chen, Z.; Fang, J. Robust interference alignment over correlated channels with imperfect CSI. In Proceedings of the 2013 IEEE 78th Vehicular Technology Conference (VTC Fall), Las Vegas, NV, USA, 2–5 September 2013.
21. Xuan, G.; Conggai, L.; Feng, L. Low complexity robust linear transceiver design for MIMO interference channel. *J. China Univ. Posts Telecommun.* **2016**, *23*, 76–81. [CrossRef]
22. Han, H.; Cai, Y.; Zhao, M.M.; Shi, Q. Robust transceiver design based on switched preprocessing for K-pair MIMO interference channels. *IET Commun.* **2020**, *14*, 300–312. [CrossRef]
23. Stoica, P.; Ganesan, G. Maximum-SNR spatial-temporal formatting designs for MIMO channels. *IEEE Trans. Signal Process.* **2002**, *50*, 3036–3042. [CrossRef]
24. Biglieri, E.; Calderbank, R.; Constantinides, A.; Goldsmith, A.; Paulraj, A.; Poor, H.V. *MIMO Wireless Communications*; Cambridge University Press: Cambridge, UK , 2007.
25. Katselis, D.; Kofidis, E.; Theodoridis, S. On training optimization for estimation of correlated MIMO channels in the presence of multiuser interference. *IEEE Trans. Signal Process.* **2008**, *56*, 4892–4904. [CrossRef]
26. Musavian, L.; Nakhai, M.R.; Dohler, M.; Aghvami, A.H. Effect of channel uncertainty on the mutual information of MIMO fading channels. *IEEE Trans. Veh. Technol.* **2007**, *56*, 2798–2806. [CrossRef]
27. Gupta, A.; Nagar, D. *Matrix Variate Distributions*; Chapman & Hall/CRC: Abingdon, UK, 2000.
28. Wang, S.; Ma, S.; Xing, C.; Gong, S.; An, J.; Poor, H.V. Optimal training design for MIMO systems with general power constraints. *IEEE Trans. Signal Process.* **2018**, *66*, 3649–3664. [CrossRef]
29. Rong, B.; Zhang, Z.; Zhao, X.; Yu, X. Robust superimposed training designs for MIMO relaying systems under general power constraints. *IEEE Access* **2019**, *7*, 80404–80420. [CrossRef]
30. Christensen, S.S.; Agarwal, R.; De Carvalho, E.; Cioffi, J.M. Weighted sum-rate maximization using weighted MMSE for MIMO-BC beamforming design. *IEEE Trans. Wirel. Commun.* **2008**, *7*, 4792–4799. [CrossRef]
31. Boyd, S.; Vandenberghe, L. *Convex Optimization*; Cambridge University Press: Cambridge, UK, 2004.
32. Bibby, J. Axiomatisations of the average and a further generalisation of monotonic sequences. *Glasg. Math. J.* **1974**, *15*, 63–65. [CrossRef]
33. Loyka, S.L. Channel capacity of MIMO architecture using the exponential correlation matrix. *IEEE Commun. Lett.* **2001**, *5*, 369–371. [CrossRef]
34. Golub, G.H.; Loan, C.F.V. An analysis of the total least-squares problem. *SIAM J. Numer. Anal.* **1980**, *17*, 883–893. [CrossRef]

mathematics

MDPI

Article

Study on Selection Diversity for MIMO 3-User Interference Channel with Interference Alignment

Dongsup Jin [1] and Xianglan Jin [2,*]

1 Department of IT Convergence, University of Ulsan, 93, Daehak-ro, Nam-gu, Ulsan 44610, Republic of Korea; dsjin@ulsan.ac.kr
2 Division of Electronic Engineering, IT Convergence Research Center, Jeonbuk National University, Jeonju 54896, Republic of Korea
* Correspondence: jinxl77@jbnu.ac.kr

Abstract: This paper explores selection diversity in the context of MIMO 3-user interference channels with interference alignment (IA), focusing on enhancing reliability through diversity order (DO) analysis. While degrees of freedom (DoF) have traditionally been emphasized for throughput optimization in IA, limited research has addressed diversity order to improve error performance. In this work, we define a conditional DO and propose a beamforming vector selection method that achieves a conditional DO of $M^2/2$, where M is the number of antennas per transceiver. The proposed scheme employs a two-stage decoding approach, combining zero-forcing for interference cancellation with maximum likelihood (ML) decoding for desired signal recovery, which is augmented by orthogonalization techniques. Simulations demonstrate the superiority of the proposed scheme in both error probability and conditional DO values compared to conventional IA methods, particularly in scenarios with higher antenna counts. These results provide insights into optimizing IA for enhanced reliability and form the foundation for future exploration of advanced decoding and beamforming strategies.

Keywords: degrees of freedom (DoF); diversity order (DO); interference alignment (IA); MIMO interference channel; selection of beamforming vector; selection diversity

MSC: 94A05; 94A13; 94A14

Citation: Jin, D.; Jin, X. Study on Selection Diversity for MIMO 3-User Interference Channel with Interference Alignment. *Mathematics* **2024**, *12*, 3877. https://doi.org/10.3390/math12243877

Academic Editors: Gurami Tsitsiashvili and Alexander Bochkov

Received: 20 November 2024
Revised: 7 December 2024
Accepted: 9 December 2024
Published: 10 December 2024

1. Introduction

Interference alignment (IA) remains a crucial technique in modern wireless communications, particularly as the demand for high data rates and spectrum efficiency grows. IA has emerged as a vital strategy in 5G-and-beyond networks, enabling more efficient interference management and allowing for higher throughput in interference-limited environments. IA optimizes the degrees of freedom (DoF) by aligning interference within lower-dimensional subspaces, particularly in dense multi-user networks like small cell deployments, where interference poses significant challenges [1].

IA has also demonstrated its effectiveness in cognitive radio networks (CRNs), where it is used to improve spectrum efficiency and manage the interference between primary and secondary users. Secondary users can align their interference in such a way that it minimizes the impact on primary users while maximizing their own throughput, significantly enhancing spectral efficiency in dynamic spectrum-sharing environments [2]. Furthermore, with the integration of massive multiple-input multiple-output (Massive MIMO) systems, IA can leverage spatial diversity to align interference across multiple users, further boosting spectral efficiency in dense networks [1].

The development of IA has focused heavily on optimizing the DoF, especially in scenarios like the K-user interference channel and the $M \times N$ X channel, where IA schemes can asymptotically achieve the maximum DoF [3,4]. While DoF are a crucial measure for spectral efficiency, the reliability of the communication system, often quantified by the diversity order (DO) introduced in [5], is also important. Research on the DO of IA has been limited, with most efforts focusing on maximizing throughput rather than improving error performance. An algorithm that applies the K-user interference alignment technique using time division has also been introduced as part of this effort, allowing the desired degrees of freedom (DoF) to be achieved with fewer antennas and low computation complexity [6].

Additionally, there have been studies exploring the use of relays to assist in IA. Relays can help to more flexibly adjust interference and enhance IA performance, particularly in scenarios with limited channel state information. For example, research on quasi-static X channels has shown that relays can significantly improve the IA's effectiveness in such environments [7], and in the opposite directional interference alignment scenario, it was proven that interference signals are aligned using relays to enhance system performance [8].

Moreover, opportunistic interference alignment (OIA) has emerged as a promising approach in scenarios where perfect channel state information (CSI) is not available at all nodes. OIA leverages multi-user diversity by opportunistically selecting users who can align interference effectively, thus further improving spectral efficiency without the need for full CSI at all transmitters [9–11].

Although beamforming selection schemes based on maximizing the sum rate have been proposed [12], they may not be optimal for minimizing the error rate. In peer-to-peer (P2P) multi-input and multi-output (MIMO) communication systems, various selection schemes, such as transmit antenna selection or equivalent path selection, are based on the signal-to-noise ratio (SNR) at the receiver. These selection methods, however, often assume a zero-forcing linear receiver, which can degrade error performance. In contrast, maximum likelihood (ML) decoding offers better error performance, although at the cost of increased computational complexity [13].

Recent advancements in interference alignment (IA) techniques have increasingly leveraged machine learning, particularly reinforcement learning and deep learning. These approaches have demonstrated significant potential in improving IA performance, especially in dynamic and complex wireless environments. Reinforcement learning (RL) has been employed to optimize IA in various scenarios by adapting to dynamic channel conditions and resource management challenges.

A deep Q-network (DQN)-based approach was proposed to optimize user selection policies in cache-enabled IA networks, where the wireless environment was modeled as a finite-state Markov channel, achieving an improved sum rate and energy efficiency [14]. A deep deterministic policy gradient (DDPG) algorithm was developed for jointly optimizing IA and power control in dense networks. This method effectively managed computational complexity and enhanced spectral efficiency [15]. Additionally, a deep RL framework for edge caching and user selection in heterogeneous networks was introduced, where edge caches were dynamically updated to optimize spectral and energy efficiency in opportunistic IA scenarios [16].

Pure deep learning (DL) techniques have been focused on simplifying IA by using neural networks to model and predict IA strategies. An autoencoder-based transmission strategy was utilized for distributed IA, enabling effective alignment with only local channel state information [17]. Deep learning was applied to enhance IA for discrete constellations, where encoder and decoder functions were designed to maximize the sum rate in interference channels [18]. A neural network framework was developed to predict optimal precoding matrices for IA in MIMO systems, demonstrating improved IA performance [19]. Finally, a hybrid approach was proposed, combining traditional signal processing with deep learning for IA and channel estimation in 6G heterogeneous network environments [20].

Machine learning-based algorithms can address problems that are difficult to model mathematically; however, they require learning the optimal weights under varying channel conditions and often involve high relative complexity, making their practical application in real communication environments challenging. Therefore, this paper aims to explore scenarios where both the transmitter and receiver use M identical antennas, focusing on mathematical beamforming and interference alignment.

Previous works without using machine learning technology have attempted to address this gap by incorporating space–time block codes (STBCs) into IA schemes to enhance the error performance, particularly for systems like the 2×2 X channel. In such systems, each transmitter uses only the CSI of its connected links, allowing for improved error performance through the use of STBCs [21,22]. Under this assumption, global CSI is assumed to be available at both the transmitter and receiver, making beamforming-based approaches for selecting matrices a more effective method from the perspective of throughput [23].

The key contributions of this paper can be summarized as follows:

1. Analysis of Diversity Order and Error Performance with and without Selection: This study compares scenarios with and without selection, clearly analyzing the performance improvement achievable through selection from the viewpoints of diversity order. Research on systematically deriving the diversity order for interference alignment (IA) with multiple antennas has been limited. This paper quantitatively analyzes and verifies that selection diversity, when using M antennas, can enhance the performance of IA from the perspective of error probability. Without selection, the conditional DO is limited to $M/2$, while selection improves it to $M^2/2$, resulting in significant error probability reduction at high SNR levels.
2. Improvement of Conditional Diversity Order (DO) through Beamforming Selection: This paper proposes a novel beamforming vector selection method in the interference alignment (IA) environment, achieving a conditional DO of $M^2/2$. This ensures higher reliability and improved error performance compared to existing methods [23].
3. Proposed Two-Stage Decoding Approach: The proposed decoding procedure removes interference signals using zero-forcing and then recovers the desired signal with maximum likelihood (ML) decoding. This significantly reduces error probability compared to conventional zero-forcing-based decoding methods.
4. Utilization of Orthogonalization Techniques: The paper optimizes the design of beamforming matrices using QR factorization and singular-value decomposition (SVD)-based orthogonalization techniques. In particular, the SVD-based approach demonstrates better Symbol Error Rate (SER) performance compared to QR factorization, improving signal quality in multi-antenna environments.

2. Characteristic Function of Multivariate Rayleigh Random Variables

Usually, the characteristic function of multivariate Rayleigh random variables is used for analysis of the error probability. Therefore, in this section, we introduce the results from [24].

Consider $L \times 1$ complex Gaussian random vector $\mathbf{x}$ given by

$$\begin{aligned} \mathbf{x} &= \mathbf{x}_c + j\mathbf{x}_s \\ &= \begin{bmatrix} x_{c_1} \\ \vdots \\ x_{c_L} \end{bmatrix} + j \begin{bmatrix} x_{s_1} \\ \vdots \\ X_{s_L} \end{bmatrix} \end{aligned} \quad (1)$$

with covariance matrix $\mathbf{K}_{cc}$ and $\mathbf{K}_{ss}$ and cross-covariance matrix $\mathbf{K}_{cs}$ such that $E[x_{c_i} x_{s_i}] = (\mathbf{K}_{cs})_{ii} = 0$.

Let $\gamma_i = |x_{c_i} + jx_{s_i}|^2$ and $\gamma = (\gamma_i, \cdots, \gamma_L)^T$. Then, the characteristic function $\Psi_\gamma(j\omega)$ of γ is given as [24]

$$\begin{aligned}\Psi_\gamma(j\omega) &= E[\exp(j\omega\gamma)] \\ &= (\det(\mathbf{I}_L - 2j\,\mathrm{diag}(\omega)\mathbf{K}_{ss}))^{-\frac{1}{2}} \\ &\times \Bigg(\det\Big(\mathbf{I}_L - 2j\,\mathrm{diag}(\omega)\mathbf{K}_{cc} \\ &+ 4\,\mathrm{diag}(\omega)\mathbf{K}_{cs}[I_L - 2j\,\mathrm{diag}(\omega)\mathbf{K}_{ss}]^{-1} \\ &\times \mathrm{diag}(\omega)\mathbf{K}_{cs}^T\Big)\Bigg)^{-\frac{1}{2}}. \end{aligned} \tag{2}$$

If $\mathbf{X}$ is a circularly symmetric complex Gaussian random vector, that is

$$\mathbf{K}_{cc} = \mathbf{K}_{ss}, \mathbf{K}_{cs}^T = -\mathbf{K}_{cs},$$

(2) is simplified as

$$\begin{aligned}\Psi_\gamma(j\omega) &= \det(\mathbf{I}_L - 2j\,\mathrm{diag}(\omega)[\mathbf{K}_{cc} + j\mathbf{K}_{cs}]) \\ &\times \det(\mathbf{I}_L - 2j\,\mathrm{diag}(\omega)[\mathbf{K}_{cc} - j\mathbf{K}_{cs}]) \\ &\times \det(\mathbf{I}_L - j\,\mathrm{diag}(\omega)\mathbf{K})^{-\frac{1}{2}} \\ &\times \det(\mathbf{I}_L - j\,\mathrm{diag}(\omega)\mathbf{K}^\dagger)^{-\frac{1}{2}}. \end{aligned} \tag{3}$$

where $\mathbf{K}$ is the covariance matrix of $\mathbf{x}$.

In general, for analysis on the error performance, (2) and (3) are important formulas, because the error probability is given as a Q function, which is upper-bounded by the exponential function. Let $\omega = j(\rho, \rho, \cdots, \rho)^T$, and then (2) is given as

$$\begin{aligned}\Psi_\gamma(-\rho) &= (\det(\mathbf{I}_L + 2\rho\mathbf{K}_{ss}))^{-\frac{1}{2}} \\ &\times \left(\det([\mathbf{I}_L + 2\rho\mathbf{K}_{cc}] + 4\rho^2\mathbf{K}_{cs}[\mathbf{I}_L + 2\rho\mathbf{K}_{ss}]^{-1}\mathbf{K}_{cs}^T)\right)^{-\frac{1}{2}}\end{aligned}$$

and (3) is also given as

$$\Psi_\gamma(-\rho) = \det(\mathbf{I}_L + \rho\mathbf{K})^{-\frac{1}{2}} \times \det(\mathbf{I}_L + \rho\mathbf{K}^\dagger)^{-\frac{1}{2}}$$

In a high SNR region, where $\rho \to \infty$, we have

$$\begin{aligned}\Psi_\gamma(-\rho) &\approx (2\rho)^{-\frac{\mathrm{rank}(\mathbf{K}_{SS}) + \mathrm{rank}(\mathbf{K}_{cc} + \mathbf{K}_{cs}\mathbf{K}_{ss}^{-1}\mathbf{K}_{cs}^T)}{2}} \\ &\times \prod \lambda_{K_{ss}} \prod \lambda_{\mathbf{K}_{cc} + \mathbf{K}_{cs}\mathbf{K}_{ss}^{-1}\mathbf{K}_{cs}^T}.\end{aligned}$$

where λ_* denotes the eigenvalues of the matrix. If there is no deterministic relation among the element of $\mathbf{X}$, $\mathbf{K}$ is a full-rank matrix.

For (3), we have

$$\Psi_\gamma(-\rho) = \rho^{-\mathrm{rank}(\mathbf{K})} |\prod \lambda_{\mathbf{K}}|^2. \tag{4}$$

From this result, the following lemma can be obtained.

Lemma 1. *Consider that the received SNR is expressed as $\gamma = \rho|\mathbf{x}|^2$, where $\mathbf{x}$ is a circularly symmetric Gaussian random vector. The DO of the decoder is the number of the random variables with non-deterministic relation in $\mathbf{x}$.*

3. Selection of Beamforming Matrices for 3-User MIMO Interference Channel

3.1. System Model and Interference Alignment for 3-User MIMO Interference Channel

In this subsection, the system model for 3-user MIMO interference channel and IA is explained [3]. And then, the proposed scheme, the selection scheme of beamforming matrices for IA, is introduced. Here is a system model for the 3-user interference channel with M antennas.

- For even M, we have

$$\mathbf{Y}_{e,k} = \sqrt{\rho}\sum_{i=1}^{3}\mathbf{H}_{ki}\mathbf{V}_i\mathbf{d}_k + \mathbf{n}_k$$

  where $\mathbf{H}_{ki}$ and $\mathbf{V}_i$ denote an $M \times M$ channel matrix between transmitter i and receiver k and $M \times \frac{M}{2}$ beamforming matrix of transmitter i, respectively. $\mathbf{d}_k$ is an $\frac{M}{2} \times 1$ data stream vector, and ρ is the parameter linearly proportional to the average transmit SNR. It is assumed that all channel coefficients are independent and identically distributed with the Rayleigh distribution.
- For odd M, we have

$$\mathbf{Y}_{o,k} = \sqrt{\rho}\sum_{i=1}^{3}\mathbf{H}_{o,ki}\mathbf{V}_i\mathbf{d}_k + \mathbf{n}_k$$

  where $\mathbf{H}_{o,ki}$ is given as

$$\mathbf{H}_{o,ki} = \begin{bmatrix}\mathbf{H}_{ki} & \mathbf{0}\\ \mathbf{0} & \mathbf{H}_{ki}\end{bmatrix}$$

  and $\mathbf{V}_i$ denotes a $2M \times M$ beamforming matrix, respectively. $\mathbf{d}_k$ is an $M \times 1$ data stream vector.

We can generate the beamforming matrices for IA as follows. From the result in [3], the beamforming matrices for IA are obtained by the following relation:

$$\begin{aligned}\text{span}(\mathbf{H}_{12}\mathbf{V}_2) &= \text{span}(\mathbf{H}_{13}\mathbf{V}_3)\\ \mathbf{H}_{21}\mathbf{V}_1 &= \mathbf{H}_{23}\mathbf{V}_3\\ \mathbf{H}_{31}\mathbf{V}_1 &= \mathbf{H}_{32}\mathbf{V}_2.\end{aligned} \tag{5}$$

It is rewritten as

$$\begin{aligned}\text{span}(\mathbf{V}_1) &= \text{span}(\mathbf{H}_{31}^{-1}\mathbf{H}_{32}\mathbf{H}_{12}^{-1}\mathbf{H}_{13}\mathbf{H}_{23}^{-1}\mathbf{H}_{21}\mathbf{V}_1)\\ \mathbf{V}_2 &= \mathbf{H}_{32}^{-1}\mathbf{H}_{31}\mathbf{V}_1\\ \mathbf{V}_3 &= \mathbf{H}_{23}^{-1}\mathbf{H}_{21}\mathbf{V}_1.\end{aligned} \tag{6}$$

Therefore, $\mathbf{V_1}$ consists of the eigenvectors of

$$\mathbf{H}_{31}^{-1}\mathbf{H}_{32}\mathbf{H}_{12}^{-1}\mathbf{H}_{13}\mathbf{H}_{23}^{-1}\mathbf{H}_{21}(=\mathbf{A}) \tag{7}$$

and $\mathbf{V}_2$ and $\mathbf{V}_3$ can be obtained from (6).

- For even M, we have

$$\mathbf{V}_1 = \left[\mathbf{b}_1\mathbf{b}_2\cdots\mathbf{b}_{\frac{M}{2}}\right] \tag{8}$$

  where $\mathbf{b}_i$ is an eigenvector of $\mathbf{A}$.
- For odd M, we have

$$\mathbf{V}_1 = \begin{bmatrix}\mathbf{b}_1 & 0 & \mathbf{b}_3 & 0\cdots & 0 & \mathbf{b}_M\\ 0 & \mathbf{b}_2 & 0 & \mathbf{b}_4\cdots & \mathbf{b}_{M-1} & 0\end{bmatrix}. \tag{9}$$

In this way, at each receiver, interference signal vectors are aligned as described in Figure 1. In Figure 1, the desired signal of transmitter 1 is represented in red, the desired signal of transmitter 2 is in green, and the desired signal of transmitter 3 is in black. Each receiver 1, 2, and 3 must receive the corresponding transmitter's signal without interference. On the right side of Figure 1, it can be observed that the desired and undesired signals at each receiver are linearly independent and aligned. This represents the core result of interference alignment (IA).

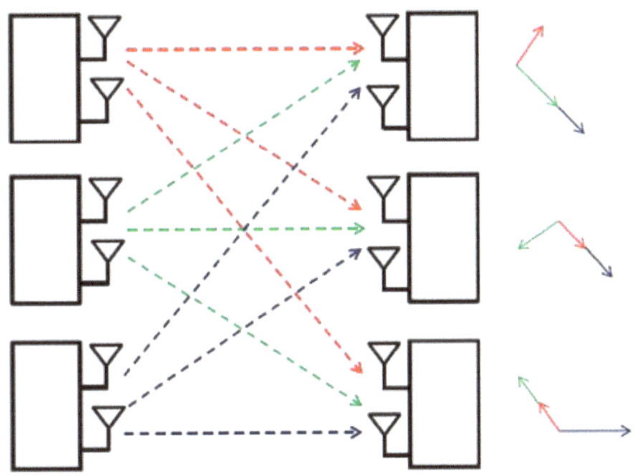

Figure 1. IA for 3-user interference channel.

In fact, for odd M, all eigenvectors of $\mathbf{A}$ should be used as in (9), and thus, the selection of beamforming matrices is complicated, and the IA should be modified for selection. Therefore, for simplicity, we consider the case that M is even in this paper.

3.2. Orthogonalization of Beamforming Matrices

In general, the orthogonalization of transmit signals can achieve the improvement in performance. Equation (6) guarantees the perfect IA but does not guarantee the orthogonalization between beamforming vectors at each transmitter. Therefore, we should modify the design of the beamforming matrices.

In [12], the QR factorization is used for the orthogonalization of beamforming matrices, which is identical to the Gram–Schmidt orthogonalization. In the Gram–Schmidt orthogonalization, there exists a critical drawback. Let us orthogonalize two beamforming matrices by using the Gram–Schmidt orthogonalization as follows:

$$\begin{aligned}&\text{Beamforming matrix 1:}(\mathbf{b_1 b_2}) = (\mathbf{b_1 b_2}')\\&\text{Beamforming matrix 2:}(\mathbf{b_1 b_3}) = (\mathbf{b_1 b_3}')\end{aligned} \tag{10}$$

In (10), it can be seen that $\mathbf{b_1}$ is the beamforming vector for the beamforming matrix 1 and 2. If the data symbol is (x,0), $(\mathbf{b_1 b_2})(x,0)^T = (\mathbf{b_1 b_3})(x,0)^T = x\mathbf{b_1}$. If the error rate becomes high for beamforming matrix 1 for $(x,0)$, the error rate also becomes high for beamforming matrix 2. In this case, we cannot use the advantage of the selection of beamforming vectors. In [12], this case does not occur almost surely, because it is focused on the sum rate where the Gaussian source is assumed. However, the practical communication system uses the quadrature amplitude modulation (QAM) or phase shift keying modulation (PSK). Therefore, the weakness of this scheme can be exposed in the practical communication system.

In this paper, instead of Gram–Schmidt orthogonalization, we use SVD for the orthogonalization of beamforming matrices as follows:

$$\text{span}(\mathbf{V}_i) = \text{span}(\mathbf{U}_i \Sigma \mathcal{V}^\dagger) = \text{span}(\mathbf{U}_i(\frac{M}{2}))$$

where $\mathbf{U}_i$, Σ, and $\mathcal{V}^\dagger$ can be obtained by SVD, and $\mathbf{U}_i(\frac{M}{2})$ is the submatrix consisting of the first $M/2$ columns of $\mathbf{U}_i$. In this way, the problem of Gram–Schmidt orthogonalization is overcome, and therefore, we can use $\mathbf{U}_i(\frac{M}{2})$ as a beamforming matrix for the transmitter i in order to improve the performance.

3.3. *Two-Stage Decoding and Selection of Beamforming Matrices*

In [3], each transmitter can send $M/2$ data streams, and each receiver can detect them without interference. In general, each receiver decodes the desired signal by only zero-forcing because of its complexity. However, zero-forcing can make error performance worse. Therefore, in this paper, we consider two cascaded decoding procedures which consist of zero-forcing for the interference signal and ML decoding for the desired signal However, in the case of ML decoding, the decoding complexity increases with M_{QAM}^d as the degrees of freedom and the number of QAM symbols (M_{QAM}) increase. In contrast, zero-forcing has a complexity proportional to M_{QAM}, resulting in higher complexity for ML decoding. This decoding procedure is described in Figure 2. In an MIMO communication system, given a channel matrix $\mathbf{H}$, the pairwise error probability (PEP) is given as

$$\text{P}_{\text{PEP}} = \Pr(\mathbf{x} \to \hat{\mathbf{x}}|\mathbf{H}) = \text{Q}(|\mathbf{H}(\mathbf{x} - \hat{\mathbf{x}})|). \tag{11}$$

Considering M-QAM, the symbol error probability (SER) can be obtained approximately by the union bound of the PEP, thus lowering the PEP is related to lowering the SER directly. If the error probability of all pair-wise error patterns can be reduced, for M-QAM, the SER can be reduced. Therefore, we can select the eigenvectors for $\mathbf{V}_1$ as follows:

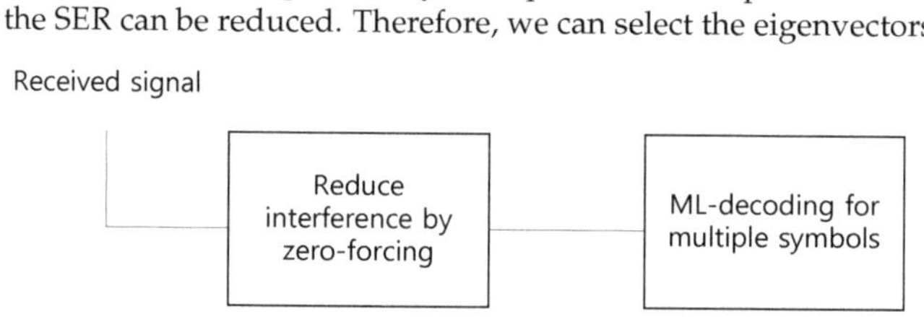

Figure 2. Two-stage decoding procedure at each receiver in MIMO interference channel.

$$\begin{aligned}\mathbf{V}_{1,\text{sel}} = \arg \max_{\mathbf{V}_1 \in \text{eig}(\mathbf{A})} \min \Big(&|\mathbf{R}_1^\dagger \mathbf{H}_{11} \mathbf{V}_1 \tilde{\mathbf{d}}_1|, \\ &|\mathbf{R}_2^\dagger \mathbf{H}_{22} \mathbf{V}_2 \tilde{\mathbf{d}}_2|, |\mathbf{R}_3^\dagger \mathbf{H}_{33} \mathbf{V}_3 \tilde{\mathbf{d}}_3| \Big).\end{aligned} \tag{12}$$

where $\tilde{\mathbf{d}}_i = \mathbf{x}_i - \hat{\mathbf{x}}_i$, which can be various vectors, because there are various pair-wise error pattens for M-QAM.

Definition 1. *In general, the DO is defined as* $DO = \lim_{SNR\to\infty} \frac{\log P_P EP(\text{SNR})}{\log \text{SNR}}$. *In this paper, we define the contional DO for simplicity as follows:*

$$DO|\mathbf{A} = \lim_{SNR\to\infty} \frac{\log P_{PEP|\mathbf{A}}(\text{SNR})}{\log \text{SNR}}. \tag{13}$$

where $\mathbf{A}$ *is defined in (6). The contional DO does not describe the error performance in the fading environment directly like the DO. However, it can make the analysis for the selection of beamforming vectors for the interference channel as that of that for the P2P MIMO channel, and we can expect that as the contitional DO become higher, the error performance become better.*

4. Diversity Analysis

In this section, the conditional DO of the proposed scheme is derived. In fact, it is too complicated to derive the exact DO, and thus, we analyze the contional DO of the proposed scheme. First, we derive the DO of the case without the selection of beamforming matrices and then analyze the conditional DO of the two-stage decoding with the selection of beamforming matrices.

4.1. Diversity Analysis for the Case Without Selection

In this subsection, we analyze the DO of IA without selection.

$$
\begin{aligned}
&E_{\mathbf{H}_{11},\mathbf{H}_{22},\mathbf{H}_{33}} \Pr(\mathbf{x} \rightarrow \hat{\mathbf{x}}|\mathbf{A}, \mathbf{H}_{11}, \mathbf{H}_{22}, \mathbf{H}_{33}) \\
&\quad = \frac{1}{3} E_{\mathbf{H}_{11},\mathbf{H}_{22},\mathbf{H}_{33}|\mathbf{A}} \left(\Pr(\mathbf{x} \rightarrow \hat{\mathbf{x}}|\mathbf{A}, \mathbf{H}_{11}) + \Pr(\mathbf{x} \rightarrow \hat{\mathbf{x}}|\mathbf{A}, \mathbf{H}_{22}) + \Pr(\mathbf{x} \rightarrow \hat{\mathbf{x}}|\mathbf{A}, \mathbf{H}_{33}) \right) \\
&\quad = \sum_{i=1}^{3} E_{\mathbf{H}_{ii}} Q\left(\sqrt{\rho} \left| \mathbf{R}_i^{\dagger} \mathbf{H}_{ii} \mathbf{V}_i \tilde{\mathbf{d}}_i \right|^2 \right) \\
&\quad \leq \sum_{i=1}^{3} E_{\mathbf{H}_{ii}} \exp\left(-\frac{\rho}{2} \left| \mathbf{R}_i^{\dagger} \mathbf{H}_{ii} \mathbf{V}_i \tilde{\mathbf{d}}_i \right|^2 \right).
\end{aligned} \tag{14}
$$

We have

$$
|\mathbf{R}_i^{\dagger} \mathbf{H}_{ii} \mathbf{V}_i \tilde{\mathbf{d}}_i|^2 = \sum_{k=1}^{\frac{M}{2}} |\mathbf{r}_k^{\dagger} \mathbf{H}_{ii} \sum_{l=1}^{\frac{M}{2}} \mathbf{v}_{i,l} \tilde{d}_l|^2 \tag{15}
$$

where $\tilde{d}_l$ is the lth element of $\tilde{\mathbf{d}}_i$.

$\mathbf{R}_i$ and $\mathbf{V}_i$ consist of orthonormal columns, and thus, each $\mathbf{r}_k^{\dagger} \mathbf{H}_{ii} \sum_{l=1} \mathbf{u}_k \tilde{d}_l$ is a complex Gaussian random variable and independent from the others. $E\{|\mathbf{r}_k^{\dagger} \mathbf{H}_{ii} \sum_{l=1} \mathbf{v}_{i,k} \tilde{d}_l|^2\} \geq E\{|\mathbf{r}_k^{\dagger} \mathbf{H}_{ii} \mathbf{v}_{l'} \tilde{d}_{l'}|^2\} = |\tilde{d}_{l'}|^2 \neq 0$. Therefore,

$$
\begin{aligned}
&\sum_{i=1,2,3} E_{\mathbf{H}_{ii}} \exp(-\frac{\rho}{2} |\mathbf{R}_i^{\dagger} \mathbf{H}_{ii} \mathbf{V}_i \tilde{\mathbf{d}}|^2) \\
&\quad = 3 E_{\mathbf{H}_{ii}} \prod_{k=1}^{\frac{M}{2}} \exp(-\frac{\rho}{2} |\mathbf{r}_k^{\dagger} \mathbf{H}_{ii} \sum_{l=1} \mathbf{v}_{i,l} \tilde{d}_l|^2) \\
&\quad \leq 3 \prod_{k=1}^{\frac{M}{2}} \left(1 + \frac{\rho |\tilde{d}_l'|^2}{2} \right)^{-1}
\end{aligned} \tag{16}
$$

each $E_{\mathbf{H}_{ii}} Q(\sqrt{\rho} |\mathbf{R}_i^{\dagger} \mathbf{H}_{ii} \mathbf{V}_i \tilde{\mathbf{d}}|^2)$ has a DO of $M/2$ regardless of $\mathbf{A}$, which means that the conditional DO is also $M/2$. Finally, without selection, IA and two-stage decoding can achieve a DO of $M/2$.

4.2. Diversity Analysis for the Case with Selection

The DO depends on the beamforming vector corresponding to the worst SNR; therefore, the analysis for the worst case is enough for diversity analysis. In fact, there are $\binom{M}{M/2}$ choices for the selection of beamforming matrices. Let each set containing $M/2$

beamforming vectors be S_k and S be a set consisting of S_k. Thus, $|S| = \binom{M}{M/2}$, and we let be $|S|$. From the selection criterion, the following PEP analysis is possible for a given $\mathbf{A}$

$$\begin{aligned}
&E_{\mathbf{H}_{11},\mathbf{H}_{22},\mathbf{H}_{33}|\mathbf{A}} \Pr(\mathbf{x} \to \hat{\mathbf{x}}|\mathbf{A}, \mathbf{H}_{11}, \mathbf{H}_{22}, \mathbf{H}_{33}) \\
&= \frac{1}{3} E_{\mathbf{H}_{11},\mathbf{H}_{22},\mathbf{H}_{33}|\mathbf{A}} \Big(\Pr(\mathbf{x} \to \hat{\mathbf{x}}|\mathbf{A}, \mathbf{H}_{11}) + \Pr(\mathbf{x} \to \hat{\mathbf{x}}|\mathbf{A}, \mathbf{H}_{22}) + \Pr(\mathbf{x} \to \hat{\mathbf{x}}|\mathbf{A}, \mathbf{H}_{33}) \Big) \\
&= \frac{1}{3} \sum_{i=1}^{3} E_{\mathbf{H}_{ii}|\mathbf{A}} Q\left(\sqrt{\rho} \left| \mathbf{R}_{i,\mathrm{sel}}^{\dagger} \mathbf{H}_{ii} \mathbf{V}_{i,\mathrm{sel}} \tilde{\mathbf{d}} \right|^2 \right) \\
&\leq E_{\mathbf{H}_{11},\mathbf{H}_{22},\mathbf{H}_{33}|\mathbf{A}} Q\left(\sqrt{\rho} \max_{k \in \{1,\ldots,L\}} \min_{i_k \in \{1,2,3\}} \left| \mathbf{R}_{i_k,k}^{\dagger} \mathbf{H}_{i_k i_k} \mathbf{V}_{i_k,k} \tilde{\mathbf{d}} \right|^2 \right) \\
&\leq E_{\mathbf{H}_{11},\mathbf{H}_{22},\mathbf{H}_{33}|\mathbf{A}} \exp\left(-\frac{\rho}{2} \max_{k \in \{1,\ldots,L\}} \min_{i_k \in \{1,2,3\}} \left| \mathbf{R}_{i_k,k}^{\dagger} \mathbf{H}_{i_k i_k} \mathbf{V}_{i_k,k} \tilde{\mathbf{d}} \right|^2 \right) \\
&\leq E_{\mathbf{H}_{11},\mathbf{H}_{22},\mathbf{H}_{33}|\mathbf{A}} \exp\left(-\frac{\rho}{2L} \sum_{k=1}^{L} \min_{i_k \in \{1,2,3\}} \left| \mathbf{R}_{i_k,k}^{\dagger} \mathbf{H}_{i_k i_k} \mathbf{V}_{i_k,k} \tilde{\mathbf{d}} \right|^2 \right) \\
&\leq E_{\mathbf{H}_{11},\mathbf{H}_{22},\mathbf{H}_{33}|\mathbf{A}} \max_{\{i_1,\ldots,i_L\} \in \{1,2,3\}^L} \exp\left(-\frac{\rho}{2L} \sum_{k=1}^{L} \left| \mathbf{R}_{i_k,k}^{\dagger} \mathbf{H}_{i_k i_k} \mathbf{V}_{i_k,k} \tilde{\mathbf{d}} \right|^2 \right) \\
&\leq E_{\mathbf{H}_{11},\mathbf{H}_{22},\mathbf{H}_{33}|\mathbf{A}} \sum_{\{i_1,\ldots,i_L\} \in \{1,2,3\}^L} \exp\left(-\frac{\rho}{2L} \sum_{k=1}^{L} \left| \mathbf{R}_{i_k,k}^{\dagger} \mathbf{H}_{i_k i_k} \mathbf{V}_{i_k,k} \tilde{\mathbf{d}} \right|^2 \right)
\end{aligned} \tag{17}$$

When $M = 4$, we have six choices for beamforming matrices. From (17),

$$\begin{aligned}
&E_{\mathbf{H}_{11},\mathbf{H}_{22},\mathbf{H}_{33}|\mathbf{A}} Pr(\mathbf{x} \to \hat{\mathbf{x}}|\mathbf{A}, \mathbf{H}_{11}, \mathbf{H}_{22}, \mathbf{H}_{33}) \\
&\leq E_{\mathbf{H}_{11},\mathbf{H}_{22},\mathbf{H}_{33}|\mathbf{A}} \sum_{\{i_1,\ldots,i_6\} \in \{1,2,3\}^6, \tilde{\mathbf{d}}} \exp\left(-\frac{\rho}{6} \sum_{k=1}^{6} |\mathbf{R}_{i_k,k}^{\dagger} \mathbf{H}_{i_k i_k} \mathbf{V}_{i_k,k} \tilde{\mathbf{d}}|^2 \right)
\end{aligned} \tag{18}$$

From Equation (18), we are able to independently analyze the diversity order for each transmitter–receiver pair. In other words, if we analyze the diversity order for one transmitter–receiver pair, the result will be the same for all pairs. Therefore, we will analyze the diversity order for the first pair $i_k = 1$.

Let $\tilde{\mathbf{d}} = (d_1, d_2)$, $\mathbf{V}_{1,1} = [\mathbf{u}_{1,1}\mathbf{u}_{1,2}]$ and $\mathbf{R}_{1,1} = [\mathbf{r}_{1,11}, \mathbf{r}_{1,12}]$ be the zero-forcing matrix for $\mathbf{V}_{1,1}$. Then,

$$|\mathbf{R}_{1,1}^{\dagger} \mathbf{H}_{11} \mathbf{V}_{1,1} \tilde{\mathbf{d}}|^2 = \underbrace{|\, \mathbf{r}_{1,11}^{\dagger} \mathbf{H}_{11} \sum_{i=1}^{2} \mathbf{u}_{1,i} d_i \,|^2}_{\tilde{h}_1} + \underbrace{|\, \mathbf{r}_{1,12}^{\dagger} \mathbf{H}_{11} \sum_{i=1}^{2} \mathbf{u}_{1,i} d_i \,|^2}_{\tilde{h}_2} \tag{19}$$

Similarly, define the other $\mathbf{V}_{1,k}$s and $\mathbf{R}_{1,k}$; then,

$$|\mathbf{R}_{1,k}^{\dagger} \mathbf{H}_{11} \mathbf{V}_{1,k} \tilde{\mathbf{d}}|^2 = \underbrace{|\, \mathbf{r}_{1,k1}^{\dagger} \mathbf{H}_{11} \sum_{i=1}^{2} \mathbf{u}_{k,i} d_i \,|^2}_{\tilde{h}_{2k+1}} + \underbrace{|\, \mathbf{r}_{1,k2}^{\dagger} \mathbf{H}_{11} \sum_{i=1}^{2} \mathbf{u}_{k,i} d_i \,|^2}_{\tilde{h}_{2k+2}} \tag{20}$$

From (5), it can be seen that $\sum_{i=1}^{2} \mathbf{u}_{1,i} d_i$, $\sum_{i=1}^{2} \mathbf{u}_{2,i} d_i$, $\sum_{i=1}^{2} \mathbf{u}_{3,i} d_i$, and $\sum_{i=1}^{2} \mathbf{u}_{4,i} d_i$ are randomly generated from $\mathrm{span}(\mathbf{b}_1\ \mathbf{b}_2)$, $\mathrm{span}(\mathbf{b}_1\ \mathbf{b}_3)$, $\mathrm{span}(\mathbf{b}_1\ \mathbf{b}_4)$, and $\mathrm{span}(\mathbf{b}_2\ \mathbf{b}_3)$, respectively. Since the $\sum_{i=1}^{2} \mathbf{u}_{k,i} d_i$s are 4×1 vectors, we can obtain only four linearly independent vectors among the $\sum_{i=1}^{2} \mathbf{u}_{k,i} d_i$s. Similar to this, we can also obtain the linearly independent vector among $\mathbf{r}_{1,lm}$

For non-zero vectors $\mathbf{r}_{1,lm}$ and $\mathbf{r}_{1,l'm'}$, even though $\mathbf{r}_{1,lm} = \mathbf{r}_{1,l'm'}$, if $\sum_{i=1}^{2} \mathbf{u}_{1,i} d_i$ and $\sum_{i'=1}^{2} \mathbf{u}_{1,i'} d_i'$ are linearly independent, $\mathbf{r}_{1,lm} \mathbf{H}_{11} \sum_{i'=1}^{2} \mathbf{u}_{1,i'} d_{i'}$ and $\mathbf{r}_{1,l'm'} \mathbf{H}_{11} \sum_{i=1}^{2} \mathbf{u}_{1,i} d_i$ cannot have a deterministic relation. Therefore, at least eight $\tilde{h}_i'$s does not have non-deterministic relations. From Lemma 1, the eighth conditional DO is guaranteed by the proposed scheme.

Theorem 1. *When M is an even number, IA with the selection criterion in (12) and the decoding scheme with zero-forcing and multi-symbol decoding can achieve the $M^2/2$ conditional DO at least.*

Proof. When each node has M antennas, the $M/2$ DoFs are achieved and thus in (17). Therefore, we can obtain the following relation like (19) and (20):

$$\begin{aligned} |\mathbf{R}_{1,1}^{*}\mathbf{H}_{11}\mathbf{V}_{1,1}\mathbf{d}|^2 &= |\mathbf{r}_{11}^{*}\mathbf{H}_{11}\sum_{i=1}^{M/2}\mathbf{u}_{1,i}d_i|^2 + |\mathbf{r}_{12}^{*}\mathbf{H}_{11}\sum_{i=1}^{M/2}\mathbf{u}_{1,i}d_i|^2 + \cdots + |\mathbf{r}_{1\frac{M}{2}}^{*}\mathbf{H}_{11}\sum_{i=1}^{M/2}\mathbf{u}_{1,i}d_i|^2 \\ |\mathbf{R}_{1,L}^{*}\mathbf{H}_{11}\mathbf{V}_{1,L}\mathbf{d}|^2 &= |\mathbf{r}_{L1}^{*}\mathbf{H}_{11}\sum_{i=1}^{M/2}\mathbf{u}_{L,i}d_i|^2 + |\mathbf{r}_{L2}^{*}\mathbf{H}_{11}\sum_{i=1}^{M/2}\mathbf{u}_{L,i}d_i|^2 + \cdots + |\mathbf{r}_{L\frac{M}{2}}^{*}\mathbf{H}_{11}\sum_{i=1}^{M/2}\mathbf{u}_{L,i}d_i|^2 \end{aligned} \quad (21)$$

where $L = \binom{M}{\frac{M}{2}}$. Similar to the four-antenna case, we can obtain M linearly independent vectors among the $\sum_{i=1}^{M/2}\mathbf{u}_{k,i}d_i$s and M linearly independent zero-forcing vectors. In (21), since each equation has $M/2$ terms, we have $M^2/2$ terms, which do not have the deterministic relation. Therefore, from lemma 1, the proposed scheme can achieve the $M^2/2$ conditional DO. □

4.3. Expected Diversity Order

In the previous subsection, we only achieved the conditional DO $M^2/2$, which is not the DO. From (18) and Theorem 1, it can be seen that the average PEP is bounded as

$$\text{PEP} \le E_{A}c \cdot \rho^{-\frac{M^2}{2}} = E_c c \cdot \rho^{-\frac{M^2}{2}} \quad (22)$$

In fact, c is a variable related to the $\mathbf{A}$, and it belongs to $(0, \infty)$. In order to achieve the DO $M^2/2$, a requirement should be added to the selection criterion, which is the minimum value of $|R_{i,\text{sel}}^{\dagger}H_{ii}V_{i,\text{sel}}\tilde{\mathbf{d}}|$. If the selected beamforming vector and the corresponding zero-forcing vector cannot satisfy this value, the data are not transferred through them. In [25], this requirement can be satisfied numerically. Therefore, we can achieve the $M^2/2$ DO.

5. Simulation Results

In this section, the simulation results for the proposed scheme are presented and analyzed in detail. Each transmitter used BPSK modulation, and it was assumed that all channel coefficients followed identical distributions. Each channel coefficient was generated from a complex Gaussian distribution, with the mean and variance of the real and imaginary parts set to 0 and $\frac{1}{2}$, respectively. The error probabilities ranged from 10^{-1} to 10^{-7}, covering both low and high signal-to-noise ratio (SNR) regimes.

Figure 3 compares the symbol error probability (SER) performance of various schemes for $M = 2$ and $M = 4$. For $M = 2$, the two-stage decoding scheme behaved identically to zero-forcing decoding, as orthogonalization of beamforming matrices is not required when transmitting a single desired symbol. However, for $M = 4$, the two schemes diverged significantly. The consequently arrived at the following observations:

- Observation 1: Impact of diversity order
  It can be observed that there was a significant difference in SER performance between cases with and without selection. In particular, when selection was not applied, the beamforming vector was not chosen as an optimal vector from the perspective of error probability. Interestingly, for $M = 4$, the SER was observed to increase compared to $M = 2$. As the DoF increased in the MIMO interference channel, the number of transmission streams grew. This deteriorated the condition number of the system's transmit–receive matrix, since the power needed to be divided among each stream.

A poor condition number amplified small noise significantly during the zero-forcing process, leading to a higher probability of errors.
It can be observed that a significant reduction in error probability was achieved in the case of $M = 4$ compared to $M = 2$. This is because the diversity order is 2 for $M = 2$ whereas it is 8 for $M = 4$, indicating that the difference in diversity order leads to a larger decrease in error probability. As the number of antennas increased, the number of eigenvector options for selection also grew, which aligns with the analysis presented in the previous section.

- Observation 2: Impact of orthogonalization
Even in scenarios where selection was not applied and only two-stage decoding was used, the effect of orthogonalizing the beamforming matrix was clearly evident. Without orthogonalization, the SER for $M = 4$ in the absence of selection was approximately 3.0×10^{-3} for an SNR of 12 db, whereas with orthogonalization, it improved to around 1.2×10^{-3}. For $M = 4$, the benefits of using QR factorization versus SVD for selection could also be observed from the perspective of the SER. Ultimately, it can be concluded that optimizing the beamforming matrix using an appropriate orthogonalization technique is essential.
- Observation 3: Two-stage decoding performance
In this paper, it was shown that through two-stage decoding, interference was first eliminated, allowing each transmitter–receiver pair to form an independent MIMO channel. While using zero-forcing in an independent MIMO channel did not provide any diversity order, performing ML decoding, as proven in Section 4, offered an additional diversity order, leading to further performance gains from an error probability perspective. Simulations also confirmed that when $M = 4$, the performance gap between zero-forcing decoding without selection and two-stage ML decoding became increasingly pronounced as the SNR increased.

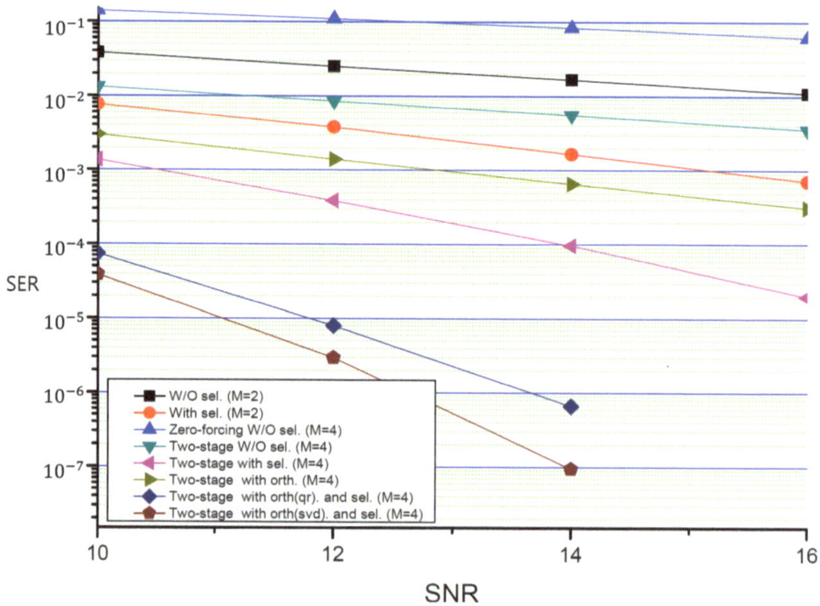

Figure 3. SER performance comparison across decoding and beamforming techniques.

In conclusion, the simulation results presented in this section clearly demonstrate the advantages of selection and the two-stage decoding scheme in various scenarios. By leveraging diversity order and optimizing beamforming through orthogonalization, significant improvements in error probability were achieved compared to conventional zero-forcing approaches. These results emphasize the importance of proper selection and orthogonalization techniques, particularly as the number of antennas increases. These findings validate the theoretical analyses presented earlier in this paper and lay the groundwork for further exploration of advanced decoding schemes.

6. Conclusions

In this paper, a selection method of beamforming matrices for IA was proposed and its DO was analyzed for a three-user interference channel, where each node had even M antennas. For the error performance enhancement, two-stage decoding was considered, which consisted of the zero-forcing for mitigation of interference signals and the ML decoding for the desired signals, and the orthogonalization of beamforming matrices was also used.

First, by using the two-stage decoding without selection of beamforming matrices, it was confirmed that a DO of $M/2$ could be achieved, and then by using the selection scheme of beamforming matrices and two-stage decoding, it was proved that an $M^2/2$ conditional DO could be guaranteed. Also, the $M^2/2$ DO could be achieved by using an additional requirement. Through the simulation, it was also shown that the performance of the proposed scheme was superior to that of the conventional IA and the $M^2/2$ is achieved.

As a further work, we analyzed the DO for odd M values. In addition, the proposed scheme can be easily extended to the other wireless communication environment where IA can be used. Therefore, the analysis on the selection diversity for the other channel models is considered to be a good further work.

Author Contributions: Conceptualization, D.J.; methodology, D.J.; software, D.J.; validation, X.J.; formal analysis, X.J.; investigation, X.J.; resources, D.J.; data curation, D.J.; writing—original draft preparation, D.J. and X.J.; writing—review and editing, D.J. and X.J.; visualization, D.J.; supervision, X.J.; project administration, X.J.; funding acquisition, D.J. All authors have read and agreed to the published version of the manuscript.

Funding: This result was supported by the "Regional Innovation Strategy (RIS)" through the National Research Foundation of Korea (NRF) funded by the Ministry of Education (MOE) (2021RIS-003).

Data Availability Statement: The data will be made available by the authors upon request.

Conflicts of Interest: The authors declare no conflicts of interest.

References

1. Trabelsi, N.; Fourati, L.C.; Chen, C.S. Interference Management in 5G and Beyond Networks. *arXiv* **2024**, arXiv:2401.01608.
2. Abdulkadir, Y.; Simpson, O.; Sun, Y. Interference Alignment for Cognitive Radio Communications and Networks: A Survey. *J. Sens. Actuator Netw.* **2019**, *8*, 50. [CrossRef]
3. Cadambe, V.; Jafar, S. Interference alignment and degrees of freedom of the *K*-user interference channel. *IEEE Trans. Inf. Theory* **2008**, *54*, 3425–3441. [CrossRef]
4. Cadambe, V.; Jafar, S. Interference alignment and the degrees of freedom of wireless X networks. *IEEE Trans. Inf. Theory* **2009**, *55*, 3893–3908. [CrossRef]
5. Tarokh, V.; Seshadri, N.; Calderbank, A. Space-time codes for high data rate wireless communication: Performance criterion and code construction. *IEEE Trans. Inf. Theory* **1998**, *44*, 744–765. [CrossRef]
6. Jing, X.; Mo, L.; Liu, H.; Zhang, C. Linear Space-Time Interference Alignment for *K*-User MIMO Interference Channels. *IEEE Access* **2018**, *6*, 3085–3095. [CrossRef]
7. Jin, D.S.; No, J.S.; Shin, D.J. Interference alignment aided by relays for the quasi-static X channel. In Proceedings of the 2011 IEEE International Symposium on Information Theory Proceedings, St. Petersburg, Russia, 31 July–5 August 2011; pp. 2637–2641. [CrossRef]
8. Liu, Z.; Sun, D. Relay-Assisted Opposite Directional Interference Alignment: Feasibility Condition and Achievable Degrees of Freedom. *IEEE Commun. Lett.* **2015**, *19*, 66–69. [CrossRef]
9. Lee, N.; Tse, D.; Heath, R.W. Opportunistic Interference Alignment for the MIMO Interference Channel. *IEEE Trans. Wirel. Commun.* **2014**, *12*, 3192–3202.
10. Sung, H.; Lee, S.R.; Lee, I. Opportunistic Interference Alignment for Two-Cell MIMO Interference Channels. *IEEE Trans. Commun.* **2014**, *62*, 3043–3053.

11. Guo, Y.; Ge, J.; Li, J. Opportunistic Interference Alignment for MIMO Interference Channels With Limited Feedback. *IEEE Trans. Veh. Technol.* **2016**, *65*, 5179–5191.
12. Sung, H.; Park, S.; Lee, K.; Lee, I. Linear Precoder Designs for *K*-user Interference Channels. *IEEE Trans. Wirel. Commun.* **2010**, *9*, 291–301. [CrossRef]
13. Sezgin, A.; Jafar, S.A.; Jafarkhani, H. Optimal Use of Antennas in Interference Networks: A Tradeoff between Rate, Diversity and Interference Alignment. In Proceedings of the Proceedings IEEE GlobeCom'09, Honolulu, HI, USA, 30 November–4 December 2009.
14. Yu, F.R.; He, Y. Deep Reinforcement Learning for Interference Alignment Wireless Networks. In *Deep Reinforcement Learning for Wireless Networks*; Springer: Berlin/Heidelberg, Germany, 2019; pp. 21–44.
15. Wang, C.; Deng, D.; Xu, L.; Wang, W.; Gao, F. Joint Interference Alignment and Power Control for Dense Networks via Deep Reinforcement Learning. *IEEE Trans. Wirel. Commun.* **2021**, *19*, 966–970. [CrossRef]
16. Zheng, J.; Gao, L.; Wang, H.; Niu, J.; Ren, J.; Guo, H.; Yang, X.; Liu, Y. Smart Edge Caching-Aided Partial Opportunistic Interference Alignment in HetNets. *Mob. Networks Appl.* **2020**, *25*, 1842–1850. [CrossRef]
17. Mishra, R.K.; Chahine, K.; Kim, H.; Jafar, S.; Vishwanath, S. Distributed Interference Alignment for K-user Interference Channels via Deep Learning. In Proceedings of the 2021 IEEE International Symposium on Information Theory Proceedings, Los Angeles, CA, USA, 12–20 July 2021; pp. 2614–2619.
18. Mishra, R.; Jafar, S.; Vishwanath, S.; Kim, H. Enhancing K-user Interference Alignment for Discrete Constellations via Learning. *arXiv* **2024**, arXiv:2407.15054.
19. Zhang, X.; Luo, Z.; Xiao, W.; Feng, L. Deep Learning-Based Modulation Recognition for MIMO Systems: Fundamental, Methods, Challenges. *IEEE Access* **2024**, *12*, 112558–112575. [CrossRef]
20. Subramanian, R.; Jayarin, J.; Chandrasekar, A. Hybrid Deep Learning Approach for 6G MIMO Channel Estimation and Interference Alignment in HetNet Environments. *Trait. Du Signal* **2022**, *40*, 1231–1240. [CrossRef]
21. Li, L.; Jafarkhani, H.; Jafar, S.A. When Alamouti codes meet interference alignment: Transmission schemes for two-user X channel. In Proceedings of the 2021 IEEE International Symposium on Information Theory Proceedings, St. Petersburg, Russia, 31 July–5 August 2011.
22. Li, F.; Jafarkhani, H. Space-time processing for X channels using precoders. *IEEE Trans. Signal Process.* **2012**, *60*, 1849–1861. [CrossRef]
23. Ordonez, L.; Palomar, D.; Pages-Zamora, A.; Fonollosa, J. High-SNR analytical performance of spatial multiplexing MIMO systems with CSI. *IEEE Trans. Signal Process.* **2007**, *55*, 5447–5463. [CrossRef]
24. Malik, R.K. On multivariate Rayleigh and exponential distributions. *IEEE Trans. Inf. Theory* **2003**, *49*, 1499–1515. [CrossRef]
25. Li, L.; Jafarkhani, H.; Jafar, S.A. Selection diversity for interference alignment systems. In Proceedings of the 2013 IEEE Global Communications Conference (GLOBECOM), Atlanta, GA, USA, 9–13 December 2013; pp. 1773–1778. [CrossRef]

MDPI AG
Grosspeteranlage 5
4052 Basel
Switzerland
Tel.: +41 61 683 77 34

Mathematics Editorial Office
E-mail: mathematics@mdpi.com
www.mdpi.com/journal/mathematics

www.ingramcontent.com/pod-product-compliance
Lightning Source LLC
Chambersburg PA
CBHW041409210926
51850CB00013B/3224
* 9 7 8 3 7 2 5 8 3 6 6 3 5 *